Time: Limits and Constraints

The Study of Time

Founding Editor

J. T. Fraser

VOLUME 13

Time: Limits and Constraints

Edited by

Jo Alyson Parker
Paul A. Harris
and
Christian Steineck

BRILL

LEIDEN • BOSTON
2010

This book is printed on acid-free paper.

Library of Congress Cataloging-in-Publication Data

International Society for the Study of Time. Conference (13th : 2007 : Asilomar Conference Center, Pacific Grove, Calif.)
Time : limits and constraints / edited by Jo Alyson Parker, Paul A. Harris, and Christian Steineck.
p. cm. — (The study of time ; v. 13)
Includes index.
ISBN 978-90-04-18575-3 (hardback : alk. paper) 1. Time—Congresses. I. Parker, Jo Alyson, 1954- II. Harris, Paul (Paul A.) III. Steineck, Christian. IV. Title. V. Series.

BD638.I72 2007
115—dc22

2010018541

ISSN 0170-9704
ISBN 978-90-04-18575-3

To Julius and Jane Fraser

CONTENTS

INTRODUCTORY ESSAYS

SECTION I

THEORY AND EMPIRIE

SECTION II

THE LIMITS OF DURATION

SECTION III

CREATIVE CONSTRAINTS

SECTION IV

FINAL QUESTIONS

LIST OF CONTRIBUTORS

MARC BOTHA has, since 2006, been a doctoral fellow in the Department of English Studies at Durham University where he is in the final stages of his Ph.D. research, *The Persistence of Minimalism: An Atopian Poetics*. Prior to this, he worked as a classical saxophonist, teacher, and lecturer. His research interests span several disciplines, including literature, the visual and performing arts, cultural theory, and philosophy, with particular interests in minimalist aesthetics, the philosophy of Agamben and Badiou, and concrete poetry (especially Ian Hamilton Finlay). His work has recently been published in *Postmodern Culture* (Johns Hopkins), *Oxford Literary Review* (Edinburgh), *TkH: Teorija koja Hoda* (Belgrade) and *Kaleidoscope* (Durham).

CAROL A. FISCHER has worked in technical theatre arenas for 20 years, assuming the various roles of costume, light, sound designer, stage and production manager, technical trainer and supervisor, and director. She received a Master's degree in Theatre Arts from San Jose State University, and she is finishing a dissertation that connects quantum physics theory and critical analysis of dramatic works to fulfill requirements for a Ph.D. in Theatre Studies from the University of California at Santa Barbara.

J. T. FRASER is the author of six volumes, editor of eleven, and author of more than seventy articles in professional periodicals on themes related to the integrated study of time. He is Founder of the International Society for the Study of Time and Founding Editor of *KronoScope: Journal for the Study of Time*. He has lectured extensively at home and abroad about themes related to the study of time.

PETER A. HANCOCK, D.Sc., Ph.D., is the University Pegasus Professor and Provost Distinguished Research Professor in the Department of Psychology and the Institute for Simulation and Training, as well as at the Department of Civil and Environmental Engineering at the University of Central Florida. He is a Fellow and past President of the Human Factors and Ergonomics Society, and he is also a Fellow and current President of the Division of Applied Experiment and Engineering Psychology of the American Psychological Association.

Paul A. Harris is Professor of English at Loyola Marymount University. His teaching and research explore interdisciplinary connections among literature, physics, philosophy, mathematics, and theology. He co-edits the literary theory journal *SubStance* and has served as ISST's President from 2004–2010.

Florian Klapproth is Associate Professor of Psychology at the Institute of Psychology and Working Sciences, Technical University of Berlin. He got his Diploma degree in 1999 and his Ph.D. in 2003. His main research areas are psychology of time, development of psychometric tests, and decision-making. He has written a book on temporal memory and a book on waiting, and he has also written several articles published in scientific journals. He is member of the Deutsche Gesellschaft für Psychologie (DGPs) and member of the International Society for the Study of Time.

Heike Klippel is Professor for Film Studies at the Hochschule für Bildende Künste Braunschweig (Braunschweig School of Fine Arts). She is co-editor of the feminist film journal *Frauen und Film*, has edited *"The Art of Programming." Film, Programm und Kontext*, (Münster, Lit Verlag, 2008), and has recently published *Zeit ohne Ende. Essays über Zeit, Frauen und Kino* (Frankfurt/Basel, Stroemfeld 2009).

William R. LaFleur is the E. Dale Saunders Professor in Japanese Studies at the University of Pennsylvania, having taught previously at Princeton and UCLA. His Ph.D. is from the University of Chicago. Major books are *The Karma of Words: Buddhism and the Literature Arts in Medieval Japan* and *Liquid Life: Abortion and Buddhism in Japan*. He edited *Zen and Western Thought: Essays by Masao Abe*; *Dôgen Studies*; and, with Gernot Böhme and Susumu Shimazono, co-edited *Dark Medicine: Rationalizing Unethical Medical Research.* A Senior Fellow at Penn's Center for Bioethics, he is currently completing a book on Japanese critics of American bioethics, although his *Awesome Nightfall: The Life, Times, and Poetry of Saigyô* recently returned him to literary studies of Japan's medieval period.

Robin Lucy is an African Americanist educated in Canada and Jamaica. She is an Associate Professor at Eastern Michigan University in Ypsilanti. Her research interests include African American women's writing, Black writers of World War II, African American folklore, and the cultures of the diaspora. She is currently working on a book about African American writing of the Cold War period.

Deirdre H. McMahon has written on a number of topics in critical race studies, particularly in relation to the structural co-dependence of race and sex in the nineteenth-century transatlantic imagination. Recent publications include analyses of black doctress Mary Seacole in *Other Mothers: Beyond the Maternal Ideal*, ed. Ellen Rosenman and Claudia Klaver, and of imperialist adventure writer G. A. Henty in *Studies in the Novel*. Currently she is co-editing a volume on domesticity and material culture in imperial Britain. She is an Assistant Teaching Professor in the Department of English and Philosophy at Drexel University.

Carlos Montemayor is Assistant Professor of Philosophy at San Francisco State University. He graduated from Rutgers University with a Ph.D. in Philosophy and a Certificate in Cognitive Science. His research focuses on issues in philosophy of mind and cognitive science. Aside from his areas of specialization, Montemayor is conversant with scholarship in a number of disciplines and fields, including psychology, phenomenology, and political philosophy. He received a J.D. from Universidad Nacional Autónoma de México and an M.A. from the New School for Social Research.

Tyler T. Ochoa is a Professor and former Academic Director of the High Technology Law Institute at Santa Clara University School of Law. He is a co-author, with Craig Joyce, Marshall Leaffer, and Peter Jaszi, of *Copyright Law* (LexisNexis 7th Ed. 2006), the best-selling copyright casebook in the United States. He has published numerous articles concerning copyright and intellectual property law, one of which was cited by the United States Supreme Court in *Eldred v. Ashcroft*, 537 U.S. 186 (2003). Prior to joining the faculty at Santa Clara, he was a Professor and Co-Director of the Center for Intellectual Property Law at Whittier Law School in Costa Mesa, California.

Steven T. Ostovich is Professor of Philosophy at the College of St. Scholastica in Duluth, Minnesota, USA. He recently published a general readership book on *The Courage of Faith* (Liturgical Press) and has a text co-authored with a historian, a Germanist, and another philosopher entitled "The Happy Burden of History, or the German

Sisyphus" under review. His current research project is on eschatological time and political theology.

Jo Alyson Parker is Professor and Chair of English at Saint Joseph's University in Philadelphia, where she teaches courses in the eighteenth- and early nineteenth-century novel, narrative, and literary theory. She is the author of *Narrative Form and Chaos Theory in Sterne, Proust, Woolf, and Faulkner* (2007), *The Author's Inheritance: Henry Fielding, Jane Austen, and the Establishment of the Novel* (1998), and essays dealing with time and narrative. With Michael Crawford and Paul Harris, she co-edited *Time and Memory: The Study of Time XII* (2007).

Laura Pattillo is Assistant Professor of English and Co-Artistic Director of Cap and Bells Dramatic Arts Society at Saint Joseph's University in Philadelphia. Her research deals with both dramatic literature and Southern literature, with a special interest in representations of Appalachia on the American stage.

Helen Sills is a professional musician, lecturer, examiner, and instrumental teacher in Sussex, U.K. She gained her Doctorate in 1992 on the spirituality and temporal characteristics of the music of Igor Stravinsky and continues to research into the perception of time, movement, and structure in music.

Christian Steineck is Professor of Japanology at Zurich University. He earned an M.A. in japanology and a Dr. phil. in philosophy from Bonn University. His research interests combine medieval and contemporary history of ideas in Japan and philosophy of culture, with a special interest in the body-mind problem, and time.

John Streamas is an Associate Professor of Comparative Ethnic Studies at Washington State University. In addition to critical and scholarly works. he writes stories, poems, and informal essays. In 2004, his play *The Serving Class* was selected for performance in the DNA Festival of very short plays.

Jonathan Tallant is a lecturer in Philosophy at the University of Nottingham. His research is primarily focused on metaphysics, in particular the metaphysics of time.

Frederick Turner, Founders Professor of Arts and Humanities at the University of Texas at Dallas, was educated at Oxford University. A poet, critic, translator, philosopher, and former editor of *The Kenyon Review*, he has authored 28 books, including *Shakespeare and the Nature of Time, Natural Classicism, The Culture of Hope, Genesis, Hadean Eclogues, Shakespeare's Twenty-First Century Economics, Paradise*, and *Natural Religion*.

Friedl Weinert is professor of philosophy at the University of Bradford in the United Kingdom. In his research he focuses on the history and philosophy of science, both the natural and social sciences. He is editor of the interdisciplinary volume *Laws of Nature* (1995) and co-editor of the *Compendium of Quantum Physics* (2009). He is author of *Traditionen, Diskurse, Argumente* (1985), *The Scientist as Philosopher* (2004) and *Copernicus, Darwin & Freud* (2009). He is currently working on a new book, which will trace changing conceptions of time in the light of scientific discoveries.

Katherine Weiss is an Assistant Professor of English at East Tennessee State University where she teaches drama. Along with Seán Kennedy, she is co-editor of *Samuel Beckett: History, Memory, Archive* (Palgrave Macmillan, forthcoming). Currently, Professor Weiss is working on a student edition of *Sweet Bird of Youth* to be published by Methuen Press.

PREFACE

Jo Alyson Parker, Paul A. Harris, and Christian Steineck

From July 28 through August 3, 2007, nearly one hundred time scholars gathered at the Asilomar Conference Center in Pacific Grove, California, for the 13th triennial conference of the International Society for the Study of Time. Nestled between white sand beaches bounded by the Pacific Ocean and a towering redwood forest, Asilomar provided a fertile site for exploring the conference theme, "Time: Limits and Constraints." As the initial call for papers stated, "Though seemingly synonymous with limitation and constriction, limits and constraints also serve as boundary conditions that enable or induce change and novelty. Limits and constraints of temporal processes are as important and powerful for what they produce as what they prevent; they demarcate the parameters of emergence as much as endings." Over the course of six days, scholars explored emergence and endings in a variety of fields, including astronomy, film, history, literature, music, philosophy, physics, psychology, religion, and sociology. Stimulating talks during the day, many presented in the Craftsman-style chapel designed by famed architect Julia Morgan, gave way to lively exchanges in the evenings. Once again, as it has done since its inception in 1966, the ISST provided a forum for work that breaches disciplinary constraints and enables participants to expand the limits of their understanding of time.

The seventeen essays appearing in this volume are revised and expanded versions of some of the presentations given at Asilomar. Approaching the issue of temporal limits and constraints from a variety of disciplines, speaking both within and across these disciplines, they add important new insights to the ISST's continuing work in the interdisciplinary study of time. In the previous volume, *Time and Memory: The Study of Time XII*, in an effort to expand the conversation beyond the ISST, reviewers were invited to provide responses to the papers that they had reviewed. We issued a similar invitation to reviewers for this volume, and several of them have done so, offering thoughtful and original commentaries that often begin to take us in new directions.

Following our usual practice, this ISST volume begins with the President's Address to the conference, followed by the Founder's Lecture. The President's Address is reproduced here as originally delivered the first evening of the conference (rather than as a formally revised essay for publication), preserving the convivial spirit of the occasion. In The Founder's Lecture, J. T. Fraser brings his hierarchical theory of time to bear on the conference theme in masterful fashion. This volume offers a particularly rich exploration of Fraser's theory: in addition to the Founder's Lecture, the volume concludes with Christian Steineck's astute investigation of fundamental questions and premises informing Fraser's work, as well as prefatory remarks from Fraser about this paper.

Section I: Theory and Empirie explores fundamental conceptions of time, pairing philosophical and empirical presentations. The first two papers look into different forms of temporal organization present in the human experience of time, and their relation to each other, exploring the constraints their coexistence means for each. They also demonstrate the liberating effects these mutual constraints have for the behavior of the human individual as a whole.

"Time: Biological, Intentional and Cultural," by Carlos Montemayor, which won the ISST's Founder's Prize for New Scholars, explores three essential modes of time representation. Drawing on pertinent empirical research, Montemayor delineates the potential and the limits of these levels of time organization in relation to the various internal "clockworks" involved in their making. He focuses especially on "motor-intentional time," the temporal organization necessary to coordinate locomotion with the environment, as an important intermediary stage between periodical rhythmization and conceptual time.

A different but equally triadic model is used in Peter Hancock's paper "The Battle for Time in the Brain." Hancock focuses on the parts of the brain involved in shaping sleep and explains why sleep is the outcome of their conflict and why it is so essential to human life. Once again, circadian activities, motor intentionality, and conceptual behavior come into play. The paper demonstrates how the inhibition of actual movement enacted in sleep is an important precondition for motor as well as conceptual learning and adaptation. Thus it is precisely the constraints that interaction between these structural levels imposes on the overall behavior of the organism that make these forms of learning and adaptation possible. As Montemayor emphasizes, the limitations each subsystem imposes on the others have liberating effects on the whole.

While the previous two papers concentrate on the differences between pre-conceptual and conceptual forms of temporal organization and their interactions, the next two papers look into purely physical time and the possible constraints placed on our understanding of time by the acceptance of certain models developed in theoretical physics.

Jonathan Tallant's "Memory, Anticipation and the (Un)reality of the Past and Future" explores the choice between two reflective conceptualizations, famously termed by the philosopher McTaggart the "A" and "B" theories of time. He argues, from a philosophical point of view, against the common assumption that science and especially the theory of relativity would predispose our choice of theory towards viewing time in terms of a homogeneous system (B-theory), and against the distinction of perspective enacted by structuring it into present, past, and future (A-theory).

Friedel Weinert delves deeper into the physicist's side of this argument by looking closely into the implications of the theory of relativity. This theory clearly restrains our possible choices in interpreting time insofar as it does not allow for a (universal) "privileged now." However, Weinert argues that it would be consistent with the notion of becoming and therefore with a dynamic conception of time if "becoming" were understood in terms of a transition from an indeterminate to a determinate state.

Both papers may be read as advocating caution in restricting our conceptualizations of time based on generalizations from abstract models of physics—and clearly, the first two papers of this section make a strong case in favor of accepting the reality of emerging new forms and levels of temporal organization.

In Section II: The Limits of Duration, the papers move the focus to the temporal structures employed in human society. These three papers explore the limits of acceptable duration in divergent but equally important social settings.

In her paper "Technical Reproduction and the Question of Material Duration," Heike Klippel draws on theories of reproduction by Marx and Benjamin in order to examine the limits of duration in film. Providing a brief but suggestive historical overview, Klippel demonstrates how the medium intrinsically problematizes terms such as "original" and "identical." Although it is no longer the case that several different versions of a film are produced (as occurred in the early years of the medium), films may be re-cut or restored and still lay claim to being the same film. One need only think, for example, of the seven different cuts of Ridley Scott's landmark science-fiction film *Blade Runner* or the pristine restoration of Jean Renoir's 1937 masterpiece *La Grande Illusion.* As Klippel concludes, even apparently "loss free" digital reproduction comes up against durational limits due to ever-changing technologies.

New technologies have certainly impacted on the issue of copyright, and Tyler T. Ochoa's essay "Limitations on the Duration of Copyright: Theories and Practice" takes us through a history of copyright law from its early implementations to the present day. In a comparative historical essay, he discusses the temporal side of intellectual property rights, demonstrating the different strategies employed in case law and code law and examining how their outcomes tend to converge over time. Copyright was introduced as a kind of necessary evil in the Anglo-American world and is therefore quite limited. In contrast, continental European legal thought tended to view intellectual property right as a "natural" right. In Ochoa's view, the global tendency to extend limits of copyright duration can be explained by the political economy involved in the decision-making process: since authors and publishers have much to gain by extension, while the general population does not lose as much, lobbying for extension is usually successful.

It remains to be seen whether society is really ready to wait longer and longer for free access to authored works or whether this tendency of legislation will in the end be undercut by technical means. Florian Klapproth explores the other side of waiting in an empirical study concerning short waits, typical for situations like waiting for a computer application to start up before the action can begin. His study shows that accurate information on the expected waiting time greatly relieves the emotional tension that usually is part of waiting, but that short waits may also be pleasant if they are interruptions of tedious activity.

In Section II: Creativity and Constraints, essayists explore the various ways in which the arts address temporal issues, focusing particularly upon constraints. Whether these constraints are the ticks of a clock or the beats in a measure, phase shifting or chronological disordering, endings deferred or the end of all things, they serve as spurs to creative enterprise, enabling imaginative and innovative solutions to what are often self-imposed problems. A common thread running through the first three papers is the clock and the forward arrow of time, in the second timing, in all the integral role that temporality plays in artistic endeavor.

In "David Mitchell's *Cloud Atlas* of Narrative Constraints and Environmental Limits," Jo Alyson Parker focuses on Mitchell's 2004 temporally innovative novel *Cloud Atlas*, arguing that its self-imposed narrative constraints enable Mitchell to demonstrate how past and present action might lead to a horrific future—an end to civilization. She suggests that Mitchell's themes resonate with the current debate over climate change. For Parker, Mitchell's novel serves as the literary equivalent of the Clock of the Long Now, a project designed to encourage long-term thinking and planning for the good of the planet.

In John Streamas's contribution "Closure and 'Colored People's Time,'" a different, yet related clock plays a role: the Doomsday Clock, which was created in 1947 to track the urgency of the nuclear threat and which is now tracking the urgency of climate change. According to Streamas, the Doomsday Clock is "the critical timekeeper of its apocalypse narrative"—a narrative that is predicated on the notion of individual redemption occurring after the near-triumph of evil. To people of color, for whom the near-triumph of evil persists, such an end-time narrative is at odds with the freedom-and-justice narrative, which is predicated on the notion of communal triumph occurring after struggle. In a wide-ranging argument, Streamas explains how Colored People's Time might be regarded as the very time-sense of that struggle. Indeed, Streamas envisions a resetting of the Doomsday Clock according to Colored People's Time, a resetting that would transform witnessing to analysis and dream. Both Streamas's and Parker's papers explore narratives that present alternative possibilities—what Streamas, drawing on Gary Morson, calls "sideshadowing." Narrative time thus becomes a means of re-envisioning the present reality to change the future. Respondent Robin Lucy extrapolates from Streamas's points, providing examples of Colored People's Time in African American texts. Respondent Deirdre McMahon points toward future work that might be done in exploring Colored People's Time as a strategy of resistance.

Whereas the clocks that Parker and Streamas discuss serve as tropes for the narrative issues under discussion, Carol Fischer discusses actual clocks that make an appearance on the stage. In her essay "Dramatic Time: Phenomena and Dilemmas," Fischer explains that a clock on stage measuring out the time passing on stage fits awkwardly with the real time being experienced by the audience. Ranging through a variety of plays, Fischer demonstrates that the on-stage clock is only one of many ways that contemporary playwrights foreground matters of time and explore temporal constraints. Playwrights write dialogue that questions its own temporal placement, they take the end of time as their subject, they reverse time's arrow, and they "wrinkle" time by having separate historical worlds communicate. Furthermore, updates and revivals call attention to the disparate times of their original and current performances. Drawing upon the work of Michel Serres, Fischer deals with the inherent temporal tension of live performance, wherein "the static nature of the text [is] reiterated within the dynamical event of live actors in front of a live audience." Respondent Laura Pattillo points out that the time that constrains performers might also be taken into consideration, and she draws our attention to the economic constraints that have led to an increased reliance on revivals. Respondent Kathleen Weiss finds that Fischer's essay illuminates the work of not only contemporary playwrights but those of early eras as well, pointing toward other provocative temporality-skewing plays.

The temporal tension inherent in live performance also figures in Marc Botha's essay "Evental Distension: Restless Simultaneity in Steve Reich's *Piano Phase*—Toward a Rehabilitation of the Real." In Reich's germinal minimalist composition, a simple melody played on two pianos simultaneously is shifted out of phase when one pianist begins to accelerate—performers thus observing a constraint of playing the same piece but varying the timing. The listener is thereby plunged into an "upset temporal world." *Piano Phase*, according to Botha, serves as an exemplar of art that foregrounds "questions of temporality as their principle substance." Such art enables a recuperation of the Real, and Botha elegantly demonstrates how this singular piece may have universal philosophical implications.

Just as Botha focuses upon *Piano Phase* in order to explore a philosophical theme, Helen Sills draws upon a particular musical composition—in this case, Igor Stravinsky's *Requiem Canticles*—as the focal point for a discussion of humanity's developing sense of time through the auditory system. Sills takes the reader through the stages of auditory development, demonstrating how at each stage "limits and constraints gives rise…to a qualitatively new level of temporal experience." For Sills, Stravinsky's composition marks a culmination of the process. Through a close reading of various passages, she demonstrates how Stravinsky imposes temporal constraints in order to give his listeners a sense of timelessness.

The essays in Section III: Final Questions probe the limits and boundary conditions of time as it is lived and thought about. These essays serve as a series of thought-experiments that demonstrate how our thinking about a particular limit of time may be profoundly changed by examining it from a different, pointedly unexpected point of view. These thought-experimental essays give the "final questions" they scrutinize (the end of time, death, and the basis of our knowledge of time in the first place) a new turn by posing provocative questions: What happens when we think about eschatology as a political theology and mode of lived temporality rather than a prophetic study in the end of time? What happens if we think about death as something invented by the evolution of life rather than the mark of human mortality? What emerges when the limits of naturalistic onto-epistemology are subjected to rigorous transcendental critique?

In "Pauline Eschatology: Thinking and Acting in the Time that Remains," Steven Ostovich develops a notion of "living eschatologically" and conceptualizes it as a specific form of communal or "socio-" temporality. Ostovich, by distinguishing a theological way of thinking within a theology without God, shifts the temporal grounds of eschatology from an emphasis on the (imminent) end of time to a non-linear notion of time framed as disruption or interruption. In this reading, Pauline eschatology becomes a political theology founded on the promise of justice. The temporal paradox of this eschatological political theology is that it is not defined by a concrete future event, a promise of justice to come (that is, the second coming), but can only be lived as an open time, in which one lives justly, while expecting or accepting the failure of justice to arrive. Pauline eschatology calls us to live an ethos founded on the *hos me*, "as if not": "that is, as if this were not an unjust world, and to live justly despite our lack of success in achieving justice." In a historical context defined by global conflict, depleted resources, and an uncertain future, Ostovich's essay challenges us to think deeply about our ethico-political commitments and to calls us to live life in the messianic mode, in a time that remains.

Just as Ostovich's essay shifts the treatment of a temporal limit (the eschaton) from theology to politics, from the religious to the secular, Frederick Turner's essay moves the discussion of death (the ur-limit of time) from existential-religious grounds to an evolutionary perspective. In "How Death Was Invented and What It Is For," Turner provides a naturalistic account of the emergence of death in different guises in tandem with developments in the nature of life, offering fascinating formulations of paradoxes that present themselves along the way. Turner notes that with the move from asexual to sexual reproduction, death ceases to be "something that happens to living things, but becomes something they do." Ultimately, Turner's interest in the evolutionary purpose of death emerges from a cultural context where the desire to reverse aging and overcome death is at a peak. From this standpoint, the biological event of death designed by evolution serves a 'higher' purpose: we need what Turner calls "the numinous force of death" to confer meaning on our lives. In typical fashion, Turner transcends polarized ways of thinking—here, framing death in evolutionary rather than existential or theological terms leads us to encounter and appreciate death in a whole new light, restoring and reinventing its spiritual power. Considering Turner's argument from the perspective of Daoism and Dôgen, respondent William LaFleur questions the notion of death's purposiveness, thus encouraging us to move beyond the boundaries of Western thinking about death.

The volume's final essay returns us to ISST's origins—the work of J. T. Fraser. Christian Steineck's forceful assessment of Fraser's theory functions as a sort of Neo-Kantian 'critique of Fraser's judgment,' in which he tests the "conceptual limits and methodological constraints" of Fraser's thought. Steineck focuses on the "extended umwelt principle" Fraser uses to circumscribe his notions of truth and reality within a comprehensive onto-epistemology. Here, Steineck uncovers fundamental tensions underlying Fraser's conflation—or at times confusion?—of epistemology and ontology within the hierarchical theory. Specifically, Steineck diagnoses a discrepancy between the hierarchical theory's allowance for the ideal to engage and change the real and the naturalistic or evolutionary epistemology inherent in the extended umwelt principle. Steineck's own resolution to the impasse presents an interesting distinction between the closed Umwelt (world as it is for a creature) and the open Welt (world as such). Included alongside Steineck's critical essay on Fraser are Fraser's prefatory remarks, in which Steineck's Neo-Kantian epistemology is treated from Fraser's naturalistic perspective. Together, these pieces point towards future lines of discussion, about the limits of time and what constrains our knowledge of them.

ACKNOWLEDGEMENTS

We owe profound thanks for the many reviewers who took time out of their busy schedules to provide incisive evaluations of the essays submitted to the volume. They include David Burrows, J. T. Fraser, Lolly Gasaway, Richard Hancox, William LaFleur, Carmen Leccardi, E. J. Lowe, Robin Lucy, Deirdre McMahon, Laura Pattillo, Anthony Reese, Gert van Tonder, Kathleen Weiss, and others who have chosen to remain anonymous. We feel particularly grateful that we were able to benefit from the insights of Professor LaFleur, whose untimely passing took place while this volume was in press. Thanks are due as well to Thomas Weissert, who created and maintained the volume website, thus greatly facilitating the work of us editors, and to Mary and John Schmelzer, who provided a congenial meeting-place in which we three editors could gather. Brill has provided vital support, enabling the dissemination of the ISST's ongoing work. And, finally, J. T. Fraser continues to inspire our efforts as an interdisciplinary society dedicated to the study of time.

INTRODUCTORY ESSAYS

PRESIDENT'S ADDRESS
COSMIC EPICS AND GLOBAL ETHICS

PAUL A. HARRIS

I realize that it's just after your first dinner of this conference—perhaps this was even your first ISST dinner—and that just when you were beginning to meet new people and/or renew old friendships and simply enjoy yourselves, you have been called to chapel to submit yourselves to the President's sermon. By the time I'm through with you, you'll miss my wonderful presidential predecessor Remy Lestienne dearly, for in the best scientific mode, he would sketch a grand vision of our conference theme in a few short moments in the dinner hall. In fact, you should be aware that this morning a family member called me "ambidextrously verbose." But please know that I, too, miss Remy Lestienne's Presidency—I always was more comfortable with the Vice-President title, as it connoted being a Master of Vices—which comes more naturally to me than virtuous presidential preaching—I prefer exceeding limits and shredding constraints, but since I'm up here and you're all out there, I'll try to respect the constraint I imposed on myself, namely to limit myself to talking for no more than an hour or two.

In all seriousness, I do hate to interrupt the flow of after-dinner conversation, because so much of what makes ISST distinctive happens over meals and after hours. I remember three years ago, just before dinner on our first evening in Cambridge, Tom Weissert came bouncing up to me and burst out, "It's working! It's happening!" "It"—that elusive affective and intellectual pleasure ride—has a way of happening at ISST, because of—or perhaps in spite of—everyone's best efforts. We also owe the advent of this conference, and the continuing intellectual exchange which distinguishes the ISST, to the persistent, inspiring presence of our Founder, J. T. Fraser.

Even though it is after dinner, I feel that my main task this evening is to set the table: to lay things out in a hospitable manner that encourages animated, engaged exchanges to follow. If, in lighting the candles on this table, I manage to kindle a fire or two, so much the better. My talk has three parts. Part One, "Cosmic Epics," presents a short version

Jo Alyson Parker, Paul A. Harris and Christian Steineck (Eds), Time: Limits and Constraints, pp. 3–17

of a grand, sweeping story of time that takes us to the limits of our temporal knowledge. Cosmic Epics integrate scientific cosmological research and the history of terrestrial life into a tale of evolving complexity where human beings constitute a tiny blip at the end—but, as it turns out, we're also a lippy blip of unprecedented complexity, poised at a critical juncture in our development. Part Two, "Global Ethics," considers different ways to understand the evolution of complexity and connect it to an ethos that might lead us to contemplate and create sustainable futures for humanity. Part Three, "Whence ISST?," asks how Cosmic Epics and Global Ethics might influence how we conceive of our work in the Society and takes note of recent discussions relevant to these concerns.

1. Cosmic Epics

Time, limits and constraints: from a human perspective, our conference theme immediately points us towards chronometry, in that what we define as temporal limits or constraints are a function of the reach of our instruments for measuring time. To take one example, consider how people have imagined the origin of time. For around 1500 years, many human beings accepted the Bible as the most reliable tool for calculating the date of God's week of creation—scholarly debate and counting generations named in the Old Testament dated the event around 4000 B.C. The Heavens, earth, and humans were all born within days of one another. Now, cosmologists date the birth of the universe at about 13.7 billion years ago. Our sun and earth appear 9 billion years later, and the anthropological record states that human history in the form of *Homo sapiens* only begins 250,000 years ago. Now while I suspect that there are few card-carrying creationists among us this evening, I would also suspect that many of us are much more familiar with the Biblical dating of creation than with separate scientific calculations of the age of the universe, solar system, and human species.

You all have a basic idea of how creation and the populations of the earth unfold in the Book of Genesis. But in case you don't know the basic picture of cosmic, terrestrial, and human history as scientific knowledge depicts it today, here is a timeline that highlights a few major events.

A TIMELINE OF COSMIC HISTORY

- About 13.7 billion years ago, the universe, contracted at a single point, exploded. In a few seconds, it expands to the size of a galaxy.
- 300,000 years later, the universe has cooled to a few thousand degrees Celsius, and electrons are captured by protons to form the first electrically neutral atoms of hydrogen and helium.
- Starting about 12 billion years ago, the first stars begin to shine; billions of stars cluster into galaxies, and all the elements up to iron form in the interior of stars, and all the elements up to uranium are forged in supernovae, huge explosions of dying stars.
- Fast-forward several eons. About 4.6 billion years ago, clouds of stardust, the debris of older stars, form the sun, earth, and solar system.
- About 3.5 billion years ago, the first living organisms appear on earth.
- 1.5 billion years ago, first complex cells and multicellular creatures form.
- 600 million years ago, Cambrian era creatures appear and ozone layer forms.
- 65 million years ago, dinosaur extinction and predominance of mammals begins.
- 7 million years ago—the first hominoids.
- 2–1.5 million years ago, *Homo habilis*, first member of our genus.
- 250,000 years ago, *Homo sapiens*.
- 100,000–13,000 years ago, hominization occurs as humans migrate to different parts of globe.
- 5,000 years ago, first cities and agrarian civilizations.
- 500 years ago, global system of exchange emerges.
- 200 years ago, Industrial Revolution.

The story in Genesis, though sketchy on details and not exactly forthcoming in explaining exactly how it all worked, has a nice ring to it and appeals to our minds on cognitive and perhaps spiritual levels. By contrast, the scientific timeline resists any real understanding. Trying to wrap one's head around the temporal spans on this timeline is—well, it just draws a blank, doesn't it? Two characteristics of our plot outline are immediately apparent though: this is a logarithmic tale, whose highlights appear in ever shorter spans, and the details get less fuzzy as we progress.

It would be hard to condense the scientific story into six days of creation. But in order to make it at least a bit more manageable mentally, let's reduce the timeline by a factor of a billion and retell the tale. This is David Christian's heuristic strategy; I take this timeline from his book *Maps of Time: An Introduction to Big History*. Now the story goes:

COSMIC TIMELINE REFIGURED:
ONE "YEAR" = ONE BILLION COSMIC YEARS

- 13 years ago, Big Bang
- 12 years ago, stars and galaxies
- 4.5 years ago, solar system (sun, earth)
- 4 years ago, first organisms
- 7 months ago, first multi-celled organisms
- 3 weeks ago, dinosaur extinction
- 3 days ago, first hominoids
- 50 minutes ago, *Homo sapiens*
- 5 minutes ago, agricultural communities
- 3 minutes ago, urban communities and literacy
- 15 seconds ago, global system of exchange
- 6 seconds ago, Industrial Revolution

Even this try leaves me a bit cold, though. Would it help me if I calibrate this timeline to fit the 45 years of my life? You can try this at home, using your own age as a yardstick! Taking this tack, my birth coincides with the Big Bang—how's that for cosmic narcissism? I was four years old when the first stars and galaxies evolved, and the solar system appeared when I was about 15 and a half; when I was 43, multicellular life began stirring, and just the middle of this past May, the dinosaurs died. This evening, while you were eating dinner, *Homo sapiens* appeared—we did wonder who those other creatures were in OUR dining room, didn't we?—and while you shuffled into this room and found seats, human societies formed; and since I started reading this paragraph, global systems of exchange and the Industrial Revolution have occurred. As of now—well, in the time it takes to say now, someone playing Pong on an Atari system suddenly found themselves with a machine in their lap enveloped in a system of instantaneous global communication that encircles the earth. (A large portion of this latter transformation was engineered just over a few hills from here, in Silicon Valley.)

Obviously, a merely factual account is as dry and dead as dust. In order to be fully compelling and do justice to the majestic scope encompassed by these facts, the bare bones of this plot need to be arranged into a coherent skeleton, which in turn gets fleshed out by an engaged, hopefully enthralled, human listener or reader. Clearly, the alchemical operation needed to bring about such a scenario exceeds the limits and constraints of my time with you this evening. We are not on a vision quest, though the Californian coastal setting does tempt one in this direction. Thankfully, the scientific story has been compellingly told for a wider audience by several people over the past few decades, including physicist Brian Swimme writing with ecological theologian Thomas Berry, and historians Fred Spier and David Christian. There is the priceless parody by cosmologist Eric Schulman, *A Briefer History of Time: From the Big Bang to the Big Mac.* And of course this story is found throughout the voluminous writings of our founder, J. T. Fraser. These authors have penned what I think could be justly termed Cosmic Epics: narratives that encompass the astrophysical description of the universe's history, the biochemical depiction of life's evolution, and human history.

I would like to highlight briefly some guiding insights gleaned from two recent Cosmic Epics. The first is Joel Primack and Nancy Ellen Abrams's *The View from the Center of the Universe.* I assure you that even if they were not present and going to give a presentation here tomorrow evening, I would have singled out their work. Since they are here though, I will offer only a very pared-down version of their vision. They begin their story bemoaning the fact that most contemporary educated humans envision themselves as tiny points in an indifferent, amorphous, unknown universe. "Cosmically homeless," they write, "our culture over the centuries downgraded the importance of having a cosmic home."[1] Furthermore, they assert that "this widespread cultural *indifference to the universe* is a staggering reality of our time—and possibly our biggest mental handicap in solving global problems" (6).

Primack and Abrams present scientific cosmology not simply to educate readers but to provoke a genuinely spiritual awareness in us of the universe and our place in it. Careful to keep the spiritual

[1] Joel R. Primack and Nancy Ellen Abrams, *The View from the Center of the Universe* (New York: Penguin, 2006), 4. Subsequent references to this edition are included in the text.

distinct from religions and specific belief systems, they define the spiritual as simply "*the relationship between a conscious mind and the cosmos*" (286; original italics). An attuned spiritual awareness would "think of reality as funneling from great galaxy clusters into us and spreading cell to cell, then soaring inward to the molecular level, the atomic, quantum levels—with our humanness the fulcrum at the center of the entire process." Such thought-experiments are important because they enable us "to experience the universe from the *inside*.... Until we find our symbolic place in the universe, we will always misinterpret ourselves, feeling as though we are outside..." (288). Joel and Nancy's mantra is that humans need to "think cosmically, act globally," because, as they quote Einstein as saying, "Problems cannot be solved at the same level of awareness that created them" (11).

2. Global Ethics

Joel and Nancy's call for humans to "think cosmically, act globally" exemplifies the way in which Cosmic Epics pursue an ambitious purpose: to pave the way to thinking about Global Ethics. The second Cosmic Epic I would like to bring to your attention, Eric Chaisson's *Epic of Evolution* (2006), makes the turn from cosmic history to global ethics very differently, in a way that I find telling for several reasons. Chaisson, a cosmologist, bases his book on an interdisciplinary course he has taught at Harvard since the 1970's. The body chapters in *Epic of Evolution* divide cosmic history into seven epochs, whose progression you could imagine traversing the timeline I presented earlier: Particle, Galactic, Stellar, Planetary, Chemical, Biological, and Cultural.

For our purposes, the most important part of Chaisson's book is his conclusion. There, Chaisson parses the periods of evolution of complexity in a much more pointed way, which situates humanity at a crucial cosmic crossroads. He discerns three essential Eras in cosmic history that mark changes in how energy has been propagated and utilized: the Radiation Era, the Matter Era, and the Life Era. During the Radiation Era, radiation dominated matter: "Vast amounts of radiant energy produced a spectacularly bright fireball inside which no atoms, stars, or ordered structure of any kind could have formed."[2] The Mat-

[2] Eric Chaisson, *Epic of Evolution: Seven Ages of the Cosmos* (New York: Columbia University Press, 2005), 434. Subsequent references to this edition are included in the text.

ter Era began "when matter first began coalescing a few thousand centuries" after the Big Bang—"and matter has dominated radiation ever since, successively and successfully forming galaxies, stars, planets and life" (434). In other words, the Matter Era comprises nearly the whole of cosmic history.

But now, Chaisson believes, we are in the midst of a transition into what he calls the Life Era. The Life Era marks a shift where living beings gain the ability to manipulate matter and thus change the way that biological evolution evolves: "The physicist unleashes the forces of Nature; the biologist experiments with the maps of genomes; the psychologist influences behavior with drugs. We are, in fact, forcing a change in the ways things change" (435). The Life Era thus "coincides not with the origin of life itself, not even with the emergence of humanity or consciousness," but rather emerges when at least some portion of humanity is able to control genes and the environment, to "tamper with matter" and "manipulate life itself" (435). Situated in the onset of the Life Era, we humans "stand on the verge of slowly becoming a meaningful factor in the future evolution of the Universe. This is a transition of astronomical significance—a fundamental change from matter's dominance to life's dominance—indeed the dawn of the second great transformation in all of history" (436).

These are heady words, but Chaisson does not see humanity necessarily headed in the right direction. He acknowledges that overpopulation, nuclear warfare, genetic degeneration and environmental pollution, among many other ills, constitute global problems unprecedented in human history. It remains "unclear," he writes, "if we humans have the wisdom to achieve and sustain th[e] lofty plateau [of the Life Era] in the cosmic hierarchy" (440). Such wisdom will only come through the formation of a "worldly ethic," which would impose a constraining "mandate for society to embrace global morality and planetary citizenship as a means to survival." Such an ethical mandate would emerge, and allow us to move towards an Ethical Epoch, in Chaisson's view, "only when we harmonize the agendas of science, philosophy, and religion… [in a] cosmic-evolutionary scenario" (440).

Well, if that's all it takes… we ought to be able to tidy that little matter up by the end of the week. Californian kharma is inherently collective—how many Californians does it take to change a lightbulb? 100—1 to do the job, and 99 to share the experience. Seriously, the clash between scientific and religious worldviews may well constitute what J. T. Fraser would term an "irresolvable conflict" facing global humanity. In Fraser's view, irresolvable conflicts that define one temporal scale cannot be worked out on that scale, but must be contained by a further level that contains without resolving the conflict—his nuanced version of Einstein's dictum that "Problems cannot be solved at the same level of awareness that created them." But humans cannot simply leap or log into some higher level of awareness to work out the conflict between secular, scientific thinking and spiritual, religious belief. Instead, we have to negotiate passages between them as a means to work towards some emergent synthesis. Perhaps, as Ken Wilber, the transdisciplinary integral philosopher, succinctly states: "The most pressing political issue of the day, both in America and abroad, is a way to integrate the tradition of liberalism with a genuine spirituality."[3] Wilber asserts that in championing an individualism built on the pursuit of economic and political freedoms, liberalism ends up imposing an "economic tyranny" in which "anything spiritual or religious tends to be quite embarrassing" (xv). Conversely, Wilber describes Conservatism as founded on civic humanism, communal standards, and religious values, but imposing a cultural tyranny of its own. How, then, could these camps be integrated? Rephrasing the question, how can scientific rationalism and individualism be reconciled with a shared spiritual vision?

Cosmic Epics, as we saw in the case of Primack and Abrams, explicitly seek to reconcile scientific and spiritual outlooks. But I think Chaisson's conclusion underscores a different problem these narratives run into in attempting to create such a reconciliation: the cosmic story, and indeed the ability to tell it at all, is completely science-driven and depends on the successes of techno-science—whose very successes play an integral role in our current global crisis. In his account, spirituality and ethics arise after the fact, as a coda of sorts—Chaisson's tale concludes with humanity at the dawn of the Life Era; an Ethical Epoch remains only a fuzzy formation somewhere at or beyond the temporal horizon. It's as if in telling a terrific success story, one reaches the end and suddenly realizes that success has some terribly unwanted consequences, and that a new story has to begin, with a very different starting point. I'm reminded of the scene in *Dr. Strangelove* when Peter Sellers as the President asks George C. Scott's General Buck Turginson if the plane carrying the bomb can make it through Russian

---

[3] Ken Wilber, *The Eye of Spirit* (Boston: Shambala, 1997). Subsequent references to this edition are included in the text.

defenses.... As Chaisson fully realizes, humanity stands in dreadful need of evolutionary transformation: we need to develop a collective will to think and act cooperatively rather than competitively, ethically rather than greedily.

What steps may be taken now to work towards such a lofty, but absolutely vital goal? I believe that for the cosmic epic narrative to awaken a spiritual and/or globally ethical perspective in readers, a critical first step is to modify how we conceptualize the evolution of complexity, how we understand it on the level of our terrestrial, socio-cultural lives. It is indisputable that, empirically speaking, cosmic history charts a path of evolving complexity. But the evolution of complexity, as it operates in socio-cultural domain, may be given different inflections and emphases, which imply very different ethics.

Currently, the evolution of complexity is understood in predominantly neo-Darwinian terms, under the aegis of what is known as the modern evolutionary synthesis. The Darwinian evolutionary model centers on natural selection, which involves the transmission of genes from one generation to the next. Over the long haul, the most successful adaptations have the best chance to survive—hence the Social Darwinist phrase "survival of the fittest." These days, the evolutionary model merges with socio-economic thinking through complexity theory and the study of self-organizing systems. Evolution and economics are both seen in terms of free-markets in which self-interested, rational individuals competing for resources and against one another, pursuing immediate goals, all self-organize into orderly, complex systems. The invisible hand of Adam Smith is replaced by self-organization; the economy of globalization is said to provide optimum utilization and distribution of resources. In this understanding of how complexity operates and evolves, the cooperative and one might say ethical dimension is reassuringly taken care of, provided for by an opaque mechanism embedded in the models of the process itself. Chaisson's Life Era seems the natural outcome and product of this process, where the evolution of complexity, fuelled by agents seeking to transform resources and each other for their own gain, culminates in Life's ability to directly manipulate matter.

But competition is of course only one side of the story of the evolution of complexity. Cooperation is the other side—complex systems evolve through intense interdependence and exchange among their components or members. We need to complement our neo-Darwinian picture of evolution with a more ecological vision and find a way the two could be integrated. The "survival of the fittest" mantra could be countered by Gregory Bateson's ecological axiom that "the unit of survival is the organism in its environment." Seen in this light, the evolution of complexity unfolds as an interlocking relationship between agents and environments embedded in feedback processes. The headlong rush forward of technologically driven evolution takes shape as a positive feedback loop going out of control; the climate crisis emerges as a negative feedback process produced by humanity's self-serving successes.

Signs of a shift towards more ecologically-based thinking are visible in debates within evolutionary biology. In a recent *New York Times* Science column, Douglas H. Erwin, a senior scientist at the Smithsonian's Museum of Natural History, speculates that the "new field of developmental evolutionary biology," dubbed "evo-devo," may mark a paradigm shift in evolutionary theory.[4] Evo-devo argues that major genetic changes in evolutionary history are driven less by natural selection than by mutations. In this view, mutations occur as species transform their environments and are in turn transformed by them. Evolution proceeds not only through internal adaptive innovations in organisms, but by evolving the possibility spaces of evolution to operate in—an interlocking, constant feedback process of mutual transformation. Ultimately, perhaps the evo-devo model will move us towards a model of being transformed rather than transforming something else, of cooperative rather than competitive complexity.

Let me stress here that it is not a question of which account of the evolution of complexity is more true—evolution is surely both competitive and cooperative—but how these views contribute to our cultural evolution. If the human species is going to shift the course of the evolution of complexity, we need to define ourselves as evolutionary creatures less in terms of our ability to affect and transform environments and others and more as creatures distinguished by their capacity to be affected and transformed. This sounds idealistic, but only humans can engineer such changes in themselves simply by adopting new ideals. The capacity of creatures to be affected is of course a function of their umwelt. Unlike other biological entities, humans have the capacity to be transformed not only by environmental factors but

[4] Douglas H. Erwin, "Darwin Still Rules, but some Biologists Dream of a Paradigm Shift." *New York Times* June 26, 2007.

symbolic expressions—the most effective of which, in the domain of collective learning, is an overarching narrative within which people situate themselves and on whose basis people shape their actions. To instantiate oneself in a narrative and then assume the role it designates is something of a spiritual act—it is in one sense how religions function in the minds of believers.

And I think that, to move towards a global ethics, this process of transforming oneself by placing oneself in a story must inform our notion of ethics as well. Generally speaking, ethics involves recognized rules of right conduct in given contexts. The trouble is that ethical debates all too often arise belatedly, as attempts to devise corrective measures to be taken in relation to a situation already established, if not out of control. People need to engage ethics not in reaction to a particular debate, but actively assume it as a vocation, an integral part of their lives. Ethical subjects are not born but made, or, dare I say, reborn: they only become ethical subjects by undergoing a transformation and championing with selfless enthusiasm a universally applicable truth. This is the rather radical position taken by philosopher Alain Badiou, who defines truth as an event marking a break from a given situation, and the ethical subject as one who displays fidelity to that truth. In so doing, the subject becomes "in excess of him[or her]self, as the fidelity passes through them."[5] Ethical subjects are different from individuals driven by selfish genetic motivations because the purpose or cause they serve takes them outside their narrow interests. Badiou argues that ethics must proceed from universally applicable premises of the True and the Good, rather than from one's trying to right some perceived wrong. While many prominent ethical paradigms presume that ethics must stave off existing inequalities or correct injustices, and hence posit an impoverished or inferior 'other' in relation to whom one is ethical, Badiou calls for ethics to proceed from a positive passion to uphold a truth.

While unquestioningly embracing Badiou's project might take us in any number of sketchy directions, I find it worth attending to in this context because it turns our eye towards developing genuinely global ethical perspectives. It might be interesting to recast Badiou's

[5] Alain Badiou, *Ethics: An Essay on the Understanding of Evil*, trans. Peter Hallward (New York and London: Verso, 2006), 45. Subsequent references to this edition are included in the text.

premise that ethics must start from *universally* applicable truths in cosmic terms—that is, to think of 'universally' not as connoting timeless metaphysical truth but a principle in harmony with a perspective informed by an informed view of the universe. Think cosmically, act globally.

3. Whence ISST?

What do the foregoing remarks on cosmic epics and global ethics imply on the comparatively humble, but perhaps increasingly significant level of human intellectual exchange—the umwelt we will inhabit this week? If, in order to evolve in an ethical direction, humans need to awaken and affirm their capacity to be transformed by others, then communities explicitly designed for truly interdisciplinary exchange assume a privileged place. Since my first experience of the ISST in 1992, I have felt deeply attracted by its collective ethos—scholars pursue and present their specialized findings, but do so in order to enter into interdisciplinary dialogue that sharpens and perhaps questions the assumptions at play in their papers. In my welcome letter to new members, I invite people not only to think of what they can contribute to the Society, but also to attend to how their work may be changed by participating in it. I envision the ISST as a cooperative rather than competitive learning community, whose interdisciplinary nature ensures that its goals and encompassing vision keep evolving. I would further propose that as Timesmiths it behooves us to become familiar with the Cosmic Epic vision of time, which provides a supple top-down dynamic which may inform our individual work, which in turn will stimulate new ideas about the grand view.

Before continuing, let me interject an aside of a more concrete nature. I would remind members that Editor Marlene Soulsby called in the recent issue of *Kronoscope* for submissions of a pedagogical nature, in which members share ideas and practices concerning what defines Time Studies. Such contributions could comprise a significant, bottom-up means to consider what our larger ends might be now and in the future.

How does its interdisciplinary nature mold ISST's identity as a cooperative rather than competitive intellectual system of exchange? An excellent example is found in the discussion about interdisciplinarity on ISST's listserv, in its affiliated journal *Kronoscope*, and in recent issues of its newsletter *Time's News*. Discussions at the 2004 Cambridge conference generated a lively, frank exchange of views on our listserv concerning disciplinary specialty and the privileging of different disciplinary models for synthesizing thought on time. In *Time's News* 36 (February 2005), an invited essay by Nicholas Tresilian offered eloquent ruminations on the "interdisciplinary question," which in turn provoked rich responses in *Time's News* 37 (February 2006) from ISST members Frederick Turner, Troy Camplin and Steven Ostovich. The top-down, global visions articulated by Tresilian, Turner, and Camplin were countered by Ostovich's plea that a top-down or what he called "hard" interdisciplinarity be complemented by a "soft interdisciplinarity [which] is literally utopian—that is, no place and every place, not a proper space at all but between spaces, where scholars speak from their disciplines but listen across boundaries (and where translation is the task, not developing a new language)" (9).

Time's News 38 (February 2007) recapitulated this dynamic interplay between hard and soft interdisciplinarity. J. T. Fraser provided what could be termed an updated vision statement of ISST in his Founder's Essay "The Integrated Study of Time: A Call for Reciprocal Literacy." There Fraser presciently observed that, "If a significant portion of the earth's societies cross a certain threshold of complexity in their communications systems and in their interdependence, the global socialization and evaluation of time will subsume the office of the person as the primary agent in controlling his or her time" (14). Negotiation of this collective, integrated study of time necessitates "reciprocal literacy" among its members—a characteristic that begins to develop precisely through the "soft" interdisciplinary exchanges marked by Ostovich. The overall view of ISST that emerges here was summarized by Tresilian in his 2006 "Postscript" to "The Interdisciplinary Question," where he argued that "the relationship between interdisciplinary and specialist thought needs to be seen not as a matter of hierarchy but a sequence of feedback loops, in which each successively fertilizes the other…in a virtuously iterative cycle" (10).

In his Response to the Founder's Column in *Time's News* 38, Tresilian identified especially fruitful ways that ISST could aim to function in order to intensify the productivity of such feedback loops. Tresilian urged us to consider that "if ISST is going successfully *to carry on the work which J. T. Fraser began*, then it needs to become more pro-active in seeking out contexts especially favourable to the interdisciplinary study of time." He further specified that "ISSTers will need to become more acute at identifying *time's trouble-spots*—regions of our collective experience where conventional object-based models of the universe break down, opportunities for the emergence of new models and meta-languages based on a perception of the world as a mutually implicated process" (6–7; original italics). Work that identifies time's trouble-spots needs to be both critically analytic and comparatively synthetic—a mode of thinking and writing that puts "soft interdisciplinarity" into active practice. An especially salutary instance of such work may be found, I believe, in Mark Aultman's contribution in the latest issue of *Kronoscope*, where he situates a confrontation between Fraser's integrative theory of time as conflict and Turner's vision expressed in his 2006 book *Natural Religion* within an imagined exchange between a human thinker and God.

I would like to conclude these remarks by extending an open invitation for contributions to this kind of exchange and work among ISST members. Such contributions might materialize in messages to the listserv, submissions to *Time's News*, or proposed essays for *Kronoscope* or possibly volume XIII of *The Study of Time*, the published proceedings of this conference. Coeditor Jo Alyson Parker and myself encourage you to listen to papers presented this week with an integrative ear. We will consider publishing writings that synthesize work presented here in order to raise questions and issues implicit in diverse papers, but explicitly identified in none of them. We have no hard and fast stipulations for the form such writings might take—such is the nature of utopian undertakings, which do not nestle themselves in a preconceived place! The idea of making room for such contributions builds on an innovation introduced by coeditors Parker and Michael Crawford in the proceedings from Cambridge, *Time and Memory: The Study of Time Volume XII*, where the remarks of some outside reviewers of submitted essays were published in order to extend the discussion of ideas expressed by ISST presenters.

It is tempting to conclude this avowedly idealistic, perhaps overreaching address by asking that in a moment of silence we all pause to reflect on our aspirations for this conference—what outcomes might we hope for, both subjectively and collectively? Don't worry, I won't put us in such a self-conscious and potentially embarrassing position—but I would recommend that you carve out small increments of time this week to contemplate these questions. In the meantime, we can all return to making "it" happen—that affectively joyful conversation that enlivens the lifeblood of the International Society for the Study of Time.

Thank you.

References

Badiou, Alain. *Ethics: An Essay on the Understanding of Evil.* Trans. Peter Hallward. New York and London: Verso, 2006.
Chaisson, Eric. *Epic of Evolution: Seven Ages of the Cosmos.* New York: Columbia University Press, 2005.
Erwin, Douglas H. "Darwin Still Rules, but some Biologists Dream of a Paradigm Shift." *NewYork Times* June 26, 2007.
Primack, Joel R. and Nancy Ellen Abrams. *The View from the Center of the Universe.* New York: Penguin, 2006.
Wilber, Ken. *The Eye of Spirit.* Boston: Shambala, 1997.

FOUNDER'S ADDRESS
CONSTRAINING CHAOS*

J. T. Fraser

Summary

With the help of the integrated study of time, this paper traces the evolution of causation and of time. Then, based on what has been learned and assisted by a critique of the idea that our experience of passage is rooted in cosmic entropy increase, it suggests that the human experience of time derives from the constitutive conflicts of matter, life, and personal and collective human identities. It then shows that all of these conflicts derive from the evolutionary constraints upon the primeval chaos.

1. The Evolution of Causation and of Time

Causal relations are constraints that govern the manner in which events may be connected. This section traces the evolution of causation-as-constraint and identifies the corresponding stages in the evolution of time. We begin—in the beginning.

Contemporary understanding associates the oldest form of causation with the origin of the universe: that is, with a very big bang of a very small and very hot dot. That dot was small enough to have passed through a contemporary atom, yet its mass is believed to have been the same as that of the universe today, which is a hefty 1054 grams, its temperature 1032 degrees centigrade. That rare object fits physical cosmology no less the mystical visions of William Blake who reminded us that "All deities reside in the human breast."[1]

I propose to think of that massive hot dot as an object of absolute chaos, of pure becoming, in which time, space, electromagnetism, and gravity were merged into what John Wheeler called a quantum foam.[2] The dot cosmos is an idea to which we assign past reality because it

---

* An earlier version of this paper appeared in *KronoScope* 7:2 (2007): 121–35.

[1] "Marriage of Heaven and Hell," in *The Portable Blake*, ed. Alfred Kazin (New York: Viking Press, 1969), 256.

[2] John Archibald Wheeler and Kenneth Ford, *Geons, Black Holes and Quantum Foam* (New York: Norton, 1998).

Jo Alyson Parker, Paul A. Harris and Christian Steineck (Eds), Time: Limits and Constraints, pp. 19–36

fits our current model of the universe.[3] And, for reasons to be given later, I postulate that the constitutive features of this thought-object—specifically, its state of pure becoming—remains the permanent substratum of the universe. In other words, pure becoming thus lurks beneath—or in—all natural phenomena and serves as the primeval pool of all potentialities. I imagine it as the state of the universe before about 10^{-45} sec. after the theoretical zero point of time. This very short period is known as Planck Time.[4] I postulate that that the primeval chaos filters through all processes, taking different forms in the worlds of matter, life, and humanity. It is the unpredictable, the creative element in nature. It should be recognized as creative, whether it leads to changes judged constructive or destructive from the human point of view.

The absolute chaos I am talking about must be distinguished from the chaos of chaos theory. Namely, the glory of chaos theory is that it can show how a deterministic, totally ordered, and predictable system can give rise to totally unpredictable conditions. But the dot universe was not preceded by a world of any form of causation or causations that were then followed by a deterioration of their order. The history of the cosmos is along the opposite direction. It began with pure becoming or absolute chaos out of which it evolved increasingly more complex forms of permanence or being.

How may this awesome drama be interpreted through the methods of physics and described in its language?

First, it is necessary to agree that speaking of the temporal locations of events so close to cosmogenesis makes sense, as is the use of units of time derived from the astronomical processes of mother earth: years, days, hours, minutes, and seconds. This way to look at the microscopic periods of time is to acknowledge that they are artifices which permit us to accommodate, within our noetic reality, conditions that are knowable only through abstract reasoning, that are unmeasurable but not uncalculable. Once these artifices are accepted, one may enter

[3] In the words of the cosmologist Abbé Georges-Henri Lemaitre, inventor of the Big Bang theory, "the beginning of the world happened a little before the beginning of space and time." See "The Beginning of the World from the Point of View of Quantum Theory," *Nature* 127 (1931): 706.

[4] This is the lower boundary of time, beneath which a theory of quantum gravitation—yet to be formulated—has to replace all theories of the macroscopic physical universe.

and admire physical cosmogenesis. Not surprisingly, that process is divided into periods for which different laws of nature apply. Thus quantum cosmology is appropriate for the cosmos between the end of the Planck period I mentioned: that is, from about 10^{-45} sec. to about 10^{-11} sec. after the theoretical zero point of time. Between 10^{-11} sec. and about one hundredth of a second, particle cosmogony applies. Thereafter standard physical cosmology applies. Temporal positions of events during this history are identified with the help of imagined "presents" or "nows," with respect to which futures and pasts may acquire meaning. Again, the time units employed are those of mother earth: years, days, hours, minutes, seconds. This artifice accommodates, within our noetic reality, conditions that are knowable only through abstract reasoning and that are alien to our senses.

Beginning perhaps 10^{-35} sec after the imagined zero point of time, the universe became populated by objects we know as elementary particles. I prefer to call them particle-waves. They formed distinct families, such that each member of any of the families was indistinguishably identical to all other particle waves of its kind. The only way the behavior of sets of identical objects may be described is statistically, that is, probabilistically. It follows that the earliest, the most primitive constraints upon the manner in which events could be connected had to be *probabilistic causation.*

Einstein could never accept the probabilistic nature of causation in the atomic world, a view represented by his often quoted but hard to verify utterance that "God does not play dice with the world." I have no way verifying whether God does or does not do so. But quantum theory demonstrates that nature certainly does. In the evolution of causation, whose steps I am tracing, probability is the oldest, the most primitive form of constraint applied to the primeval chaos. It is the only form of causation in the quantum world, and it remained one of the forms of constraints-as-causation in all later, more complex integrative levels, to wit, those of solid matter, life, human minding, and human society. One may say that probability remained the surviving ancestor of all later causal connectedness.

With the expansion and hence cooling of the universe, immense collections of elementary objects froze out and gathered into the early galaxies. There, by the law of large numbers formulated by Swiss mathematician Jakob Bernoulli, the probabilistic laws of the particulate world gave rise to the *deterministic causation* of ponderable matter.[5] The new form of causation did not replace its probabilistic ancestor but was added to it as a new constraint. Now there were two constraints upon the ever-present primeval chaos, and, because of the nested hierarchical organization of nature, there may be no deterministic causation among events without subsuming probabilistic causation as well as chaos.

By 12.5 billion years after the Big Bang the hot dot became a very cold immensity and gave rise to a yet newer way of connecting causes with their effects. It was *organic intentionality*, a feature and hallmark of the life process. How did life and organic intentionality come about?

There are good reasons to believe that our nearest nonliving ancestors were crystals. The idea may be traced at least to the Russian biochemist A. I. Oparin and the English scholar Joseph Needham. But the first detailed formulation of that idea, acceptable to contemporary biology, is due to A. G. Cairns-Smith.[6] See also my review paper of this subject.[7] Let me now sum up this view of biogenesis with brutal brevity.

Convincing reasons may be adduced in support of the idea that, some four billion years ago, certain crystalline structures, submersed as they were in the rich, oscillatory spectrum of the earth, became subject to natural selection and that such selection favored systems that could maintain their growth and decay processes in stable equilibrium and, through that balance, protect them from internal and external perturbations. Such systems could display self-directed purposes in terms of their needs and, hence, could provide distinctions between future and past with respect to a now, where the "now" could be defined by the instant-to-instant maintenance of balancing of their equilibrium. Organic intentionality was then favored because it helped satisfy the needs of the system by making it possible for it to provide

---

[5] Jakob Bernoulli, *Ars Conjectandi* (1713), reprinted in translation as "The Law of Large Numbers," in James R. Newman, *The World of Mathematics* (London: Allen and Unwin, 1956), 1455.

[6] See his "Beginning of Organic Evolution," in *The Study of Time IV*, ed. J. T. Fraser, N. Lawrence, and D. Park (New York: Springer Verlag, 1981), 15-63. See also his *Genetic Takeover and the Mineral Origin of Life* (Cambridge University Press, 1982) and his *Seven Clues to the Origin of Life* (Cambridge: Cambridge University Press, 1985).

[7] "Time and the Origin of Life," *Time and Time Again: Reports from a Boundary of the Universe* (Leiden: Brill, 2007), 135–50, especially "The evolution of the primeval clockshop," 40.

for contingencies beyond the present instant. Organic intentionality, as a form of causation, did not replace probabilistic and deterministic causations but was added to them as a third type of constraint upon pure becoming.

During the eons between about two million and about 40,000 years ago, a fourth constraint emerged, one that is unique to the human mind. "Mind" is a noun and hence it suggests an object in space. Instead of implying such an object, I prefer to think of "minding." The word "minding," is traceable to the tenth century, when it pertained to remembering, thinking, attending, intending. One must have a human brain to do minding, but it is no more necessary to have a body part called "mind" for remembering or intending, than a body part named "quilt" for quilting or "wink" for winking.

I would like to represent the constraint unique to minding may by a line from *Troilus and Cressida*: "This is the monstruosity in love, lady, that the will is infinite, and the execution confined, that the desire is boundless, and the act slave to limit."[8] The causation appropriate to minding is *noetic intentionality*. Through the symbolic transformation of experience, noetic intentionality expands the temporal world of organic intentionality. It makes it possible for humans to perceive the world, with themselves in it, in terms of open-ended futures and pasts. The nested hierarchical organization of nature still holds, of course. There may be no noetic intentionality without the presence of organic intentionality plus deterministic causation, probabilistic causation and—the ever present—pure becoming,

The next step in the evolution of causation was *collective human intentionality*. It functions through religious, political and cultural means. As one would expect, there can be no collective intentionality without the presence of its ancestral forms of causation, identified.

In his *Space, Time, and Spacetime*, Lawrence Sklar, distinguished philosopher of science, judged doubtful the whole point of searching for a physical theory of time different in detail from a causal theory of time.[9] The hierarchical theory of time shares this view. Specifically, by recognizing a nested hierarchy of qualitatively different causations,

---

[8] William Shakespeare, *Troilus and Cressida*, 3.2.85–89.

[9] Lawrence Sklar, *Space, Time, and Spacetime* (Berkeley: University of California Press, 1974), 356.

as just detailed, it identifies a nested hierarchy of qualitatively distinct temporalities. Let me describe them, starting from the top.

The time of collective intentionality is *sociotemporality.* It is the way a culture experiences, thinks of, and represents time. It is created through a dialectic between consensus and dissent pertaining to those events that secure or demolish the identity of the group. It concerns social futures and pasts, with respect to a collectively created and maintained *social present.*

The time of noetic intentionality is *nootemporality*, the temporal world of mature humans of historical times. Within its horizons a person may establish his or her expectations and memories with reference to a *mental present.* The mental present is the phenomenal expression of the population dynamics of the neurons of the human brain, as that dynamics is understood in the theory of neural Darwinism proposed and developed by Gerald Edelman.[10]

The time of organic intentionality is *biotemporality*, the temporal reality of all life forms. Its horizons, compared with those of nootemporality, are limited. Its reference or anchor is the organic present, created and maintained by the synchronized organic oscillatory complements of living organisms.

The time of the world of deterministic causation is *eotemporality*, the time of the physicist's "t" with all its usefulness and problems. The time of the world of particle-waves is *prototemporality*, the time of quantum physics. The time of pure becoming is *atemporality*. It is useful to stress that what is novel in this identification of the different temporalities and different causations-as-constraints is the broad perspective that it offers. The different causations and temporalities, considered separately, have been recognized, but not as parts of a nested hierarchical system. As an example—standing *pars pro toto*—let me quote from a 1966 paper on "Time and Quantum Theory" by E. J. Zimmerman. Continuous and directed time, with its well defined instants, if needed as a reference to quantum theory, must be imported into quantum theory from the macroscopic world:

[10] An entry to neural Darwinism may be had through Gerald Edelman's *The Remembered Present: A Biological Theory of Consciousness* (New York: Basic Books, 1989).

> "time" which appears in the equations is not a quantum mechanical observable... but rather a parameter external to the microscopic system. This "time" does not refer to something internal to the quantum system, but to something measured by a laboratory device—a macroscopic clock.[11]

In what Zimmerman asserts, the hierarchical theory of time identifies the simultaneous presence of qualitatively different temporalities: specifically, prototemporality for the quantum system, eotemporality for the clock on the wall, bio-, noo- and sociotemporality for the experimenter.

Having considered the temporalities of society, man, life, matter and particle-waves, let us turn to the time of absolute chaos or pure becoming. That world is *atemporal.* This term does *not* signify the theological timelessness of God or of anything else. Nor does atemporality signify the human experience often, and uncritically, described as that of timelessness. Rather, it signifies conditions to which none of our notions pertaining to time apply.[12]

Next, let me collect the different causations-as-constraints. They are as follows: (1) absolute chaos or pure becoming, at the foundation of the universe, (2) probabilistic causation, (3) deterministic causation, (4) organic intentionality, (5) noetic intentionality, (6) and collective intentionality. These are the *canonical forms of time.* The term "canonical" is borrowed from mathematics. It signifies the simplest forms of causations-as-constraints to which all more complex kinds of causation may be reduced.

Earlier I postulated that the absolute chaos or pure becoming of the dot universe remained a permanent substratum of the universe and that it keeps on serving as the source of the unpredictable, the unexpectable. What has been said about the canonical forms of causation and time supports that postulate. Namely, none of the emerging forms of causation replaced its evolutionarily simpler form or forms. Instead, each was added to the earlier forms. It is reasonable to maintain, therefore, that the primeval world of pure becoming or absolute chaos also

---

[11] E. J. Zimmerman, "Time and Quantum Theory," in *The Voices of Time: A Cooperative Survey of Man's Views of Time as Expressed by the Sciences and Humanties*, 2nd ed., ed. J. T. Fraser (Amherst: University of Massachusetts Press, 1981), 492.

[12] For an entry to these ideas, see J. T. Fraser, "A Nested Hierarchy of Causations, Languages, Temporalities and Conflicts," in *Time, Conflict, and Human Values* (Chicago: University of Illinois Press), 26–43.

was not replaced by later forms of causation but remained present in all processes of nature.

I also suggested that the atemporal chaos is the primeval pool of all potentialities, that it is the source of becoming, of the emergence of the unpredictably new, of whatever we may describe as creative—whether or not it is desirable. Let me, therefore, turn to issues of the emergence of the unpredictably new in nature and man.

2. Creativeness in Nature and Humanity

To trace the path from the ever-present pure becoming or absolute chaos at the foundations of the cosmos, to creativeness in nature and man, we start with a critique of a broadly held belief about the thermodynamic roots of time.

Arthur Stanley Eddington, distinguished physical cosmologist, was convinced that anything as important as time's passage must necessarily derive from, and must be explicable in terms of, physical processes. In a 1927 lecture titled "The Running Down of the Universe," he remarked that the direction of time "makes no appearance in physical science except in the study of organisation of a number of individuals." He added that "randomness is the only thing that cannot be undone."[13] This statistical irreversiblity reminded him of the irreversibility of time. In turn, the similarities made him remark that, to find the roots of our experience of passage, we should explore the "increase of the random element" in the physical world. He even coined a telling term, "time's arrow," to describe this one-way property of increasing randomness. The increase in randomness, he believed, is "vividly recognized by consciousness."[14]

Eddington's model of time's arrow has been responsible for eighty years of extensive work built on the dogmatic belief that time's passage is a human recognition of the thermodynamic "running down" of the universe, of the increase in the entropy of the cosmos. I propose

[13] Reprinted in Arthur Eddington, *The Nature of the Physical World* (Ann Arbor: University of Michigan Press, 1958), 63–86. The quote is from p. 69.

[14] Eddington, 69.

to cast a very critical eye upon this claim and recommend a way of repairing it.[15]

The increase of entropy to which Eddington appealed is predicated on the Second Law of Thermodynamics. That law of nature asserts that the entropy of a thermodynamically closed system, in the long run, may only increase or remain constant but cannot decrease. This is true enough. But, deriving the passage of time from an application of the Second Law to the cosmos has (a) a profound problem concerning its validity for the universe at large and, more importantly, (b) its reasoning has a fatal mistake.

With regard to (a)—it is not at all certain that that the Second Law of Thermodynamics holds for the universe at large, for the following three reasons:

(1) It is not at all certain that the universe may be regarded as thermodynamically closed. Namely, a closed system must have something external to it, something to which it could be open or else closed. But by definition there is nothing external to the universe.
(2) Entropy is a mathematical measure of the disorganization of a system.[16] It needs a reference level of organization. Was there an incredible degree of orderliness in the Big Bang, one that will need billions of years to get disorganized? No. The universe is believed to have originated not from some absolute order but from absolute chaos.
(3) The cosmologist Edward Harrison has given convincing reasons is support of the idea that the total entropy of the universe remains constant.[17]

Let me assume that, in spite of these serious doubts, the universe is thermodynamically closed and its entropy increases, and attend to the fatal mistake in the reasoning that leads to the suggestion that our experience of passage is a recognition of the increasing disorder of the cosmos.

---

[15] First sketched in the section "Two Arrows for the Price of One," Parts I and II, in J. T. Fraser, *Time as Conflict: A Scientific and Humanistic Study* (Basel and Stuttgart: Birkhäuser Verlag, 1978), 62–69.

[16] See G. J. Whitrow's superb summary s. v. "Entropy," in *The Encyclopedia of Philosophy*, ed. Paul Edwards (New York and London: Macmillan, 1972), 2: 526–29.

[17] E. R. Harrison, *Cosmology: The Science of the Universe* (Cambridge: Cambridge University Press, 1986), 354–60.

Consider that we, the formulators of questions about time and the thermodynamics of the universe, are living organisms. As such, we systematically and steadily self-organize. Until the day we die, our bodies oppose the cosmic "running down." The sense of entropy change of that self-organization, of the entropy decrease—the growing of the body, the healing of wounds—proceeds along time's experienced arrow. It parallels our sense of time's passage, as does the direction of the cosmic entropy increase. The direction of entropy decrease, due to the self-organizing processes of our bodies, points along time's experiential arrow just as does the direction of the cosmic entropy increase. Obviously, deciding to attribute our experience of time's passage to (a) a steady, statistical increase of cosmic disorder (increase of entropy) or else (b) to a steady, statistical increase of local, living order (decrease of entropy) is an arbitrary choice. Both selections are valid. Pick your preference.

I dare speculate that if Eddington had been a biologist, he would have asserted with conviction and authority that, for the sources of our sense of passing time, we must look to the capacity of living systems to self-organize. Eddington, the biologist, would have insisted that for the roots of our experience of passage we must look to biological processes that decrease entropy.

Since the experiential direction of time may be attached with equal justification to the entropy increase of the physical cosmos no less than to the local entropy decrease of the life process, it cannot be identified with either. Rather, time's experienced passage relates to the opposition between entropy increasing and decreasing processes. It relates to the fact that they define each other, that there is a conflict between growth and decay. The temporalities of the physical world—the older, undirected temporalities—permit the emergence of directed temporalities, but they do not demand direction for their own completeness.

To demonstrate that the two opposing arrows are co-present in the directed temporalities, I will appeal to what in the natural sciences are known as minimal principles. Namely, for physical, chemical and biological processes, it can often be shown that from among all processes that could take place (from among all processes allowed by the laws of nature), the processes that actually do occur are those for which some characteristic variable assumes a minimum value.[18] An example

[18] Examples of minimal principles include Fermat's principle of light paths with minimal travel time, Hamilton's principle of least action, LeChatelier's law of displacement against stress.

directly applicable to our theme is the principle of minimal entropy production, formulated by Ilya Prigogine. It states that when a system is in a steady state, the rate of its entropy increase is a minimum.[19] I interpret this minimalizing process as the superposition of two simultaneous, opposing trends: the general increase in disorder and the local increase in order.

Let me restate this interpretation in a form that makes it easy to illustrate it.

Although the long-term fate of the universe may well be governed by the Second Law of Thermodynamics, the final state predicted is reached along paths governed by principles that minimize the rates of entropy increase. To illustrate this grand policy, I will proceed to identify the specific oppositions between ordering and disordering associated with each form of causation-as-constraint as traced in Section 1. Then, I will give reasons why the conflicts to be identified should be regarded as constitutive of the temporalities of matter, life, humanity and society.

3. The Constitutive Conflicts of Time

We begin with sociotemporality—the time of human society—and work our way down to absolute chaos.

A *society* is usually thought of as a community of individuals working toward a shared goal. But this is only half of the story. The complete story involves the instant-by-instant balancing of the conflicts between conduct that supports and conduct that opposes whatever that goal may be. The instant-by-instant coordination of the opposing trends defines the *social present.* It is with respect to the social present that collective memories (called history) and collective intentions may acquire meaning. Such conflicts are constitutive of society because, if they cease, persons may still remain around but society will have collapsed into anarchy.[20] A group of people in anarchy cannot create, much less maintain, a social present. In the absence of a social

---

[19] See his *Introduction to the Thermodynamics of Irreversible Processes*, 3rd ed. (New York: Interscience, 1961), 83. "We may conclude that internal irreversible processes always operate in such a way that their effect is to lower the entropy production per unit time." Then consult H. E. Meijer and J. C. Edwards, "Modification of the Minimum-Entropy-Production Principle," *Physical Review A* 1:5 (May 1970): 1527–1536, and E. Ebhan's "Generalizations of the theorem of minimum entropy production to linear systems involving inertia," *Physical Review A* 32:1 (July 1985): 581–589.

[20] Fraser, *Time, Conflict, and Human Values*, *op. cit.*, 40–43.

present, no meaning may be assigned either to history or to planning and, hence, to sociotemporality.

From the constitutive conflicts of society, let me turn to the conflicts that are constitutive of personhood.

A *person* or *individual* is a member of our species, possessing stable identity: that is, stable biological and psychological functions unique to the species. That stability is created and maintained by a steady balancing of conflicts between the processes that support it and the social, mental, biological, and physical perturbations that oppose it. The instant-by-instant balancing acts define the *mental present*. It is with respect to the mental present that noetic intentionality or goal-directedness and noetic memories may acquire meaning. In their absence, noetic time cannot acquire direction. The conflicts implied are constitutive of personal identity and with it of nootemporality because, if they cease, as they do, for instance, in senility, a man or woman may well remain alive but his or her personal identity becomes ill-defined or vanishes. He or she becomes "a walking shadow…a tale / Told by an idiot, full of sound and fury / Signifying nothing."[21] Lady Macbeth, after the evil deed, ceased to live in a nootemporal reality.

The *life process* is usually thought of as one of growth through self-organization. But that is only half of the story because "from hour to hour we ripe and ripe / And from hour to hour we rot and rot / And thereby hands the tale."[22] Ripening and rotting are jointly constitutive of life and hence of biotemporality because, if either growth or decay ceases, their dialectic ceases, as does life, and, with it, organic intentionality and biotemporality vanish.

Conflicts between growth and decay are also constitutive of ponderable, *inanimate matter* in a grand style. The entropy increase of the cosmos is opposed by the creation of order brought about by its cooling. But the two trends are not and cannot be coordinated in a cosmic present because, as we know from Relativity Theory, "there is no longer world-wide simultaneity, and hence no universal common time…."[23] Presentness or nowness has only local meaning.

---

[21] William Shakespeare, *Macbeth*, 5.5.28.

[22] William Shakespeare, *As You Like It*, 2.7.26.

[23] On the absence of a cosmic present see "The determination of time at a distance," in G. J. Whitrow's magisterial *The Natural Philosophy of Time* (Oxford: Clarendon Press, 1980), 249.

We arrived at the dot universe. Good reasons may be given in support of the view that the important, creative conditions that emerged from the primeval chaos were not those of order but those of conflicts between different forms of ordering and disordering.[24] All later constitutive conflicts—those of complexifying inanimate matter, those of life, those of minding, and those of society—may then be seen as stages in the evolution of that primeval opposition.[25]

At this point of our survey, a remarkable policy of nature comes into view. It pertains to the conflicts between ordering versus disordering. Let me identify its evolutionary steps.

The most primitive ordering opposition to the "running down of the universe" is the gathering of particle-waves, governed by probabilistic laws. The next level of self-organization, that by solid matter, is governed by deterministic laws. They are steps in increasing the efficiency of ordering: from probability to certainty. But even the most efficient deterministic self-organization of inanimate matter can do no more than to decrease the rate of disorganization, the rate of entropy increase.

But life, governed by events that connect through organic intentionality, can do better. Specifically, for small regions—within the confines of a living body—it can actively, purposefully, reverse decay by self-directed growth. This active opposition to decay, by life, becomes more effective for processes whose events are connected by noetic intentionality. And it can become even more efficient in the world of collective intentionality.[26] We observe a generalized natural selection, identified in the evolution in the means that oppose an increase in cosmic disordering. It is a grand process, an immense journey along steps of constitutive conflicts between increasingly more complex forms of ripening and rotting.

4. Human Values

Constraints, as understood in this paper, may be divided into two distinct categories: the laws of nature is one, human values the other.

---

[24] Fraser, "From Chaos to Conflict," in *Time and Time Again, op. cit.*, 83–97.

[25] Fraser, *Time, Conflict, and Human Values, op. cit.*, 41.

[26] See the section "Time's Rites of Passage," in Fraser, *Time as Conflict, op. cit.*, 161–185.

I submit that there is a nested hierarchical relationship between the two—that is, there is a continuity between them. Also, while upon a superficial view, both categories of constraints appear to favor permanence, upon closer examination, they both turn out to be agents of change.

Human values are generalizations arrived at by inductive reasoning. They express in collectively intelligible forms what appear to be useful constraints upon belief: conduct and emotions. Since Greek antiquity, these constraints have been classed under the idea of the true, the good, and the beautiful. I will define each in terms of its relation to time. But first I want to explain that by constraints being "useful" is meant that they assist in maintaining the peculiarly human conflicts pertaining to beliefs, to conduct, and to emotions. These conflicts must be maintained because, for reasons to be given, if they cease their dialectic, our value judgments vanish and, with them, vanish our hallmarks as human.

As are human values, the laws of nature are also generalizations established by inductive reasoning. They express, in collectively acceptable forms, what appear to be stable—unchanging—and hence easily quantifiable relationships.[27] That there appears to be a grand natural process that favors increasing efficiency in opposing decay is prima facie evidence that the laws of nature are agents of change. Probabilistic causation was a necessary step to the emergence of deterministic causation. It was then probabilistic and deterministic causations together that made possible the emergence of organic intentionality.

It is the nested and hierarchical relationship between probabilistic and deterministic constraints upon atemporal chaos that provides the continuity from the laws of nature to life and to minding and hence to the constraints known as human values. But as we leave the physical integrative levels for those of life, minding, and society, the complexity of processes and structures increases by many orders of magnitude.[28] Quantitative methods remain important but tend to become inadequate for exploring the constraints and formulating guidelines for the life, the mental and the social processes of humans. What-

---

[27] On the nature and role of quantifiability in developing knowledge, see A. W. Crosby's *The Measure of Reality: Quantification and Western Society, 1250–1600* (Cambridge: Cambridge University Press, 1997).

[28] See Fraser, "Complexity and its measure," in *Time and Time Again, op. cit.*, 5–12.

ever regularities seem to be relevant become increasingly qualitative in their nature. It is the distinctions between the quantitative and the quantitative-plus-qualitative constraints that distinguish the laws of nature from human values.

Let me now turn to the working details of the broad claims just made. First, I will propose working definitions of the true, the good, and the beautiful in terms of their relations to time. I will then explore their roles, as constraints, that bring about changes in beliefs, conduct, and preferred emotions.[29]

Truth as a human value may be defined as the recognition of permanence in reality.[30] Let good as a human value be defined as an assertion that certain conduct, intent, or character traits will promote stable harmony in the minds and affairs of persons and societies. Finally, let beauty as a human value be defined as a quality of an object or event that engenders feelings that one would like to perpetuate. Employing these definitions, we may now explore the long-term roles of human values.

First: truth as a human value. Although searching for truth is driven by the desire for identifying permanence in humanity and nature, the historical function of seeking truth has been the creation of conflicts that, in their turn, create changes in the very principles that generated them.

The working definition of the good appeals to the desire for stability and harmony in human affairs. This goal is sometimes achieved, in the short run. But the long-term office of the good has been the opposite. Moral judgments have been creating and maintaining conflicts concerning conduct and intent and, through them, have been keeping alive a steady revolt against whatever principles happened to be guiding peoples' conduct and intent. I believe that the incredible creativeness and destructiveness of our species are not due to *not* having and following moral principles but to having and following them.

The working definition of the beautiful related it to the desire of identifying feelings that one would wish to perpetuate. Yet the

[29] The definitions of human values through their relations to time, and their respective dynamics are the theme of *Time, Conflict, and Human Values, op. cit.*

[30] But, consider that…although searching for truth is driven by the desire for permanence, the historical function of seeking truth has been the creation of conflicts which, in their turn, create changes in the very principles that generated them. See Fraser, *Time, Conflict, and Human Values, op. cit.*, 73.

evolutionary office of the aesthetic faculties has been that of generating and maintaining conflicts between the world as we find it to be and that other world of imagination to which art gives transient forms and changing names.

Let me summarize what was said about human values as constraints. It was this: they share with the laws of nature the qualities of promoting local ordering that opposes cosmic disordering. This would suggest that human values promote conservative trends, which they may do for the short run. But for the long run they promote change by creating new and more complex conflicts between local representations of the cosmic features of growth and decay, recognized in the words of the Bard, as ripening and rotting.

With an idea of the laws of nature and of human values as agents of change, together with an idea of the constitutive conflicts of matter, life, mind and society, we are ready to turn to the issue of human time—as something immensely more intricate than the recognition of the increasing entropy of the cosmos.

5. Human Time

As humans we are members of human societies. As part of our tasks, we practice minding. To be able to do so, we must be alive. To be alive, we must be made of matter in its ponderable form, while also depending on the functions of particle-waves. We are even in contact with the absolute chaos of pure becoming of the dot universe, as Chaucer well knew when he wrote that "It is good to keep one's poise and be protected / Since all day long we meet the unexpected."[31] In other words, we are parts of all integrative levels of nature.

It follows that all the causations-as-constraints that I identified in this paper and everything that I said said about time—all the conflicts between ripening and rotting—may be identified among the features of human time. So are all the laws of nature. Also, we are constrained by the family of human values serving as agents of change. Positioned as we are at the complexity boundary of the universe, we are subject

[31] Geoffrey Chaucer, "The Knight's Tale," ll. 1523–24, in *The Canterbury Tales*, trans. Nevill Coghill (London: Penguin Books, 1960), 60. English by Neville Coghill Original: "It is ful faire a man to nere him evene / For al-day meteth men at unset stevene."

to natural selection that favors increased efficiency in opposing the cosmic entropy increase.

To keep ourselves alive and our minds functional, we continuously balance or try to balance uncountably many constraints. Our situation makes King Lear's concerns a concern of the species: "O! let me not be mad, not mad, sweet heaven! / Keep me in temper; I would not be mad!"[32] I believe that our unique, species-specific state is that of madness, both the controlled and uncontrolled types—that it is a condition responsible for the immense creativity and destructiveness of men and women. Therefore, the question arises....

6. How to Manage the Constraints upon Being Human

I take my cue from all thinkers, East and West, who emphasized the importance of recognizing the unity of all things. If so, then our program should be to willingly share the dynamics of all constraints and constitutive conflicts of matter, life, minding, and society.

But where may we find instructions on how to proceed? In the literary arts. I selected a poem by William Butler Yeats called "Those Images," because it is a mapping into concise poetic language of what the theory of time as conflict says about constraints, about nature and about human values.[33]

> Seek those images
> That constitute the wild.
> The lion and the virgin,
> The harlot and the child.
>
> Find in middle air
> An eagle on the wing,
> Recognize the five
> That make the Muses sing.

References

Bernoulli, Jakob. *Ars Conjectandi* (1713). Reprinted in translation as "The Law of Large Numbers." In James R. Newman, *The World of Mathematics* (London: Allen and Unwin, 1956), 1452–55.

[32] William Shakespeare, *King Lear*, 1.5.50–51.
[33] *The Collected Poems* (New York: Macmillan, 1956), 316.

Blake, William. "Marriage of Heaven and Hell." In *The Portable Blake*, ed. Alfred Kazin. New York: Viking Press, 1969.

Cairns-Smith, A. G. "Beginning of Organic Evolution." In *The Study of Time IV*, ed. J. T. Fraser, N. Lawrence, and D. Park, 15–63. New York: Springer Verlag, 1981.

———. *Genetic Takeover and the Mineral Origin of Life*. Cambridge University Press, 1982.

———. *Seven Clues to the Origin of Life*. Cambridge: Cambridge University Press, 1985.

Chaucer, Geoffrey. "The Knight's Tale." ln *The Canterbury Tales*, trans. Nevill Coghill, 26–85. London: Penguin Books, 1960.

Crosby, A. W. *The Measure of Reality: Quantification and Western Society, 1250–1600*. Cambridge: Cambridge University Press, 1997.

Ebhan, E. "Generalizations of the theorem of minimum entropy production to linear systems involving inertia." *Physical Review A* 32:1 (July 1985): 581–589.

Eddington, Arthur. *The Nature of the Physical World*. Ann Arbor: University of Michigan Press, 1958.

Edelman, Gerald. *The Remembered Present: A Biological Theory of Consciousness*. New York: Basic Books, 1989.

Fraser, J. T. *Time and Time Again*. Leiden: Brill, 2007.

———. *Time as Conflict: A Scientific and Humanistic Study*. Basel and Stuttgart: Birkhäuser Verlag, 1978.

———. *Time, Conflict, and Human Values*. Chicago: University of Illinois Press.

Harrison, E. R. *Cosmology: The Science of the Universe*. Cambridge: Cambridge University Press, 1986.

Lemaitre, Georges-Henri. "The Beginning of the World from the Point of View of Quantum Theory." *Nature* 127 (1931): 706.

Meijer, H. E. and J. C. Edwards, "Modification of the Minimum-Entropy-Production Principle," *Physical Review A* 1:5 (May 1970): 1527–1536.

Prigogine, Ilya. *Introduction to the Thermodynamics of Irreversible Processes*, 3rd ed. New York: Interscience, 1961.

Sklar, Lawrence. *Space, Time, and Spacetime*. Berkeley: University of California Press, 1974.

Wheeler, John Archibald and Kenneth Ford. *Geons, Black Holes and Quantum Foam*. New York: Norton, 1998.

Whitrow, G. J. "Entropy." In *The Encyclopedia of Philosophy*, ed. Paul Edwards, 2: 526–29. New York and London: Macmillan, 1972.

———. The Natural Philosophy of Time. Oxford: Clarendon Press, 1980.

Yeats, William Butler. *The Collected Poems*. New York: Macmillan, 1956.

Zimmerman, E. J. "Time and Quantum Theory." In *The Voices of Time: A Cooperative Survey of Man's Views of Time as Expressed by the Sciences and Humanties*, 2nd ed., ed. J. T. Fraser, 479–99. Amherst: University of Massachusetts Press, 1981.

SECTION I

THEORY AND EMPIRIE

CHAPTER ONE

TIME: BIOLOGICAL, INTENTIONAL AND CULTURAL*

Carlos Montemayor

Summary

In this paper, I propose that time representation should be classified as agent dependent motor-intentional, agent dependent conceptual and agent independent conceptual.[1] I employ this classification to explain certain features of psychological and cultural time and discuss how biological time constrains such features. The paper argues that motor-intentional time is a crucial psychophysical link that bridges the gap between purely biochemical cycles and conceptual-intentional representations of time, and proposes that the best way to understand the transitions from biological to psychological time is to characterize them as phase transitions that emerge on the global properties of a system (in this case temporal behavior) regardless of specific aspects of the underlying processing units.

With respect to hierarchical models of time, the paper claims that although the hierarchical organization of constraints limits the number of possible instantiations of temporal information, informational phases have principles of organization of their own. Based on these proposals, I conclude that since information assembles itself into the temporal behavior of an agent, it is more accurate to call the principles governing such a behavioral unity a phase, rather than a hierarchy. To support these theses, the paper draws upon research in biology, neuroscience, philosophy (more specifically philosophy of mind and philosophy of time) and psychology.

1. Introduction

In this paper, I give an account of time representation—our experiences of, and beliefs about, time—and argue that there is a particular type of temporal representation, which I call motor-intentional time, that bridges the gap between biological and cultural time. I will refer to our representation of time as psychological time and to beliefs about time as conceptual-intentional time (a subset of representations of time). I will assume that all representations of time contain information that

* This paper was awarded the 2007 Founder's Prize for New Scholars.

[1] I would like to thank the editors of this volume and an anonymous referee for helpful comments that improved this paper.

Jo Alyson Parker, Paul A. Harris and Christian Steineck (Eds), Time: Limits and Constraints, pp. 39–63

is computed by organisms and that these computations occur in some physical medium, for example, their brain, their nervous system, etc.

I will also argue, in the concluding section of this paper, that the transition from one level of temporal organization to another (biological to intentional, or intentional to cultural) is best understood as a boundary of a phase transition (for example the phase transition from liquid to solid: water to ice). By this, I mean that the limits that delineate and constrain biological, psychological, and cultural temporal organization are best understood as points where such temporal organization undergoes an abrupt structural change.

This idea, however, needs to be complemented with the hierarchical view of time, which states that the relation between phases of temporal organizations is layered (e.g., biological time causally underlies psychological time). I explain in the conclusion why combining the idea of phase transitions with the hierarchical model provides a better framework for understanding the *structural complexity* of each level of temporal organization.[2]

In this essay, I first describe the biological basis for time representation (or biological time), emphasizing the role that complexity plays in the emergence of temporal behavior. I then focus on the representational dimension of temporal behavior, or psychological time, and classify it as motor-intentional and conceptual-intentional. I conclude by addressing the implications of delineating biological and psychological time in terms of constraints, and offer an explanation of how these constraints relate to hierarchical models of time representation and to the transition between biological and psychological time. However, before addressing these issues, I shall define the notion of representation that I will assume in this paper.

[2] I am using the term 'phase transition' as a concept, in order to refer to complex information processes that would require mathematical equations and physical models in order to be sensibly understood. My account speaks to the issue of how to understand the relation between biological, psychological, and cultural time so that one can get a better sense of how to model these transitions from a strictly conceptual perspective. I shall also clarify that I am not concerned in this paper with physical time (or time as is understood in physics, as a dimension of space-time). Anything I say in this paper is in principle compatible with whatever notion of time is favored by physicists. Thus, I am not concerned with time itself, if by 'time itself' one means physical time, which differs drastically from the understanding of time assumed in biology, psychology and the humanities.

Although all representation is information, not all information is representation.[3] Nelson Goodman (1976) noticed that representation is an asymmetric relation; for example, a toy with a hat represents Napoleon, but Napoleon does not represent a toy with a hat. This shows that representation is not the same as resemblance, which is a symmetric relation (e.g., if I resemble you, then you resemble me). The difference between representation and resemblance does not entail that resemblance plays no role in representation, which it clearly does, as in some cases of artistic representation. This difference rather entails that representation and resemblance are not the same relation. Both resemblance and representation are relations about information and determining their existence is informative. Yet, representation involves a unique type of information encoding that cannot be fully captured by other informative relations, such as resemblance. But what can this uniqueness be? The best way to describe it is by exemplification. The example I will use involves communication in animals, and it was originally presented by Philip Johnson-Laird (1990). It illustrates nicely how representation is a unique type of information encoding that detaches from immediate environmental constraints, which enables the generation of an abstract symbolic structure that needs to be interpreted.

Fire ants communicate information about food and the paths that lead to it through pheromones that they attach to the ground. The biochemical composition of the pheromones works as a code through which the ants retrieve information. On the other hand, bees communicate the same information by performing a dance whose tempo is correlated with distance and flight orientation. As Johnson-Laird says:

> Certainly, there is no need for the ant's nervous system to construct an internal representation of some rudimentary 'idea' of food that lights up in the insect's mind as soon as it smells the pheromone. Indeed, the use of a trail counts against such an internal representation: you do not need a map if you are traveling by rail. (Johnson-Laird 1990, 3)

[3] I am using the term 'representation' in a restrictive sense that refers only to mental representation. As I am about to explain, without this restricted sense of 'representation', the distinctions that need to be drawn in the analysis of temporal behavior cannot be drawn, which generates ambiguities and confusion.

Unlike the chemical guide used by ants, bees display a *symbolic* dance that serves as a map that orients them in their environment. The bees need to decode this dance by means of representations—rather than purely biochemical reactions—and they can only do that if they have a code, somehow instantiated in their nervous system, to *interpret* the dance in terms of direction and distance. In this case, there is no traveling by rail.[4]

The asymmetric nature of the 'represents' relation reveals something very important about the process of decoding spatial information from the bee's symbolic dance, that is, that representations are *arbitrary and dependent upon interpretation by the cognitive agent.* It could be objected that the pheromone map used by ants is also arbitrary and that they could have employed a different set of biochemical combinations to construct their pathways. This is true, but notice that the constraints on how ants could arbitrarily change the chemical composition of their pathways are biological, while the constraints on bee communication are symbolic and thus much more flexible.

Representations have a symbolic structure that is preserved regardless of how they are instantiated in a physical medium. This is clearly not the case with the ant's pheromone map. In order to change a message, ants need to change the biochemical code that physically instantiates it. However, bees can change a message by interpreting a dance differently, or by slightly modifying the tempo of the dance, or by orienting themselves differently, etc. Thus, they have a lot more options than ants.

Although there is arbitrariness in both systems of communication, the ant's pheromone map is much more stable and independent of the cognitive agent, while the bee's is more flexible and entirely dependent on the cognitive agent. The stability of biochemical communication generates the impossibility of misrepresenting information by employing a different interpretation. But misrepresentation must always be possible if interpretation is critical. Therefore, we can conclude that if a message *R* does not allow for misrepresentation, then *R* is not a representation. Biological constraints on how information is computed are

[4] The fact that bees manifest this sophisticated symbolic behavior does not imply that ants lack sophistication or that their pheromone-based communication is not computational. It is entirely a contingency of the kind of environment these animals move through (land versus air) that accounts for the use of pheromones instead of symbolic dances, i.e. pheromones do not create stable paths in the air.

determined by biochemistry. Constraints on how to interpret dances and establish which moves constitute an informative dance depend upon the animal's capacity to represent information, which is determined by something more than biochemistry, i.e. the internal code of the bee, used to interpret dances. Although both types of constraints concern information, the latter requires *representational* capacities, not merely bio-computational ones. As I will explain in the following pages, these biological and representational constraints on behavior highlight very important aspects of our representation of time.

2. The Biological Basis for Time Representation or Biological Time

A great achievement of modern biology was to identify the suprachiasmatic nucleus (SCN) as the locus of the master circadian clock in mammals.[5] This master clock synchronizes central and peripheral oscillations through circadian output regulation. The complexity of the bottom-up and top-down informational processes of the master clock's output regulation is astonishing and covers all levels of biological organization, from the molecular to the behavioral levels. As Panda and Hogenesch (1998) explain:

> Circadian modulation of gene transcription translates into rhythmic activity of their protein products, which in turn generates rhythmic metabolic flux in different tissues and rhythmic behavior and physiology at the organism level, thereby enabling the animal to adapt to diurnal changes in its environment. (Panda and Hogenesch 1998, 375)

Clearly, the circadian clock has an enormous impact on an animal's behavior, because it regulates feeding time, absorption of nutrients, muscle tone for motor reaction, hormonal rhythms, neuro-anatomic outputs, sleep-wake cycles and even locomotor activity (Panda and Hogenesch 1998, 376). Having such an extensive impact on an organism's daily life, circadian rhythms help shape environments and determine niche formation. In an ecological niche, the partition between

[5] The origin of this important discovery is the work of Carl Richter (1967), who demonstrated that lesions to the frontal part of the hypothalamus of rats eliminated rhythmic behaviors. The identification of the SCN as the locus of the circadian clock is due to research by Stephan and Zucker (1972), Moore and Eichler (1973) and Ralph, Foster, Davis and Menaker (1990).

diurnal and nocturnal animals maintains species diversity by predator avoidance and the lack of other competitors for food. Thus, the circadian clock's influence ranges from the molecular biochemical level to the formation of ecosystems.

Plants, insects and bacteria also have circadian clocks, which gives them an evolutionary advantage because by matching environmental cycles, their circadian rhythms allow them to optimize energy consumption. For plants it is crucial to synchronize their orientation with changes in light intensity to absorb as much energy as possible, which explains why they have a very complex system of photoreceptors that input information to the clock, particularly for phase resetting (see A. Millar, 2003). The experimental data reveal that circadian rhythms have been fundamental in the development and evolution of life. Without doubt, our behavior is deeply influenced by the master clock and so is our experience of time. We share this mechanism for synchronization with changes in the environment—a wonderful evolutionary achievement—with most life on earth.

However, as complex as the outputs and inputs of the clock are, it would be a mistake to call these biochemical processes 'representations of time'. The term used in the biological literature is 'projections' and I think it is quite appropriate to characterize the nature of biochemical information processing. The master clock located at the SCN computes information in a temporal fashion, much like the circadian clocks of plants and bacteria. But plants and bacteria are not *representing* time. I will explain what is required for time representation in the next section. For now, the most important conclusion of this brief presentation on biological time is that circadian clocks constitute an important *constraint* on time representation because of the scope of their influence on temporal behavior.

The constraint constituted by biological time could perfectly be described as a set of principles governing the self-organization of an animal at the biochemical level. Something crucial about these principles of organization is that although they ultimately depend on the information contained in the smallest units of information encoding—genes, neurons, molecules, etc.—they cannot be reduced to any such atomic unit. Rather, these principles *emerge* from the combined complexity of a) the biochemical basis that creates the rich temporal behavior that animals and plants exhibit and b) the animal's interaction with the environment. This is very important to understand the nature of the *transition* from biological to psychological time, which I discuss at the end of the paper.

3. Time Representation or Psychological Time

Time *psychology* is the study of mental representations of time (representations that I am calling psychological time) and the mechanisms that are responsible for producing these representations, which include mechanisms for temporal memory, temporal reasoning, and temporal perception. In order to determine the properties of these representations, I distinguish three types of time representation—motor-intentional, agent-dependent conceptual, and agent-independent conceptual—and argue that these three kinds of temporal representation are intentional; they are *about* time and the agent's temporal involvement in its immediate environment.[6] However, they have very different properties and cognitive implications.

The notion of 'motor-intentionality' was introduced by Maurice Merleau-Ponty (1945) to capture the embodiment and situatedness of perceptual experience.[7] Our actions and perceptions are always embedded in contexts in which our body skillfully accesses information from the environment and integrates it into a unique continuous perceptual experience. Merleau-Ponty describes this skillful access as a 'body schema' or 'habit', a concept that Taylor Carman (1999) interprets as follows: "the embodied poise or readiness, which Merleau-Ponty calls 'habit' consists in a kind of noncognitive preconceptual 'motor intentionality'." "Habit is not a function of reflective thought" (Carman 1999, 219).

Notice that if reflective thought is considered a necessary condition for intentionality, then the appropriate term to identify motor-intentional time should be '*non-intentional* time.' However, I think Merleau-Ponty's term aptly captures the *proto-intentional* qualities of these primitive temporal representations and conveys the kind of mentality that I am attributing to agents capable of representing time in this way. Thus, the term 'non-intentional' is best used in the context of strictly biological processes, where no *mental content* is involved.

---

[6] Franz Brentano (1874) introduced the term 'intentionality' to describe the property of our mental representations that makes them *about* something (even if this 'something' does not exist).

[7] I shall clarify that my brief presentation of Merleau-Ponty's ideas is not strictly exegetical in character. Thus, my aim in presenting his views on motor-intentionality is not to determine exactly what he meant by this term, or relate such notion to other aspects of his work. Rather, my aim is merely to provide some terminology for my discussion of time representation at the sensory-motor level.

Brentano's notion of intentionality applies paradigmatically to the propositional account of what a thought is—thoughts are about objects and their properties. Grasping a thought requires the conceptual capacities to articulate its content in order to determine what it refers to or what it is about. But similarly, there is an embodied kind of intentionality in actions that require specific forms of *engagement* with the environment, which our nervous system (and the nervous systems of animals) performs without propositional-thought representations. It is adequate to call this type of embodied mental content 'intentional' because it is constituted by representations that are *about* the environment and the close and constant interaction of the agent's body with the environment.

Motor-intentionality is non-propositional and pre-conceptual, and it is best described in terms of analog representation, which lacks discrete symbolic categories. But it is unsatisfactory to define it merely in terms of embodiment and environmental situatedness. More importantly, in concrete cases of animal communication, like in the case of bee communication, should we characterize the representations involved in these cognitive processes as motor-intentional or as conceptual? The latter option has the unfortunate consequence of attributing beliefs and propositional attitudes to bees. The former seems to leave the phenomenon of bee communication unexplained.

There is shared meaning in communication that requires conceptual capacities to grasp the content of the propositions (the 'objects' of propositional attitudes like beliefs) that agents attribute to each other. But communication can also be highly situated and non-propositional. A lot of the brain's capacity is devoted to motor coordination, and there are specialized neurons called 'mirror neurons' that are act-specific: they fire both when an agent observes an action performed by another agent and when the agent performs that same action. Research has shown that primates (specifically macaque monkeys) have mirror neurons that fire when they simulate and interpret the behavior of other agents (Rizzolatti 1988). These neurons have also been found in humans.[8]

[8] Alvin Goldman (2008) explores the philosophical implications of this research for theories concerning how we are capable of identifying or "mind-read" one's own and other's mental states. Admittedly, there is controversy over claims made about the functions of mirror neurons, but an interesting implication of how mirror neurons

Interestingly, mirror neurons are associated mainly with motor activity, such as grasping, and also with more refined motor skills such as gesturing and vocalizing. Simulating the behavior of other agents is a solid basis for communication even if no conceptual content is involved. Rizzolatti et al. (1996) found mirror neurons in monkeys' Broca area (the area responsible for speech in humans). Monkeys have a vast repertoire of sounds that are crucial for communication, and they learn these sounds through motor simulation. Their sounds do not "mean" something compositionally refined and articulated like "A jaguar is hiding behind the tree where Tom lives," but they *encode* a motor-reproducible signal that generates a motor-intentional representation. We could translate this signal as 'danger' and it produces in the monkey the motor-intentional command "jump back to the tree."[9]

This type of action-based communication is similar to conceptualized communication because there is a message from an agent that produces a mental state in another agent. However, in this case, *the mental state that the communicator produces in the audience is a motor-intentional one.* Thus, another key feature of motor-intentionality (besides its being pre-conceptual) is that *it produces only motor-intentional mental representations.* If an action is interpreted as a symbol or a sign that can be thought, desired, etc., then it has been conceptualized and it is not a motor-intentional representation: it produces a belief or a desire, that is, a propositional attitude. An example of this type of action-based *conceptual* communication is sign language.

Let us go back to bee communication. It seems that there is a weak sense of interpretation and attribution of mental states in their dances. However, these mental states are not propositional attitudes. Rather, they are motor-intentional representations that only produce other motor-intentional representations. Bees attend to a dance in order to motor-project their paths (not to fix the reference of the terms of propositional representations).

---

work is that the brain could dispense with "theories" as to how to interpret the mental states of other agents.

[9] Irene Pepperberg's work on Alex the parrot (2000) is also very useful here. Pepperberg demonstrated through her studies with this bird the amazing capacity these animals have to reproduce and articulate sounds and learn categories even though they lack the capacity to represent complex linguistic structures.

4. Motor-Intentional Time

We can represent time by registering moments in time and by determining the duration of intervals in a very short scale. In order to do this, a clock with a register (or accumulator) must encode and update temporal information (see Gallistel 1990, 231). However, unlike the outputs of the circadian clock, in the case of representations of time concerning duration and temporal order (motor-intentional time), the timing mechanism outputs its measurements with considerable speed, and it is finely tuned to the sense organs. This clock is, as it were, free from the biochemical cycles of inputs and outputs dictated by the circadian clock. What is crucial is that the mechanisms responsible for these motor-intentional representations of time *measure* temporal intervals regardless of what is happening in the circadian clock.[10]

Animals need to perform precise measurements of short intervals for movement coordination, action, and planning. As mentioned, these short intervals are not measured by their circadian clocks, but by a different clock, called the stopwatch.[11] This clock measures time at a scale that is much shorter than that at which the circadian clock operates, and its calibration depends on external and internal frequencies.[12] The stopwatch for short interval timing can accelerate its rate, creating the impression of time going slower, or decelerate and create the opposite effect (Cheng et al. 2006). These are representations (presumably even with phenomenal content) that inform the agent about temporal features of its body in relation to the immediate environment and produce motor-intentional responses, such as changes of orientation, gaze, attention, etc.[13] This is why these *cognitive* temporal

[10] The circadian clock may have outputs that are genuine representations of time (as in the case of the computation of the ephemeris function for navigation). I am ignoring these outputs for the sake of simplicity in presentation, in order to highlight the contrast between non-representational or strictly biological outputs of the circadian clock (like those involved in metabolic regulation processes) and the motor-intentional outputs of the interval clock for short durations, which are always representational.

[11] When you are at a red light and 'intuitively' know when to get ready to accelerate you are using the stopwatch, not your circadian clock.

[12] For a model of a clock for duration with a calibration unit that varies depending on external frequencies, rather than internal cues see Treisman 2006.

[13] The phenomenology of motor-intentional time is a touchy issue. The impression of time slowing down or accelerating clearly produces emotions of anxiety or relaxation. However, it is difficult to attribute these emotions to animals without discussing consciousness. I am not denying that motor-intentionality might be accompanied by

representations that are well beyond the scope of the circadian master clock are best described by the term 'motor-intentional'.

Short-term memory plays a major role in computing motor-intentional time. It updates the information registered by the stopwatch and helps relate this temporal information with spatial computations in order to provide animals with complex proprioceptive representations that are crucial for self-location. The most important computational characteristics of motor-intentional time are:

1. It is encapsulated or cognitively impenetrable: even if the agent possesses conceptual thoughts about time, her beliefs or other propositional attitudes do not influence motor-intentional information processing.[14]
2. The format in which information is encoded and represented is analog: it preserves metric (not conceptual) structure.
3. The type of memory it involves is mainly short-term memory: its content is lost after a brief period of time and is cognitively situated (interacting directly with the environment).

The principles governing motor-intentional time are psychophysical and dependent not only on the agent's biology and its immediate environment, but also on attention and short-term memory for action coordination. Attention is very relevant because the stopwatch, unlike the circadian clock, can be activated by the agent (for example, by attending to a particular event). The type of short-term memory used in motor-intentional time is episodic—the other two types of memories, semantic and autobiographic, are associated with conceptual content. In the transition from biological time to our temporal mental capacities, motor-intentional time is the first cognitive step towards

a rich phenomenology, but I am not going to defend this claim. For a type of consciousness that may be well suited for motor intentionality see Ned Block's (1995) characterization of phenomenal consciousness. In the conclusion, I speculate about the relationship between consciousness and temporal experience in terms of *emergent temporal behavior*.

[14] An excellent example of cognitive encapsulation is provided by perceptual illusions. For instance, in the Müller-Lyer illusion, the information you have about the difference in length of the lines in front of you, which makes you *believe correctly* that they are not the same length, does not help you change your visual perception that they *look* as having the same length.

our rich representation of time, and it bridges the gap between biological and conceptual-cultural time.

One can think of many animals with the capacity to represent motor-intentional time, including many insects, birds, mammals and reptiles. Coordinating action depends on measuring and computing time through cognitive mechanisms. The circadian clock is not designed to coordinate real time action (e.g., escaping from a predator, calculating time for collision, etc.). A cognitive clock is required to perform that job. This distinction is best understood by assessing a concrete example.

In the case of a type of fly called 'ephemeroptera', biological cycles mark crucial transitions in the insect's life and temporal behavior.[15] For instance, biorhythms determine the period in which the fly is in the 'naiad' stage—which tends to be the longest period in this animal's short life. They also mark the transition from naiad to subimago and then to imago—the mature fly. While different species of ephemeroptera remain in the naiad stage from 4 weeks to more than 3 years, some mature flies live as shortly as one day, a period of time in which they fly, copulate, and die after copulation. While they are flying, the temporal coordination of adult flies is crucial for them to navigate and find a mate, and the precision with which they perform these complex tasks signals the existence of motor-intentional time.

Compare the cycles marked by biological rhythms with the activity of flying and finding a mate. The biological transformations that ephemeroptera undergo *happen* without the agent's cognitive involvement. Flies simply change, and they do not need representations of their body and their environment to trigger these changes. However, in order to fly, navigate and find a mate, the fly needs representations of itself and its environment. Navigating and finding mates are things flies *do* and require their *cognitive involvement*.[16] Given its ephemeral life, it would be useless for ephemeroptera to access information about time symbolically for other cognitive purposes, like embedding temporal information in propositions, generating beliefs about time,

[15] There are many species of ephemeroptera, and most of them are mayflies.

[16] A similar point can be made by comparing the temporal behavior of two animals. Baby Galapagos land iguanas hatch at about the same time from a single nest, and the Galapagos hawks arrive about the same time in order to eat them. Hatching is something that *happens* to the baby iguanas, but showing up at the feast is something that Galapagos hawks *do*.

or storing such information in long-term memory. In other words, ephemeroptera flies lack conceptual time.

5. Conceptual Intentional Time

Ephemeroptera flies have a primitive mechanism for representing motor-intentional time, employed mainly for navigation. Other insects, bees for example, have a more complex temporal behavior, which includes temporal representations that optimize food collection. But how complex can animal temporal behavior be? Recent experiments have shown that some animals, particularly the higher mammals and many species of birds, can represent time in a quasi-symbolic (quasi-conceptual) fashion.

For instance, the western scrub jay displays a very complex temporal behavior. In their interactions with other scrub jays that involve hiding food, they use not only episodic, but also semantic and autobiographic memory—or at least a primitive form of it (see Clayton and Dickinson 1999a). They have a variety of food caching strategies and are able to determine which type of food decomposes first, which allows them to have a temporally organized "menu."[17] The fact that they recognize types of food and categorize them according to the time in which they decompose reveals that they are able to embed temporal information in a symbolic format, which enables the formation of proto-beliefs and other proto-propositional attitudes.

They can also determine when and where they hid food and visit these places before the food decomposes. They also change the place where they hid food if they noticed that another scrub jay was looking. They organize all these representations in a way that informs them about what they did in the past, which caches they have at a given moment, and what they need to do in the immediate future. Their use of this form of (quasi-autobiographic) memory suggests that they have a primitive notion of self—they can remember what they did and act upon it, plan strategies, etc. All these aspects of the western scrub jay's behavior seem to be signs of conceptual-intentional time, a kind of representation of time that seems similar to the one we enjoy.

[17] See Clayton and Dickinson 1998, 1999a and 1999b.

However, although the scrub jay's behavior seems to be informed by conceptual representations, I will argue that it is a sophisticated form of motor-intentional time because an essential requirement of conceptual time is that *it can be combined with other conceptual representations without producing motor-intentional mental states*. The main computational characteristics of conceptual time (which is paradigmatically intentional) are:

1. It is not encapsulated or cognitively impenetrable: beliefs or other intentional states of the agent influence information-processing. The background beliefs of the agent determine how information is interpreted.
2. The format in which information is encoded and represented is digital-conceptual, or symbolic: it preserves *conceptual*, rather than metric structure.
3. The types of memory it involves include the outputs of short-term memory, and all types of long-term memory: semantic, autobiographic and episodic.

While the cognitive basis (representations) of the temporal behavior of western scrub jays seems to be similar to ours, it is also very different, especially concerning the representations of past, present, and future. The contrast may be seen in terms of the distinction between the A and B series and how it relates to our, but not the scrub jay's, temporal behavior. The A series is tensed. It concerns the notions of past, present and future. In terms of representation, the A series locates a cognitive agent's perceptions in the present, its memories in the past, and its plans and strategies in the future. These are essential ingredients of autobiographic memory, a type of memory that updates information in a way that creates a coherent narrative of the self. Scrub jays seem to be able to represent time in this way and presumably have representations, call them 'A proto-beliefs', such as "I was at cache *A* yesterday," "I am now caching worms," and "The content of cache *B* will not decompose until tomorrow."

Undoubtedly, A beliefs and the type of conceptual intentional time captured by the A series is a crucial part of *our* temporal behavior. So what is the difference between the scrub jay's A proto-beliefs and our A-beliefs? The key to answer this question is to understand that beliefs, being conceptual representations, can be combined with other beliefs without producing motor-intentional mental states. Again, the best way to capture this distinction is by exemplification.

A-beliefs are what philosophers call *de se* beliefs. They have temporal indexical content that relates exclusively to the cognitive agent. John Perry (1979) offers an example that involves a "tardy professor" in order to describe temporal indexical content. The professor believes the proposition "The meeting starts at noon." This is a B-proposition, because its truth-value is fixed. A hallmark of B-propositions is that they have dates or fixed times of the day (the B series is not tensed and depends only on the relations 'before', 'at the same time', and 'after'). If a B-proposition is true, it *was* true yesterday, *is* true today and *will* be true tomorrow.[18] The tardy professor believes the B-proposition "The meeting starts at noon" all along while checking her email. Then, all of a sudden she realizes that the meeting starts *now*. Since she believes the proposition "The meeting starts *now*," she jumps off her seat and runs down the hallway.[19] What could be the difference between the tardy professor's A-belief and the A proto-beliefs of scrub jays?

I said that motor-intentional representations only produce other motor-intentional mental representations. In contrast, conceptual representations, like A-beliefs, produce motor-intentional representations *and* other conceptual representations. The scrub jay has, as many other animals, categorization capacities. But these "categories" are not concepts. They are what J. J. Gibson (1979) called 'affordances': their content is situated, requires the skillful access of an agent, and is determined by the goals of the agent and its immediate environment. Notice that affordances are discrete representations, but they are not yet conceptualized because they cannot be *generally* combined without producing motor-intentional representations.

I propose to define affordances as motor-intentional categories (action-oriented representations with limited categoricity). An affordance gives a cognitive agent clear advantages. The scrub jay has a temporally organized menu and each item in the menu is an affordance that *produces* certain motor-intentional representations. A bee might have affordances concerning certain flowers, other bees, etc. But

[18] To show that it will be true tomorrow, suppose this is a daily meeting.

[19] Notice that the proposition "the meeting starts now" *changes* truth-value. For the professor it is true now (assuming it is noon), but will not be true in an hour. It changes truth-value because it depends on the agent's temporal perspective.

an ephemeroptera fly, given its very short adult life, has less affordances (maybe just 'mate' and 'food'). This is what makes the temporal behavior of some animals much more complex than others. But none of these affordances is a concept and none of the representations of these animals a belief. If they were concepts and beliefs. the scrub jay would be able to combine these concepts, think thoughts about the concepts 'past' and 'present' in a non motor-intentional fashion, and *generalize* that if something is past and another event is *more* past, then the notion 'past' can be combined with the notion 'more' recursively, thereby producing a representation of the past that goes back in time, as it were, forever. As far as we know, only humans can represent the past in this way.

What scrub jays represent is a temporal affordance triggered by their memory and an accurate mechanism for estimating time. Their motor-intentional representations produce other motor-intentional representations, like "go to cache *A*." As mentioned, only humans have a concept of the past that goes as far back as the big bang (or even before if need be). The tardy professor elicited a motor-intentional response (jumping off her seat) through an A belief. But her concept NOW is much broader and powerful than the scrub jay's. She can have thoughts about the concept NOW without jumping off her seat. She understands that NOW is an *idea* that she can think about, ask questions about, and not only react upon.

However, there is something illustrative about the tardy professor's jumping off her seat. Why didn't she jump off her seat when she read "The meeting is at noon"? Only when she represented NOON as NOW did she jump. The answer, which philosophers of time have thought about in many ways, is that the A series is agent dependent and the B series is agent independent. I will not delve here into the nature of the A and B *series*. Rather, I will focus on A and B *representations*. B representations—determined by the concepts BEFORE, AT THE SAME TIME, and AFTER, concatenate dates *independently* of agent-relative temporal information, e.g. 'now', 'today', etc. It is thanks to our *conceptual* representation of time in terms of the B scale (dates or temporal marks that can be arbitrarily expanded and analyzed into smaller units) that we have calendars, birthdays, ceremonies and atomic clocks.

The cognitive power provided by our representation of time in terms of A and B representations is impressive. It allows us to date the age of the earth and to marvel about the future of our galaxy. The existence of calendars and scientific representations of time have dramatically transformed our temporal *behavior*, which we share with no animal on earth. Notice that it is the *combination* of agent relative A representations with agent independent B representations that gives us this remarkable cognitive power to represent time.[20] This can happen only because these are conceptual, as opposed to motor-intentional, representations. Therefore, I submit that there are three types of mental temporal representations:

a) Agent dependent motor-intentional temporal representations
b) Agent dependent conceptual temporal representations: A representations and
c) Agent independent conceptual temporal representations: B representations

Clearly, B representations are agent dependent in the sense that they depend on being thought by an agent. What I mean by 'agent independent' is that *their content is fixed independently of agent-dependent information*. This is why their truth-value does not change at different times. It is interesting that we have all these representations of time. How they interact (especially how motor-intentional time interacts with conceptual time) is an important research question. My guess is that motor-intentional representation plays a much bigger role than we have traditionally thought. It definitely grounds perception, gives us a skillful access to our environment, and allows us to navigate through the world without much thoughtful effort. But it is hard to overstate the importance of our conceptual representations of time. Just think of how much we depend on calendars, clocks, and all sorts of religious and scientific temporal representations. Our temporal behavior can be considered as one of the most dramatically distinctive manifestations of what it is to think like a human.

The A series presupposes the relations of the B series, but these relations have to be anchored in (and limited by) an agent relative time. It is for this reason (the agent-centered nature of A beliefs) that I conclude that *cultural time is determined by B-beliefs*. As a symbolic

[20] Clearly, quantifiers help boost these representations of the A and B series. Since quantifiers are part of the machinery underlying conceptual-linguistic cognition, I will not explain their role in any detail here because it would be distracting.

structure, the best way to characterize B representation is in terms of social conventions that transcend the agent-dependency of A representation. The study of B representations alone is extremely challenging, and a thorough overview of the many ways of representing time in terms of B representations *combined* with A representations requires serious multidisciplinary research. The key disciplines for the study of conceptual time (besides psychology and philosophy) are sociology and anthropology as they apply to scientific, religious, political and artistic conventions concerning B representation and how it influences A conceptual representation.[21]

The complexity of cultural time emerges from the lower-level temporal constraints (biological and motor-intentional). When structured in terms of phase transitions, which I describe in the next section, these constraints become *possibilities* for new representations. However, in the evolution of temporal behavior, there is a dramatically important temporal representation that makes us unique: the cultural convention we call the B series, which can truly be called mathematical time.[22]

6. Conclusion: Constraints and Transitions

I have argued that motor-intentional time is a crucial psychophysical link that bridges the gap between purely biochemical cycles and conceptual-intentional representations of time. But how are the transitions from biological to motor-intentional time, and then from motor-

[21] See Zerubavel 1979, 1981 and 2003 for work on sociology that focuses on conceptualizations of time in concrete social settings. See Gell 2001 for work on conceptual time from the perspective of anthropology that addresses the distinction between the A and B series. For work on aesthetics that emphasizes the importance of approaching the issue of conceptual time from various disciplines see Rodowick 1997.

[22] L. E. J. Brouwer founded mathematical intuitionism, inspired by how our intuition or "sensation" of time can be described in terms of the B series, particularly how the uncountable infinity of numbers correspond to the uncountable infinity of moments in which our perception of time can be decomposed. Brouwer describes the 'first act of intuitionism' as follows: "Completely separating mathematics from mathematical language and hence from the phenomena of language described by theoretical logic, recognizing that intuitionistic mathematics is an essentially languageless activity of the mind having its origin in the perception of a move of time. This perception of a move of time may be described as the falling apart of a life moment into two distinct things, one of which gives way to the other, but is retained by memory" (Brouwer 1981, 4–5). Immanuel Kant and Arthur Schopenhauer held similar views about the foundations of mathematics.

intentional time to conceptual-time, possible? In a sense, to say that biological time *constrains* motor-intentional time (and that, likewise, motor-intentional time constrains conceptual time) is informative but not explanatory. In order to give an explanation of these transitions, one needs to say more about the nature of the processes that lead to complex temporal behavior. Although it is not possible at the moment to provide a complete explanation about how life organizes itself in temporal fashion, there are critical aspects of temporal behavior that provide clues regarding the underlying mechanisms that support biological and psychological time. I will focus on the nature of the aforementioned transitions and explain how these transitions relate to hierarchical models of time.

As mentioned, our temporal behavior (and presumably the temporal behavior of many animals) is intentional in the sense that it is constituted by representations that are about time. Many philosophers have argued that there must be a fundamental cognitive basis underlying and unifying our experience of time. These authors relate our experience of time with our capacity for conscious awareness—a type of consciousness that might be described as consciousness *of* something, rather than *self*-awareness. Although this foundation for consciousness, which Immanuel Kant called the unity of apperception, is discussed in his *Critique of Pure Reason*, it was Edmund Husserl's work on time and consciousness (1893–1917) that provided a detailed account of it in phenomenological terms. At the phenomenological level, our experience and representation of time seems inseparable from our being *conscious of* objects and events. One way of supporting this claim, offered by Ernst Mach, is by contrasting spatial perception with temporal perception. As Mach said:

> Much more difficult than the investigation of space-sensation is that of time-sensation. Many sensations make their appearance with, others without, a clear sensation of space. But time-sensation accompanies every other sensation, and can be wholly separated from none. (Mach 1959, 245)

Mach's contention is that the inseparability of time-sensation from *sensation in general* suggests that time-sensation is omnipresent, and thus, *sui generis*. Husserl devoted a lot of effort to understanding the *sui generis* nature of time perception and proposed that "Time-consciousness is the original seat of the constitution of the unity of identity in general" (Husserl 1948, 73). The question that concerns us now, in the context of hierarchical models of time, is *how is it possible to experience time as a single unified representation that accompanies all sensation in spite of the fact that there are at least three levels of complexity*—biological, motor-intentional, and conceptual—that cannot be reduced to one another?

Clearly, biological constraints *determine* all the temporal behavior of an animal. However, the temporal behavior of an animal cannot be explained by, or reduced to, purely biological principles. The situation is very similar to the physics of condensed matter and its relationship to the principles governing life. The physics of condensed matter constrains and determines the way in which life organizes itself. But life cannot be explained by, or reduced to, purely condensed matter physics—other principles of organization that are well beyond the purview of physicists must do that job. This is very important to understand the nature of the transitions I discussed previously.

How does the unity of temporal experience and phenomenal consciousness relate to models of time representation? First, it is a *datum* that needs to be explained by these models. And second, it serves as a criterion to assess these models by favoring those that emphasize the *emergent* nature of temporal behavior, rather than the distinct and incompatible levels upon which different aspects of temporal behavior depend. One can focus on consciousness and the unity of time representation (or biological time or any other specific aspect of temporal behavior) and take it as the crucial issue to be explained. The ultra-specialization of the biological and social sciences actually forces researchers to take this path in order to produce results. However, one needs to go further and study the unity of the temporal behavior of an animal as a whole, because it is the only way in which the emergent nature of the transitions from the biological to the intentional can be highlighted.[23]

I said previously that motor-intentional time is a link that bridges the gap between biological and conceptual-intentional time. This suggests that there is a hierarchy of levels that needs to be accounted for in a model of time representation, and that each level is a necessary step towards more complex temporal behavior. Indeed, it is

[23] For an approach to the unity of time representation that focuses on the neurological basis as well as the phenomenology of the experience of time see Varela 1999 and references therein.

very intuitive to model time representation as a hierarchy of levels. For instance, Ernst Pöppel (1997 and 2004) has given very powerful arguments, based on experimental evidence, in favor of a hierarchical model of temporal perception. A key advantage of Pöppel's model is that it explains how temporal integration is possible and why layers of processing units are needed in order to eliminate complexity and generate the three-seconds window that frames all our experience: the phenomenal present. The hierarchy Pöppel is proposing is of *units* that generate a temporal output, which structures phenomenal content, thereby unifying time perception with consciousness.

Pöppel's hierarchical model improves our scientific understanding of the unity of temporal perception and consciousness. However, if what we are interested in is the temporal behavior of an animal as a whole, then hierarchical models fall short of explaining how temporal behavior is possible. I do not want to challenge hierarchical models of time, which are right about the *mechanisms* underlying biological time and time representation. What I want to challenge is the notion, suggested by these models, that time representation is *reducible* to the computational operations happening at each level (or unit) and that the transitions from the biological to the intentional are *steps* in a hierarchy of temporal processes, equally reducible to the component parts responsible for computing such processes. But how to make hierarchical (scientific) models of biological and psychological time compatible with the idea that the transitions between biology and psychology are not reducible to ever-smaller units, or levels of a hierarchy? I propose that the best way to solve this problem is to characterize these transitions as *phase transitions that emerge on the global properties of a system* (in this case temporal behavior) regardless of specific aspects of the underlying processing units.

I draw upon J. T. Fraser's hierarchical theory of time,[24] which is, because of its scope, compatible with the notion that transitions between organizational principles for temporal behavior are emergent properties of temporal behavior as a whole. My goal is to explain how, although different layers of complexity within the hierarchy of temporal processes *constrain* the way in which behavioral transitions are possible, these transitions occur at the temporal-behavior level, and the overall effect of the transitions is actually liberating rather than

[24] See for instance Fraser J. T. 1970.

limiting. I will briefly define what I mean by phase transition and then conclude by connecting this notion to my previous discussion on biological and psychological time.

In condensed matter physics, a phase is a state of matter governed by principles that emerge from the large aggregation of constituents (atoms) at a certain pressure and temperature. What is interesting about phases is the *exactness* of these principles (as well as the vast scope of processes they govern).[25] Transitions between phases are sharp and produce abrupt changes in organization, like the transition from liquid to solid. It would be incorrect to characterize one phase as more complex than the other, and I think the same is true about the principles governing biological and psychological time.

There is a sharp transition between temporal behavior governed by the circadian clock and the temporal behavior produced by motor-intentional time. If we think exclusively in terms of information processing, this change in behavior could be characterized as a phase transition. The principles governing biological time are constrained by biochemistry, while the principles governing motor-intentional time are constrained not only by biochemistry but also by the metric structure of analog representation. These principles are exact and the contrast between, for instance motor-intentional and conceptual-intentional temporal behavior, is as sharp and dramatic as the transitions of condensed matter physics. However, these transitions concern the organization of *behavior* and not the organization of the atomic aggregates or processing units.

Thus, I submit that it is incorrect to say that the biological basis is inferior *in complexity* or at a lower level in the suggested scale of a hierarchy. It is merely a different type of temporal behavior, or a different phase of temporal organization. The hierarchical model is accurate if we are thinking in terms of biological or computational processing units, and not in terms of the overall temporal information that constitutes the temporal behavior of an organism. Indeed, there is a hierarchy of constraints on processing units that starts with physical laws, then biological time, and finally culminates in cultural time. And it is in this sense of 'hierarchy', that is, a hierarchy of constraints, that time is best understood as the result of conflict between constraints. Thus, our biology constrains the way in which we represent time, but

[25] See Laughlin 2005, 206 and Laughlin and Pines 2000.

it allows us to share a common temporal experience with other animals, helping us understand their behavior. On the other hand, psychological time in a way liberates us from our biology, allowing us to represent time in terms of the A and B series.

If overall temporal behavior is our main subject of study, then there is a phase transition from motor-intentional to conceptual-intentional time, manifest in the behavior of the cognitive agent. However, if our main topic is the constraints on the physical instantiation of temporal information in the processing units, then psychological time can be understood as a cognitive capacity that relates the most all-encompassing temporal behavior (cultural time) to its biological basis. Therefore, although the constraints on temporal behavior are hierarchical and limiting (i.e., there is a limited set of possibilities as to which chemical combinations instantiate life and its temporal behavior), the informational phases that *emerge* from these constraints are equally complex and, in their own way, liberating. The A and B series emerge from the conceptual-intentional representation of time; metric representations of the environment emerge from motor-intentional time, and complex behavioral cycles that lead eventually to niche formation emerge from biological time.

The hierarchical organization of constraints limits the number of possible instantiations of information. However, the informational phases have principles of organization of their own, and we are just beginning to understand how these informational phases that go from the biological to the psychological transform, organize, and liberate temporal behavior from the hierarchical principles underlying its instantiation. It is because information assembles itself into the temporal behavior of an agent that it is more accurate to call the principles governing such a behavioral unity a phase, rather than a hierarchy. Understanding these phase transitions in temporal behavior might eventually open the door to the secrets of consciousness, the emergence of synchronic patterns in animal communication and many other fascinating aspects of life and its relationship to time representation and temporal organization. Each phase has, as J. T. Fraser proposed for temporalities (1993), its own complexity and degree of creative freedom.

References

Block, Ned. 1995. On a confusion about the function of consciousness. *Behavioral and Brain Sciences* 18: 227–47.

Brentano, Franz. 1995. *Psychology from an Empirical Standpoint*. London: Routledge.

Brouwer, L. E. J. 1951. Cambridge Lectures on Intuitionism. Pp. 40–110 in *Brouwer's Cambridge Lectures on Intuitionism*, ed. Dirk van Dalen. New York: Cambridge University Press.

Carman, Taylor. 1999. The Body in Husserl and Merleau-Ponty. *Philosophical Topics* 27(2): 205–26.

Cheng, R. K., MacDonald, C. J., and Meck, W. H. 2006. Differential effects of cocaine and ketamine on time estimation: Implications for neurobiological models of interval timing. *Pharmacology Biochemistry and Behavior* 85: 114–122.

Clayton, N. S., and Dickinson, A. 1998. Episodic-like memory during cache recovery by Scrub-jays. *Nature* 395: 272–74.

———. 1999a. Memory for the contents of caches by Scrub Jays (Aphelocoma coerulescens). *Journal of Experimental Psychology: Animal Behavior Processes* 25: 82–91.

———. 1999b. Scrub jays (Aphelocoma coerulescens) remember the relative time of caching as well as the location and content of their caches. *Journal of Comparative Psychology* 113: 403–16.

Fraser, J. T. 1970. Time as a Hierarchy of Creative Conflicts. *Studium Generale* 23: 597–689.

———. 1993. Change, Permanence and Human Values. Pp. 1–25 in *Time and Process: The Study of Time VII*, ed. J. T. Fraser and L. Rowell. Madison, CT: International Universities Press.

Gallistel, C. R. 1990. *The Organization of Learning*. Cambridge: MIT Press.

Gell, Alfred. 2001. *The Anthropology of Time: Cultural Constructions of Temporal Maps and Images*. Oxford: Berg.

Gibson, J. J. 1979. *The Ecological Approach to Visual Perception*. Hillsdale: Lawrence Erlbaum.

Goldman, A. I. 2008. *Simulating Minds: The Philosophy, Psychology and Neuroscience of Mindreading*. New York: Oxford University Press.

Goodman, Nelson. 1976. *Languages of Art*. Indianapolis: Hackett.

Husserl, Edmund. 1991. *On the Phenomenology of the Consciousness of Internal Time*. Boston, MA: Kluwer Academic Publishers.

———. 1997. *Experience and Judgment*. Evanston: Northwestern University Press.

Johnson-Laird, P. N. 1990 "Introduction to 'What is Communication.' " Pp. 1–13 in *Ways of Communicating*, ed. D. H. Mellor. Cambridge, UK: Cambridge University Press.

Laughlin, R. B. 2005. *A Different Universe*. Cambridge, MA: Basic Books.

———. and Pines, D. 2000. The Theory of Everything. *Proceedings of the National Academy of Sciences* 97(1): 28–31.

Mach, Ernst. 1959. *The Analysis of Sensations*. New York, NY: Dover Publications.

Merleau-Ponty, Maurice. 1962. *Phenomenology of Perception*. New York, NY: Routledge and Kegan Paul.

Millar, Andrew. 2003. A Suite of Photoreceptors Entrains the Plant Circadian Clock. *Journal of Biological Rhythms* 18 (3): 217–26.

Moore, R. Y. and Eichler, V. B. 1972. Loss of a circadian adrenal corticosterone rhythm following suprachiasmatic lesions in the rat. *Brain Research* 42: 201–06.

Panda, S. and Hogenesch, J. 1998. It's All in the Timing: Many Clocks, Many Outputs. *Journal of Biological Rhythms* 19 (5): 374–87.

Pepperberg, Irene. 2000. *The Alex Studies: Cognitive and Communicative Abilities of Grey Parrots.* Cambridge, MA: Harvard University Press.
Perry, John. 1979. The Problem of the Essential Indexical. *Nous* 13: 3–20.
Pöppel, Ernst. 1997. A hierarchical model of Time Perception. *Trends in Cognitive Sciences* 1: 56–61.
———. 2004. Lost in Time: A Historical Frame, Elementary Processing Units and the 3-Second Window. *Acta Neurobiologiae Experimentalis* 64: 295–301.
Ralph, M. R., Foster, R. G., Davis, F. C. and Menaker, M. 1990. Transplanted suprachiasmatic nucleus determines circadian period. *Science* 247 (4945): 975–78.
Richter, C. P. 1967. "Sleep and activity: Their relation to the 24 hour clock." Pp. 8–29 in *Sleep and altered states of consciousness*, ed. S. S. Kety, E. V. Evarts and H. L. Williams. Baltimore: Williams and Wilkins.
Rizzolatti, Giacomo. 1988. Functional organization of inferior area 6 in the macaque monkey. II. Area F5 and the control of distal movements. *Experimental Brain Research* 71(3): 491–507.
Rizzolatti, G., Fadiga, L., Gallese, V., & Fogassi, L. 1996. Premotor cortex and the recognition of motor actions. *Cognitive Brain Research* 3 (2): 131–41.
Rodowick, D. N. 1997. *Gilles Deleuze's Time Machine.* Durham: Duke University Press.
Stephan, F. K. and Zucker, I. 1972. Circadian rhythms in drinking behavior and locomotion activity of rats are eliminated by hypothalamic lesions. *Proceedings of the National Academy of Sciences* 69: 1583–1586.
Treisman, Michel. 2006. The Perception of Time: Philosophical Views and Psychological Evidence. Pp. 217–45 in *The Arguments of Time*, ed. Jeremy Butterfield. New York: Oxford University Press.
Varela, Francisco. 1999. The Specious Present: A Neurophenomenology of Time Consciousness. Pp. 266–316 in *Naturalizing Phenomenology*, ed. Jean Petitot, Francisco Varela, Bernard Pachoud and Jean-Michel Roy. Palo Alto: Stanford University Press.
Zerubavel, Eviatar. 1979. *Patterns of Time in Hospital Life: A Sociological Perspective.* Chicago: University of Chicago Press.
———. 1981. *Hidden Rhythms: Schedules and Calendars in Social Life.* Chicago: University of Chicago Press.
———. 2003. *Time Maps: Collective Memory and the Social Shape of the Past.* Chicago: University of Chicago Press.

CHAPTER TWO

THE BATTLE FOR TIME IN THE BRAIN

Peter A. Hancock

"Living backward!" Alice repeated in great astonishment. "I never heard of such a thing!"
"—but there's one great advantage in it, that one's memory works both ways."
"I'm sure *mine* only works one way," Alice remarked. "I can't remember things before they happen."
"It's a poor sort of memory that only works backward," the Queen remarked.
—Lewis Carroll, (1871) *Through the Looking-Glass, and What Alice Found There.*

Summary

All our conceptions of time spring from our brains. However, the brain itself is a structure that has evolved over several millions of years. In the process of evolution, nature often has to erect and integrate newer structures and functionalities on existing systems and capacities. I argue that many fundamental dimensions of human response such as memory, dreaming, and the persistence of the consciousness of self derive from the interplay attendant on the integration of these stages of brain development. I argue that time has played a central role in this progression. From the earliest notion of physiological time as simple duration through the complexities of the spatio-temporal coordination of action to the ability of the frontal cortex to anticipate future conditions a river runs through it and that river is time. Provocatively, I have termed these sequential epochs of evolutionary integration as the 'battle' for time in the brain.

1. Preamble

It is quite possible that physicists, at some future point, will solve the problem of duration. Indeed, this is one of their manifest aspirations (Hawking 1988). A closed-end description, at some level of analysis, of the interrelationship between object and object would seem to represent at least an intellectually feasible goal. However, physics (in its present incarnation at least) will never resolve the problem of time. This impasse derives from the fact that, unlike duration, time is

Jo Alyson Parker, Paul A. Harris and Christian Steineck (Eds), Time: Limits and Constraints, pp. 65–87

not a property of object-to-object relations but is rather an intrinsic property of the living observer (that is subject-object relations; and see Gibson 1975; also Russell 1915). Indeed, from the special and general theory of relativity, we understand the critical requirement for the presence of a living, intelligent observer (Einstein 1905, 1950). Consequently, if we wish to address the mystery of time, and especially if we wish to understand the subtleties and nuances of the human apperception of time, we have to look in the human brain. Even referring to this organ as *the* brain encourages us to the view that it is a single, discrete entity. Certainly this is how the brain is conceived of in the everyday world. However, the brain is a multi-leveled, multi-structured, modular assemblage. It is an evolutionary palimpsest in which newer structures have developed and have subsequently been overlaid upon older ones. Thus, progressively more complex response strategies have been super-imposed upon simpler and more primitive behavior patterns. Our own personal phenomenological experience tells us that this hybrid system "works" and indeed many would say that the brain works harmoniously. However, this perception of normality is largely an illusion of everyday experience. As with the purported unity of consciousness itself, this belief in the normality of experience is a result of habituation rather than understanding. In this chapter, I suggest that there is actually a continuing battle for control within the brain. The landscape of this conflict is time (Fraser 1966, 1). To illustrate this argument at a general level, I use a popular tripartite division of the brain. However, in actuality there are many modular elements within the brain that bid for control. In the process of human evolution, each modular advance has conferred a sequentially increasing survival advantage. Through juxtaposition of these respective brain systems, I shall look here to resolve some fundamental paradoxes and conundrums in basic behavioral phenomena such as sleep and memory, as well as to explore the further ramifications of what I have asserted for a greater understanding of time itself.

2. Introduction

All time, as we know it, is a product of the mind enacted in the brain. Our experience of time and our understanding of it result from the interaction between several, somewhat discrete brain structures. These deal with the temporal processing of changes in the pattern of environmental stimulation and engage in efforts to anticipate future conditions. Intrinsic evolutionary constraints are placed upon later brain structures because they are necessarily erected on existing components. The earliest of these mechanisms subserves a capacity that is shared by all living organisms (Schrödinger 1944, 1), namely, the ability to retain information concerning the persistence of self. As all living organisms possess both spatial and temporal extent, a vital survival requirement is that they be able to distinguish self from non-self. Spatially, this distinction is accomplished at the respective boundary layers between the organism and its ambient environment (but see Ehrsson 2007). However, in addition to spatial separation, a comparable temporal distinction must also be sustained. In humans the most evident emergent action of this imperative is found in the supra-chiasmatic nucleus (SCN) of the anterior hypothalamus. However, as a necessary character of all living things, this function of sustained spatio-temporal independence is almost certainly cellular in its most primitive form. Unfortunately, the categorical confusion between time as a subject-object relationship and duration as an object-object relationship (see Russell 1915) has led to the inappropriate naming of this mechanism in humans as—the "internal clock" (and see Hancock 1993).

First then, as a characteristic perhaps of life itself, there is a necessity for an awareness of self-persistence. However, simply sustaining some form of differentiation between the self and the rest of the environment (non-self) is to remain vulnerable to all of the vagaries of that environment. The next step of progress in securing greater adaptive capabilities is for the self to attain a degree of self-direction. This enhanced capacity for flexible, self-directed response is clearly an evolutionary advantage, and indeed most such living organisms now act with differing degrees of self-control. However, such a perception-action advantage comes at a cost since it now implies the necessary imperative to synchronize the organism's activities with the spatial and temporal constraints of the environment (Gibson 1975, 1979). To work efficiently, this perception-action system must permit the organism to achieve critical synchronous responses. Such synchronizations can be, and in their primitive forms obviously are, independent of any necessity to reference external and arbitrary time-keeping conventions (see Hancock 2005; Hancock & de Ridder 2003; Hancock & Manser

1997).[1] With the selective evolutionary advantage of social interaction, humans have developed external formalized referents such as clocks, watches and timekeepers of all forms of fabrication (Cippola 1967, 1). These creations are obviously useful tools. However, their output in the form of a train of homogenous intervals such as seconds is still incompletely integrated with the intrinsic human perception-action responses (Hancock 2002). In essence, the innate human capabilities to "time" their actions in the environment have been around for many millions of years, while the formal methods of "clocking" our world are only millennia old at best (Fraser 1987, 45). Thus, while we have the capacity to "time" our environmental actions, we do not necessarily need to do so.[2] If the lowest level function of the brain provides recognition of self-persistence in space and time and the next super-imposed perception-action system deals with the imperative of synchronized responses to the immediacy of present demands, where is evolution to go in order to improve any organism's temporal capacities? The answer nature returns in the human brain has been to find a way to go faster than so-called real time.[3] Largely centered in the frontal cortex, humans exceed the constraints of "real time" by generating "what-if" scenarios that permit the anticipation of possible future courses of events, especially under stressful and threatening conditions (Hancock and Weaver 2005). This highest level temporal mechanism, which acts to project, compare, and confirm possible courses of future action, also shows that most of memory itself is largely created by the functional requirement to anticipate the future (and see Hancock 2005; Nairne, Pandeirada, and Thompson 2008). This modular division within the brain also explains the nature and function of the types of sleep we experience and particularly how these respective modular

[1] It might well be observed that only human beings "keep" time, although almost all living organisms still have to synchronize their activities to match the external environmental conditions.

[2] For example, in athletic competition it is necessary to use an external referent to establish a world record, which record is then open to challenge by all. In contrast, the winner of the Olympic Games has to beat all of his or her competition in the Games but not necessarily get anywhere near the world record to do so. Thus "relative" activities such as head-to-head competition (for example, predator vs. prey) need no reference to any external timing system. Absolute, social actions (such as breaking world records), obviously need this common, external, and socially acceptable measure.

[3] This is not to imply that many other members of the animal kingdom do not possess anticipatory capacities; they assuredly do (see Wang and Yuwono 1995; Wilkie, Carr, Galloway, Parker, and Aiko 1997).

elements periodically form a truce in the battle for time in the brain. It is to the explicit consideration of this "battle" scenario that I now wish to turn.[4]

3. Let Battle Commence

The foregoing is essentially an introduction but what follows represents a largely non-technical account of the concerns at hand. Thus, although reference is made to various contemporary theories and notions in the cognitive and neurosciences, it is explicitly written in order to be accessible to a more general audience. Therefore, while the brain is modular in its structure, containing multiple interacting components, I here use a three-part division, or the notion of the triune brain (MacLean 1990), largely for explanatory convenience. In respect to time in the brain, the traditional and accepted account is that each part of the brain cooperates fairly seamlessly with the others in order to produce what we experience as an apparently harmonious and integrated temporal experience expressed as normal consciousness. This conception is confirmed directly in our own personal experience as an internal observer of our own being. In such experience, we do not encounter any obvious discontinuities, disruptions, or dysfunctions that might lead us to suspect otherwise. Obviously, there are neuropathologies of time in which some individuals do experience clinical problems, but these are, by definition, non-normal states (see Cohen 1967; Fischer, Griffin, and Liss 1962). Thus, disorders such as schizophrenia (Spencer et al. 2004) and Korsakov's syndrome (Mimura, Kinsbourne, and O'Connor 2000), among others, are of interest as "windows" on the way in which "normal" perception has somehow been altered or perverted.

4. The Molyneux Problem

However, there is an alternative to the idea of harmony that can perhaps be illustrated through consideration of a classic issue in perceptual psychology that is generally known as the 'Molyneux' problem

[4] The term "battle" is used expressly here to stimulate a degree of controversy. Many see the brain as working as an harmonious whole, and indeed it is this traditional perspective that I wish to challenge.

(Molyneux 1688; 1693). Understanding the Molyneux problem requires a temporary but critical excursion away from the main theme of the present work. However, it is an excursion that repays a brief interruption. The Molyneux problem is named after an Irishman, William Molyneux, who posed the following question to the philosopher John Locke in relation to some of his statements made in reference to his pivotal text *An Essay Concerning Human Understanding* (Locke 1690).[5] Molyneux asked the following question. Suppose a person had been blind from birth but knew the difference between a cube and a sphere from his or her experience of touch alone. If that individual now gained the capacity for sight, would he or she be able to tell the cube from the sphere simply by sight alone? It is a question related to what we now refer to as the "binding" problem, which asks how ongoing and accumulated sensory experience is combined into a single reality (and see Hancock 2005). What is perhaps most interesting is that one can go much further than the question that Molyneux posed. Such an extension could be stated as follows. Could an individual, denied from birth all forms of external sensory experience, actually think? In essence, what could be known from pure contemplation of self? It is an issue that philosophers will readily recognize as one of their own special conundrums (and see Kant 1781, 1). One answer is that any impoverished observer who was doomed to such a fate could possess a primitive sense of time. That is, as a living being, an individual would have access to an internal sense of the persistence of self. What they could further make of, or from, this persistence is a most interesting line of philosophical inquiry, but it is one for another occasion. However, it is this primitive persistence of self that triggers a subsequent line of thought. For example, where is this persistence encoded precisely? Further, how does this capacity for persistence function in complex, multi-cellular organisms. Finally, is this sense of persistence a fundamental characteristic of life itself (and see Schrödinger 1944, 1), as I have suggested? Indeed, from a personal perspective, it was the notion that any subsequent sensory integration would have to be erected upon this primal capacity that caused me to once again re-visit basic assumptions about the seamlessness of integration of "normal" experience. This line of exploration naturally leads to an explicit con-

[5] Further specific information on the Molyneux problem, its motivation and origin can be found at: http://plato.stanford.edu/entries/molyneux-problem/.

sideration of what it would be like if these two general brain mechanisms of temporal experience did not cooperate harmoniously? What would be the case if they actually battled for control of time in the brain?

To derive an account of what that battle might look like, we have first to examine the gross morphology of the human brain. As can be seen from Figure 1, the brain is no homogenous medium but shows indications of parsing between a number of obviously differentiated structures. Conceptually, this perspective can be conceived as an assembly of modular structures (and/or processes). Such modules and processes would provide a strong degree of response flexibility to counter the uncertainties of the future against which evolution struggles (see Barrett and Kurzban 2006). Here is not the place to go into a detailed exposition of neuro-anatomy (but see Nolte 2001). Thus the following argument is based on the general proposition that the brain can be differentiated vertically into three general regions (see Maclean 1990). First, the structure at the base of the brain is the brain stem (see Figure 1). Second, the structure that surmounts the brain stem is the limbic system. Finally, the structure that is super-imposed on the limbic system is the neo-cortex. This general differentiation has been refereed to as the 'triune' brain (MacLean 1990). This conception postulates the division into the R-Complex, the Limbic system, and the neo-cortex. For the sake of didactic simplicity I shall here refer to these as the "lower," "middle," and "upper" level of the brain, although this refers to their physical location, not necessarily their place in the hierarchy of control.

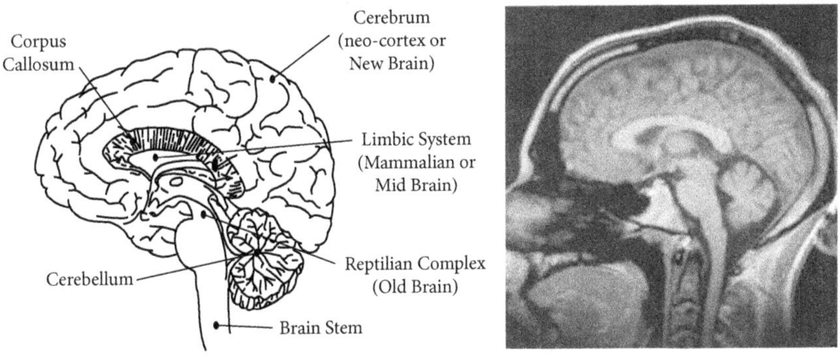

Figure 1: Gross anatomy of the human brain as shown in saggital section. The primary point of present interest is the manifest vertical differentiation of the structures shown.

5. Sleep and the Battle for Control

The brain stem, which is primarily of reptilian origin, has the advantage of being first in existence. In both the law and the brain, possession is nine-tenths of the battle. This seminal form of the brain owns the metaphorical high ground, especially in terms of the "basic" drives for sustenance, sex, or more generally, survival. The middle level, that is the limbic system, must necessarily then battle this pre-existing structure for its own periodic (and attemptedly exclusive) control of the brain. This event occurs daily but in its evolutionary origin this longer-term struggle for control of the brain is inextricably linked with the growth of thermal self-regulation, or homeothermy (and see Hancock 1981; Prosser 1991, 984). Homeothermy is the anatomical expression of the next step, or saltation, in evolution (Gould and Eldredge 1993). Not only does this capacity for homeothermy permit the general exploration of different spatial and temporal opportunities (Crompton, Taylor, and Jagger 1978), but it represents the site in the brain at which control of general circadian functioning of the organism is housed. It is little wonder that these structures and the evolutionary struggles they represent occur neuroanatomically at the interface between the reptilian brain and the limbic system. Thus the long-term integration of the brain stem and the limbic system is actually a type of warfare, while the short-term conflict represents a daily battle for control. This battle, unlike an all-out conflict, must be tempered with the needs of the organism to survive. So, in general, various levels of the brain *do* live in an uneasy alliance, but periodically (and that is, on a daily basis) each fights for and succeeds to supremacy. The middle level, limbic system begins to win its own individual battle as the lower-level reptilian brain starts to weaken.[6] This weakening occurs as the sun goes down.

The lower level brain stem can be influenced by the oscillation of the circadian cycle (Marshall and Donchin 1981). Such mutual influences show that although I have characterized this as a battle, there is

[6] The reptilian portion of the brain is of poikilothermic origin. That is, in its earlier stage of existence it derived its motive power largely from the heat of the natural environment. In its later incarnation with its homeothermic characteristic, the organism has been selected for its freedom of action. However, homeothermism, or self-generated constant body temperature, actually comes at great cost. To sustain this constant internal body-temperature the organism must be in continual search of food as the source of the calories to fuel this constant internal fire.

a strong degree of inter-dependency and indeed integration between such levels. It also illustrates how elusive the idea of "control" is itself. The limbic system starts to dominate as the circadian cycle descends toward its lowest point in the very late hours of the evening and the very early hours of the morning. The morphology of this rhythm is certainly related to the light-dark (diurnal) cycle associated with the Earth's rotation, but the organism's circadian cycle itself is not inflexibly locked to this period of rotation. Indeed, a completely inflexible association would inhibit opportunities for evolutionary adaptation of the form of spatio-temporal exploration indicated by Crompton and his colleagues (1978). Such immalleability would, for example, preclude extensive migration across multiple time zones and so curtail the opportunity to explore other diverse and accommodating ecological niches.

The ascendancy of the middle level occurs then in the very late hours of the night and the earliest hours of the morning when the ambient temperature of the surrounding environment itself reaches its lowest levels. This correlation is no coincidence and is causal. The take-over by the limbic system is, however, largely a pyrrhic victory. A degree of supremacy is only gained during the hours of sleep and quiescence. Therefore, this first characteristic of sleep could be considered as an uneasy truce between the lower and middle levels of the brain. This accord is fundamentally dictated by the issues of energy conservation and also the concurrent and critical necessity to cement memory for the learning of basic psycho-motor sequences (and see Huber et al. 2004; and see also Stickgold 2006). Many organisms exhibit such slow-wave sleep and use this self-same truce as an opportunity to enhance the permanence of learning (see Smith 1985). During sleep, each of these brain components tacitly agree to this armistice. The lower level does this by diminishing part of its influence (which to some degree is inevitable given the state of the sun as the source of energy). The middle level cooperates in this enterprise by seeking a quiet place in which there is no stimulation for the ears, closing off light from the eyes, reducing the temperature differential between the skin and the immediate thermal surround while cushioning acute tactile stimulation in order to enhance perceived comfort and to reduce the response demands of the surrounding environment (that is, making a bed, nest, etc). These pro-survival conditions optimize the facilitation of the processing of psycho-motor learning in the absence of competing "noise" from an otherwise very active central nervous system (see Stickgold

2006; Stickgold and Walker 2007). To illustrate this effect, a number of the cited studies have shown that sleep, not the passage of time per se, can greatly improve one's perceptual motor skill acquisition. Simplistically speaking, sleep, a respite in the battle for time in the brain, provides a landscape wherein the brain can concentrate mostly on learning and novel ways of adaptation (Huber, Gilardi, Massimini, and Tononi 2004). The body is not inert at this juncture but is very energy efficient, and movements here are a combination of those required for learning and those required to maintain optimal comfort for sleeping. It is at this time when some individuals "sleepwalk." In so doing, they often engage in routine behaviors. It is as though the underlying learning process is actually being played out in full orchestration, rather than the more horizontal tossing and turning of what are termed more normal individuals. Virtually all animals experience some form of sleep[7], and this, as noted, is partially the result of the compromise between intrinsic poikilothermic tendencies and the later addition of homeothermic independence. However, this low-level skirmish is not the only battle for time in the brain.

As is shown in accompanying illustrations, the three-level division of the brain can be presented either as a general descriptive relationship (Figure 2) or a more formal interconnected model (Figure 3). The uppermost level of this three-part description is here referred to as the cognitive clock. As to function, the cognitive clock is largely located in the frontal cortex. This temporal element of the brain bids for control in order to run a series of critical "what if" simulations. The running of these simulations subsequently permit a "faster than real-time" response in complex circumstances that require instantaneous response in the waking state (and see Hancock and Weaver 2005). Consciousness itself can well be characterized as a series of constant comparisons between what is currently being experienced and what has been previously anticipated by such "what-if" simulations. However, if these pattern-matching experiences do connote consciousness, it is not possible to create (and run simulations of) such scenarios on top of reality itself. If such a process were attempted, the individual would become disoriented with respect to reality and eventually show

[7] Indeed, it is most interesting that, for example, horses spend over 90% of each day standing and can sleep while standing and yet they must lie down during REM sleep (see Morrison 2003).

Cognitive Clock

(Second Order)

Future-Anticipative

Dennett Schacter

[Mismatch Comparator of
Possible vs. Actual Information]

**Sensory
Chronocomparator**

(First Order)

Fast-Responsive

Iberall Poppell

[Rate of Change of Sensor Systems]

Internal Clock

(Zero Order)

Slow-Continuous

Hoagland Hancock

[Continuous Duration Propagation]

Figure 2: Tri-level description of the model of human temporal capacities and their respective interconnections (after Hancock et al. 2005). The lowest level, the "internal clock" dispenses a continuous, analog signal whose role is the sustenance of the persistence of self. It is very rare that this capacity is suspended, that is, someone believes that no time has passed, even though a sidereal interval has actually passed. The second level, the sensory chronocomparator, provides comparison capacities across differing forms of sensory input and effector systems (Hancock 2005). That these systems *can* interact in reference to external temporal mechanisms is a tribute to their adaptive capability, not an operational necessity. The main function of this level is the synchronization of perception-action with the external spatio-temporal constraints of the world (and see Calvin 1983). Finally, the third level is the "cognitive clock." In terms of function, its role is to go "faster than real-time" by anticipating the future. It cannot do this in terms of responding but it can look to anticipate future probable events. This anticipation is only helpful if it can lead to appropriate response. The cognitive clock spends its existence searching through perception for anticipation matches. In non-real time—that is, in dreams—it tries all sorts of possibilities and unlikely courses of events. Thus the perplexing nature of dreaming. It is this "cognitive clock" and the ramifications of detailed future predictions that represent the essential characteristic of the human condition.

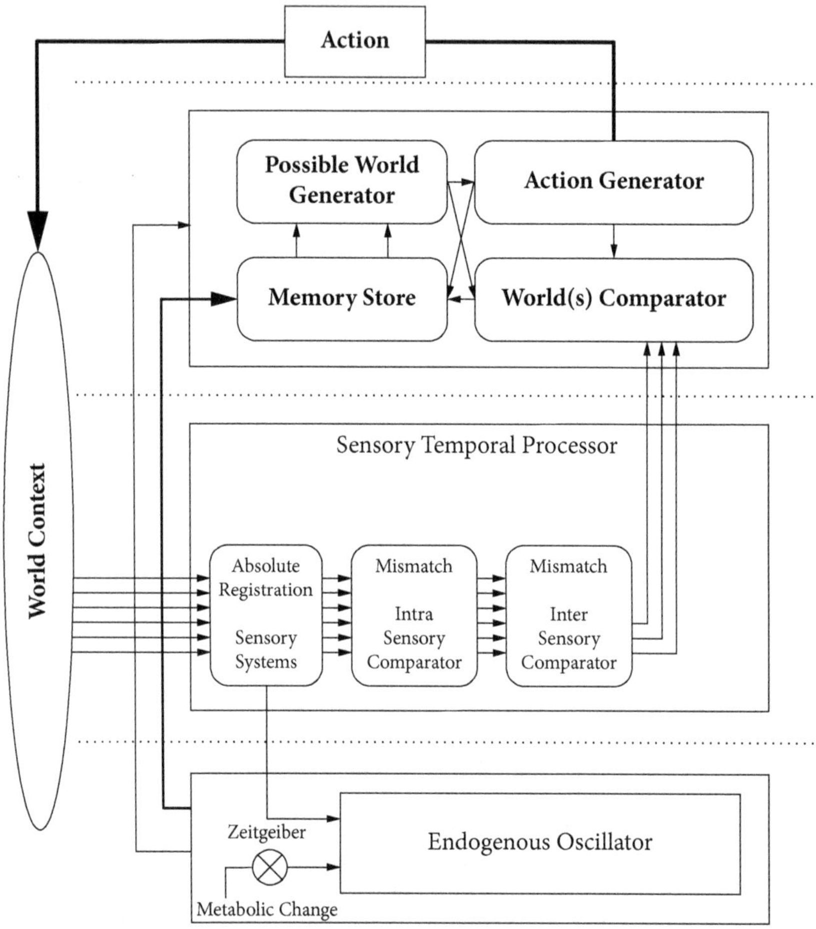

Figure 3: Model of human temporal capacities and their respective interconnections (after Hancock et al. 2005). This expresses the same basic conception as the previous descriptive illustration except that the differing levels are now articulated in more detail. The lowest level responds to internal and external zeitgebers; the middle level is now responsive to the context of the external world and through sensory integration provides perceptual displays that are used for the world comparator at the upper level. Interacting with memory stores and possible worlds, actions are resolved either by an immediate pattern match or a piecemeal solution based upon reactive response.

schizophrenic tendencies (Pearl and Berg 1963). This same form of disorientation is evident also in the behavior patterns of those who are sleep deprived.[8] Therefore, the cognitive clock must have its chance to run the more bizarre (that is, those scenarios further away from the probability of reality-matching) simulations in order to refine its arsenal of "what-if" propositions that primarily allows humans to go "faster than real-time" (and see Hobson, Pace-Schott, and Stickgold 2000). These are dreams. Thus, the cognitive clock must be refined and tuned, but it cannot do this whilst actually working on the imperative dynamics of the world itself, that is, while responding to the demands that it faces during conscious experience (Horne 2000). Thus, in most circumstances, we dream while we are asleep.

It is in the frontal cortex that the meta-levels of adaptive capability are entrenched. Hence when\does the cognitive clock supersede the two lower levels of control? The answer is: When both are at their weakest at just past the lowest point of the circadian cycle. Thus, REM begins to dominate in the latter phase of the sleep cycle. However, there is a major problem. This upper level "cognitive clock" wants to run full-scale perception-action, gross body simulations—not just the simple movement based sequences of psycho-motor learning. It is not enough to just conceive of possible future scenarios; this upper level wants to "run" these cognitive simulations in all their full, action-based glory. The upper level wants to engage the body in order to do this. However, the two lower levels are both constrained by the issue of energy usage, as well as the concerns for the actions of diverse, nocturnal predators. If, for example, in sleeping the human starts to thrash about too wildly and erratically, then not only does the individual lose precious energy but that individual also becomes an obvious target for attack by some violent predator. Despite the presence of such benign, modern, ecological niches as suburban bedrooms, the upper level still has to run its simulations but without engaging the full muscular system. The brain still engenders output signals, as it normally would, but such signals are prevented from getting to the muscles. Thus, during REM sleep, the brain is extremely active, but the body is very largely inert. This is because the required learning is largely cognitive in nature and composed of context-contingent response strategies. These are not the psycho-motor sequences of learning enacted by the lower

[8] Parenthetically, this is why sleep deprivation is a common and, unfortunately, effective form of torture (and see Hancock 2003).

levels of control and referred to previously. This explanation of events serves to address at least two of the central mysteries of sleep. As night goes on, the upper level dominates, giving over more and more time to REM (simulation testing) sleep. If disturbed during REM, a person might, on regaining full consciousness, wake during one of these dreaming episodes. At such a juncture, the person is able consciously to survey the results of one of these running simulations. Such events are experienced as dreams and nightmares.[9] Nightmares foreshadow radical survival situations. They are aversive because they represent possible future encounters with survival-threatening situations. In contrast, dreams are the happier foreshadows of desired goal fulfillment. Slow Wave Sleep is primarily an energy and low-level learning compromise while REM is the comparative compromise in terms of cognitive energy.

Thus, the battle for time in the brain explains the two primary forms of sleep, which are superimposed one upon the other. It also explains why lack of sleep itself is, in general, not fatal. One doesn't die of failure to run "what-if" scenarios or the acquisition of psycho-motor skills. Rather, an individual deprived in this manner just gets more and more confused about reality and slower in anticipating and responding to possible (and actual) future threats. The demise of an individual under these conditions does not derive from a lack of energy per se. However, in a nasty world, the progressively more enervated individual becomes progressively more vulnerable. It is the capacity of any roaming predator (such as powered vehicles in the modern world) that now exploits tired and confused individuals' failure to respond. In this way fatigue is still very much a modern killer (see Desmond and Hancock 2001). Thus, strictly speaking, sleep is not obligatory. However, if the person wishes to function effectively, he or she is well advised to get a good night's rest (Meddis 1977, 1). As far as possible, the brain will compensate for degrees of sleep loss to the extent that it can. It is only when this becomes a radical and chronic level of deprivation that behavioral disturbance become evident. The failure experienced under sleep loss has an antithesis. That is, there are occasions when the top-down "what-if" scenarios practiced by the frontal cortex during

[9] Since, these simulations must necessarily consist of attempts to integrate our recent experiences with our capacities for prospective planning they almost inevitably focus on potential survival issues. Such simulations play "what if" scenarios with some of our greatest fears and greatest aspirations so that we might survive and indeed prosper if we have to actually meet them in the waking state.

sleeping match exactly with the bottom-up stimulation of a moment of reality. These perfect matches (which statistically we are justified in expecting) are experienced by the individual as episodes of "deja-vu" (see Cleary 2008). If such "top-down" projections over-dominate then an individual might experience sensory and perceptual distortion, perhaps similar to those reported in so-called "paranormal" experiences.

As with all forms of behavior, there are large individual differences. Thus, there are people for whom each respective level of control proves either very strong or very weak. As a consequence, a percentage of the populace are insomniacs while, in contrast, others such as neonates and teenagers sleep for extended intervals. These different behavioral outcomes depend upon which of the modules of the brain are most advanced at each stage of maturation. The REM sleep of babies for example, indicates a lot of simulation "run-time" but little scenario input data. The sleep of neonates must be even more of a "*blooming, buzzing confusion*" than their waking world (James 1890, 1). Teenagers also show the effects of an increasingly powerful cognitive clock and thus upper level adaptive learning through their extensive hours of sleep.

6. The Mystery of Memory

One of the greatest of all conundrums with respect to time is the apparent memory paradox that life is lived forward but remembered backward. It is possible that an understanding of the function of the respective brain mechanisms of time that have been presented can serve to address and explain this paradox. First, it is obvious that human memory is not a complete and veridical record of all past events in the lifetime of the organism. Human memory is selective, sporadic, fallible, and evidently incomplete (Schacter 2001). Why is this so? It has been suggested that even the phenomenal storage capacities of the human brain could not contain all the information assimilated during a single lifetime. However, it is clear that memory is incomplete in rather special ways. We do not, for example, appear to forget on a consciously selective basis. Rather, what remains with us are special moments of particular pertinence and relevance.[10] It is true that we

[10] Interestingly, collective memory at a social level has the self-same principles, and we call this collective memory—history.

personally have a sense of autobiographical continuity, but when we are asked to recall our past, it is particular moments which stand out and not a detailed litany of any specific interval. Thus, we might well remember snapshots of events, such as birthdays, but not whether we had a cup of coffee on a particularly specified day. What is the purpose of this form of selectivity? The answer appears to lie in the general purpose of memory (and see Dudai and Carruthers 2005). From the perspective adopted here, the function of memory is overwhelmingly designed to anticipate the future (see Carroll 1871; Nairne and Pandeirada 2008; Schacter and Addis 2007).

Earlier it was indicated that the identification of the continuity of self in space-time was a primary function of all living systems. It is this level of functioning that underlies the autobiographical continuity of self. It allows us to continue to identify ourselves as ourselves on a moment-by-moment and day-by-day basis. This autobiographical capability supports the architecture of higher level functioning that informs our consciousness of this personal continuity in the face of the challenges of everyday life. However, memory for specific events (that is, episodic memory) is often tied to emotion. It is these emotion-contingent memories that are laid down at the crucial points in our existence. Such memories are often associated with extremes of all emotion but are often triggered at times of extreme stress (Hancock and Weaver 2005) when we are involved in battles for survival. That this stress may be either distress (as in adverse conditions) or eustress (in which something extremely pleasant is occurring) appears somewhat immaterial to the process of establishing discrete memories themselves. These occasions are often the subject of a form of after-action review in which the appraisal process is critical to the perception of stress itself and its subsequent memorial foundations. Thus our memory for specific episodes represents the way in which the survival process is gearing us toward dealing with stressful events in the future. The fact that the information so contained is exceptionally relevant to us as individuals in an autobiographical sense is immaterial to the more general progression of evolution itself. Evolution as a process only cares[11] about the past to the extent that knowledge of the past can help one anticipate and deal with the future. Thus memory can be con-

[11] In a true sense, evolution does not "care" about anything since that implies a teleology, which is almost certainly false in this case (and see Hancock 2009).

ceived of as a "string of pearls" in which the discrete events of episodic memory (the pearls) are strung out along the line of autobiographical continuity (the string). The analogy can be taken still further in that, in general, the size of the pearls (that is the intensity of the episodic memories) tend to covary with distance from the present in the same way that a string of pearls often has a large central pearl (James's 'specious present') with others of diminishing size on either side.

That memory itself is also distributed (generally) across the whole of the brain is a form of defense against discrete and drastic failure. That is, needing this information for survival to be available on demand, as future occasion requires, necessitates that it not be confined to any one single location. For if it was so confined and if this location were somehow damaged or immobilized, then one's ability to use the highest level of temporal capacities (what is referred to here as the "cognitive clock") would be largely obviated. An individual, constrained only to reactive response, would have a very limited chance of survival in a non-technical, non-supportive world. We see shades of this sad state in those individuals who suffer from Alzheimer's disease, which unfortunately and evidently influences this memory storage and retrieval function. Now the apparent paradox of memory that was noted earlier is laid bare and can be readily resolved. The fact that life is remembered backward is simply an artifact of our own personal consciousness, which recalls these events in an autobiographical trail. It is this autobiographical trail that is itself then formally elaborated into our general conception of time as composed of past, present, and future. This observation itself explains a number of allied observations in human as well as animal activity.[12] The conceptual framework that I

[12] One incident in my own life I found very instructive. I was present at the death of my Grandfather, who lived to the ripe old age of 100. I was sitting with him in the hospital room and he was murmuring to himself. My mother, never the most patient of individuals (even though a nurse herself) was desperately trying to talk with him and understand what he was saying. Presumably, she thought he was asking for something and her nursing instinct took over. Hearing him talk of some earlier part of his life, she assured me that he was 'losing it' and that this was not so unusual with terminally ill individuals. I listened carefully however and noted that he was muttering about his part in the invention of the rotary lawn-mower! Some time after his death I found out that indeed my grandfather had been employed in a carpet-weaving factory and had been involved in the development of a tool to cut and smooth the pile of each carpet, which was held vertically for this action to occur. Eventually, the horizontal version of this self-same instrument became the basis for the rotary lawn-mower. Of course, the full story is more complex than this. However, in retrospect I understood very well what was happening in that hospital room. My grandfather was dying and,

have advanced here thus helps explain certain puzzles associated with basic human processes such as memory and sleep. However, I do not wish to suggest that these are the only ramifications of the model I have put forward (cf. also, Cleeremans and Sarrazin 2007). Sleep and memory are the two aspects that have been discussed in this present paper. However, other temporal phenomena, such as decision-making, déjà vu, and dream content, can also be encompassed by the present form of explanation.

7. The Parliament of the Mind—The Congress of the Brain

The foregoing has largely represented time from a brain-based perspective. It has dealt with how some issues concerned with sleep, memory and associated phenomena may be approached by an examination of the evolution of the interaction of the modular structures in the brain. However, there is an alternative level of analysis with respect to human time as represented by the emergent property of mind. These respective levels of description (mind and brain) may be different in their form, but an integrated account of their actions must be a coherent one. The present case of sleep and time provides an interesting intersection. What has been described in the present chapter takes place largely in the absence of the conscious experience of time. So, for example, we are consciously aware when we wake up in the morning that time has passed since the onset of sleep (even in conditions where the external, environmental zeitgebers give us no clue as to the actual time of day or night; see Siffre 1965, 1). However, we are not directly aware of the content or even the duration of that passage of time (except in this very general sense). However, at some level, the

I assume, he knew it. My mother's actions were largely guided and oriented by an imperative for the future. But he had no future and he also knew that as well. In his last moments, both the future and even the present began to fade in importance and all that was left was for him to survey some of the most important moments of his past life. While this eventuality will come to us all, it is the fate of human beings, because of the functioning of the highest level cognitive clock, to know that they will individually and personally cease to exist at some point in their future. As this inevitable event approaches, the mandate of survival dissipates, and the evolution-induced by-product of adaptive living, which is autobiographical memory, assumes ascendancy in the very last vestiges of life. This selfsame process might be behind the anecdotal reports of individuals involved in highly dangerous near-miss emergencies who subsequently report that their life "flashed" before their eyes.

mind's understanding that time has passed must be informed by the continuity function (that is, the autobiographical persistence of self) discussed earlier. However, the absence of a sense of time-in-passing during sleep is due to the fact that no conscious episodic memories are being laid down. Indeed, this may be a necessity of sleep since it is the confirmation of previously learned episodic lessons that now occupy the brain. Thus, the mind's experience of time is itself largely limited to intervals of consciousness. That both the organization of the brain and the properties of the mind may each be the result of a collective emergence is an important aspect of human experience to be explored (and see Minsky 1988, 1).

8. Summary and Conclusion

The present view of time and duration is strongly influenced by Russell (1915) and more recently by the inherent challenges posed by Pirsig (1974) and Tarnas (1991). If we can, for the moment, accept Russell's perspective, then duration is a property of the relationship between object and object. It is thus quite reasonable to talk in physical terms about the properties of such duration(s) and thus believe in the proposition which distinguishes the A and B versions of time. The latter distinction has been discussed extensively (and see McTaggart 1993) and indeed is a central issue in all of the time literature. However, using Russell's resolution, the two perspectives are far from incompatible. If Russell is correct, then time itself is a property of living systems, being a relationship between subject and subject or equally between subject and object. If this is so, then it is a direct categorical error to talk about time in relation to non-living things. If it were not apparently tautological, it might be appropriate to specify time as one (if not the) characteristic of life itself. Thus time depends on an observer, whereas duration goes its merry way, independent of any need for any conscious or living entity. In this view, time is a truly multi-faceted construct since different forms of time will be experienced by different entities. I have here suggested that as evolution apparently creates ever-more complex forms of life, there is a comparable increase in the ever-more sophisticated forms of time, although this assertion lies in danger of the hubris that seems to accompany the human condition. The differentiation of time and duration, as well as the purported supremacy of human temporal experience, carries with it a natural attraction to ourselves as the ultimate reporters and auditors of knowledge (and see Hancock 2005). However, there may be important challenges to such a collective egocentric framework. In his highly popular text, Pirsig (1974) sought to challenge this division between self and other, arguing that it parsed existence inappropriately. The brisance derived from the dissolution of this division is informative but leaves behind no obvious structure by which understanding can progress. Pirsig's following text (Pirsig 1991) aspires towards this achievement of a new unity but, understandably, largely fails to reach it. In general, it seems that humans are constrained to divide experience in order to understand it. Looking to embrace the holistic life is currently much more of a spiritual journey toward progress than a scientific one. Perhaps that will change in our future. If anything can elicit that change it has to be the study of some fundamental facet of experience that ranges across the whole of human understanding. At present, the only effective topic that fulfills this criterion is time itself.

Acknowledgements

I am very grateful to the multiple agencies that have sponsored my work over the years. Details of the views expressed here are my own and do not represent those of any supporting agency so. My previous work with my colleagues Dr. James Szalma and Dr. Tal Oron-Gilad has been particularly influential on this work, and I am most grateful for their observations. The final version of this work was greatly improved by the observations of Nushien Shahnami and the editing of Gabriella Hancock. I am grateful the comments of the editors and two unknown reviewers, whose remarks were very helpful in revising and expanding the present text.

References

Atance, C. M. and D. K. O'Neill. 2001. Episodic future thinking. *TRENDS in Cognitive Science* 5 (12): 533–539.

Barrett, H. C. and R. Kurzban. 2006. Modularity in cognition: Framing the debate. *Psychological Review, 113* (3), 628–647.

Carroll, L. [Dodgson, C. L.]. 1871. *Through the Looking-Glass, and What Alice Found There*. London: Macmillan.

Calvin, W. H. 1983. A stone's throw and its launch window: Timing precision and its implications for language and hominid brains. *Journal of Theoretical Biology* 104: 121–135.

Cippola, C. M. 1967. *Clocks and Culture 1300–1700*. New York: W. W. Norton.

Cleary, A. M. 2008. Recognition memory, familiarity, and déjà vu experiences. *Current Directions in Psychological Science* 17: 353–357.
Cleeremans, A., and J. C. Sarrazin. 2007. Time, action, and consciousness. *Human Movement Science* 26: 180–202.
Cohen, J. 1967. *Psychological Time in Health and Disease.* Springfield: C. C. Thomas.
Crompton, A. W., C. R. Taylor, and J. A. Jagger. 1978. Evolution of homeothermy in mammals. *Nature* 272: 333–336.
Dennett, D. C. 1991. *Consciousness Explained.* Boston: Little, Brown & Co.
Desmond, P. A., and P. A. Hancock. 2001. Active and passive states of fatigue. Pp. 455–465 in *Stress, workload, and fatigue*, ed. P. A. Hancock and P. A. Desmond Mahwah, New Jersey: Lawrence Erlbaum.
Dudai, Y. and M. Carruthers. 2005. The Janus face of Mnemosyne. *Nature* 434: 567.
Ehrsson, H. H. 2007. The experimental induction of out-of-body experiences. *Science* 317: 1048.
Einstein, A. 1905. Zur Elektrodynamik bewegter Körper. *Annalen der Physik* 17: 891–921.
———. 1950. On the Generalized Theory of Gravitation. *Scientific American* 182 (4): 13–17.
Fraser, J. T., ed. 1966. *The Voices of Time.* Braziller: New York.
———. 1987. *Time the Familiar Stranger.* Redmond, Washington: Tempus Books.
Fischer, R., F. Griffin, and L. Liss. 1962. Biological aspects of time in relation to (model) psychoses. *Annals of the New York Academy of Sciences* 96: 44–65.
Gibson, J. J. 1975. Events are perceivable, time is not. Pp. 295–301 in *The Study of Time II*, ed J. T. Fraser and N. Lawrence, eds. New York: Springer-Verlag.
———. 1979. *The Ecological Approach to Visual Perception.* Hillsdale, N.J.: Erlbaum.
Gould, S. J., and N. Eldredge. 1993. Punctuated equilibrium comes of age. *Nature* 366 (6452): 223–227.
Hancock, P. A. 1981. The simulation of human core temperature. *International Journal of Bio-Medical Computing* 12: 59–66.
———. 1993. Body temperature influences on time perception. *Journal of General Psychology* 120 (3): 197–216.
———. 2002. The time of your life. *Kronoscope* 2 (2): 135–165.
———. 2003. The ergonomics of torture: The moral dimension of evolving human-machine technology. *Proceedings of the Human Factors and Ergonomics Society* 47: 1009–1011.
———. 2005. Time and the privileged observer. *Kronoscope* 5 (2): 176–191.
———. 2009. *Mind, machine, and morality.* Chichester, England: Ashgate.
Hancock, P. A., and S. de Ridder. 2003. Behavioral response in accident-likely situations. *Ergonomics* 46 (12): 1111–1135.
Hancock, P. A., and M. P. Manser. 1997. Time-to-contact: More than tau alone. *Ecological Psychology* 9 (4): 265–297.
Hancock, P. A., J. L. Szalma, and T. Oron-Gilad. 2005. Time, emotion, and the limits to human information processing. Pp. 157–175 in *Quantifying human information processing*, ed. D. McBride and D. Schmorrow. Boulder, CO.: Lexington Books.
Hancock, P. A., and J. L. Weaver. 2005. On time distortions under stress. *Theoretical Issues in Ergonomic Science* 6 (2): 193–211.
Hawking, S. 1988. *A Brief History of Time.* New York: Bantam Books.
Hoagland, H. 1933. The physiological control of judgments of duration: Evidence for a chemical clock. *Journal of General Psychology* 9: 267–287.
Hobson, J. A., E. F. Pace-Schott., and R. Stickgold. 2000. Dreaming and the brain: Toward a cognitive neuroscience of conscious states. *Behavioral and Brain Sciences* 23(6): 793–842.
Horne, J. A. 2000. Images of lost sleep. *Nature* 403: 605–606.

Huber, R., M. F. Ghilardi, M. Massimini, and G. Tononi. 2004. Local sleep and learning. *Nature* 430: 79–81.
Iberall, A. S. 1992. Does intention have a characteristic fast time scale. *Ecological Psychology* 4 (2): 39–61.
James, W. 1890. *Principles of Psychology*. New York: Holt.
Kant, I. 1781. *Critik der reinen Vernunft* Riga: Hartnockt.
Locke, J. 1690. *An Essay Concerning Human Understanding*. St. Paul's Churchyard, London: Edward Mory.
MacLean, P. 1990. *The Triune Brain in Evolution*. New York: Plenum Press.
Marshall, N. K., and E. Donchin. 1981. Circadian variation in the latency of brainstem responses and its relation to body temperature. *Science* 212 (4492): 356–358.
McTaggart, J. M. E. 1993. The unreality of time. Pp. 23–34 in *The Philosophy of Time*, ed. R. Le Poidevin and M. McBeath. Oxford: Oxford University Press.
Meddis, R. 1977. *The Sleep Instinct*. Boston: Routledge & Kegan Paul.
Mimura, M., M. Kinsbourne, and M. O'Connor. 2000. Time estimation by patients with frontal lesions and by Korsakoff amnesiacs. *Journal of the International Neuropsychological Society* 6 (5): 517–528.
Minsky, M. 1988. *The Society of Mind*. New York: Simon & Schuster.
Molyneux, W. [1688]. 1978. Letter to John Locke, 7 July. Vol. 3, no. 1064 in *The Correspondence of John Locke*, ed. E. S. de Beer. 9 vols. Oxford: Clarendon Press, 1978.
———. [1693]. 1979. Letter to John Locke, 2 March. Vol. 4, no. 1609 in *The Correspondence of John Locke.*, ed. E. S. de Beer. 9 vols. Oxford: Clarendon Press, 1979.
Morrison, A. R. 2003. The brain on night shift. *Cerebrum* 5 (3): 23–36.
Nairne, J. S., and J. N. S. Pandeirada. 2008. Adaptive memory: Remembering with a stone-age brain. *Current Directions in Psychological Science* 17: 239–243.
Nairne, J. S., J. N. S. Pandeirada and S. R. Thompson. 2008. Adaptive memory: The comparative value of survival processing. *Psychological Science* 19: 176–180.
Nolte, J. 2001. *The Human Brain: An Introduction to Its Functional Anatomy*. Amsterdam: Elsevier.
Pearl, D., and P. S. D. Berg. 1963. Time perception and conflict arousal in schizophrenia. *Journal of Abnormal and Social Psychology* 66 (4): 332–338.
Pirsig, R. M. 1974. *Zen and the Art of Motorcycle Maintenance: An Inquiry into Values*. New York: Morrow.
———. 1991. *Lila: An Inquiry into Morals*. New York: Bantam Books.
Pöppel, E. 1988. *Mindworks: Time and Conscious Experience*. Boston: Harcourt Brace Jovanovich.
Prosser, C. L., ed. 1991. *Environmental and metabolic animal physiology*. Fourth Edition. New York: Wiley.
Prosser, C. L. 1984. Personal communication.
Russell, B. 1915. On the experience of time. *Monist* 25: 212–233.
Servan-Schreiber, J. L. 1989. *The Art of Time*. New York: Addison-Wesley.
Schacter, D. L. 2001. *The Seven Sins of Memory: How the Mind Forgets and Remembers*. Boston: Houghton-Mifflin.
Schacter. D. L., and D. R. Addis. 2007. The ghosts of past and future. *Nature* 445: 27.
Schrödinger, E. [1944]. 1967. *What Is Life?* Cambridge: Cambridge University Press.
Siffre, M. 1965. *Beyond Time*. London: Chatto & Windus.
Smith, C. 1985. Sleep states and learning: A review of the animal literature. *Neuroscience and Biobehavioral Reviews* 9 (2): 157–168.
Spencer, K. M., P. G. Nestor, R. Perlmutter, M. A. Niznikiewicz, M. C. Klump, M. Frumin, M., M. E. Shenton, and R. W. McCarley, R. W. 2004. Neural synchrony indexes disordered perception and cognition in schizophrenia. *Proceedings of the National Academy of Sciences* 101 (49): 17288–17293.
Stickgold, R. 2006. A memory boost while you sleep. *Nature* 444:559.

Stickgold, R., and M. P. Walker. 2007. Sleep-dependent memory consolidation and reconsolidation. *Sleep Medicine* 8: 331–343.

Tarnas, R. 1991. *The Passion of the Western Mind: Understanding the Ideas that Have Shaped Our World View.* New York: Random House.

Vohs, K. D., and B. J. Schmeichel. 2003. Self-regulation and the extended now: Controlling the self alters the subjective experience of time. *Journal of Personality and Social Psychology* 85 (2): 217–230.

Wang, D., and B. Yuwono. 1995. Anticipation-based temporal pattern generation. *IEEE Transactions on Systems, man and Cybernetics* 25 (4): 615–627.

Wilkie, D. M., J. A. R. Carr, J. Galloway, K. J. Parker, and Y. A. Aiko. 1997. Conditional time-place learning. *Behavioural Processes* 40 (2): 165–170.

CHAPTER THREE

MEMORY, ANTICIPATION AND THE (UN)REALITY OF THE PAST AND FUTURE[1]

Jonathan Tallant

Summary

The chief concern of this paper is to show, contra the views of a number of philosophers, that the possibility of scientific inquiry does not constitute an independent constraint on our theory choice when it comes to time. In section 1, I introduce the debate between A- and B-theorists and identify the factors typically taken to constrain our theory choice in the philosophy of time. In section 2, I discuss some arguments intended to show that the possibility of scientific inquiry constrains our theory choice in favor of the B-theory and, in section 3, the replies to these arguments that purport to show our theory choice constrained by the same possibility of scientific inquiry in favor of the A-theory. In section 4, I then argue that both sides are wrong by showing that the possibility of scientific inquiry puts no constraints on our theory of time that are not already present in the traditional debate.

1. Introducing the Debate

Let's begin with some putative truths:

(1) The battle of Hastings is past.
(2) The battle of Hastings is earlier than the First World War.

These 'truths' are what we might call 'temporal truths'. That is, a part of what makes these sentences true is that they correctly describe the temporal positions of entities in time. By way of contrast we might say that '2+2=4' is *not* a temporal truth since no part of what makes this sentence true is the temporal position of any entity in time.[2]

---

[1] This paper has its ancestry in a chapter of my thesis. With this in mind, thanks are due to Jonathan Lowe, Robin Le Poidevin. and Robin Hendry for comments and criticism. Thanks also to members of the Durham postgraduate philosophy society for comments on an even earlier draft. Finally, thanks go to all participants of the thirteenth ISST conference in Monterey for comments and criticisms.

[2] See Lowe 1998, 96–7, for a brief exploration of this characterisation.

Jo Alyson Parker, Paul A. Harris and Christian Steineck (Eds), Time: Limits and Constraints, pp. 89–108

Now whilst philosophers are occasionally prone to denying that which we take to be obviously true,[3] it is typically a desideratum of a philosophical theory that it preserve commonsense claims about reality. So our theory of time, whatever it is, is partially constrained by the desirability of preserving the truth of sentences like (1) and (2). Of course, our theory may only be de-feasibly constrained by such considerations: if maintaining the truth of either (1) or (2) commits us to a contradiction, or if we have strong theoretical reasons to deny either (1) or (2), then we may give them up. But let us allow that preserving the truth of sentences like (1) and (2) *is* a constraint on a good theory of time.[4]

So what *are* the competing theories constrained by the intuitive need to preserve the truth of sentences like (1) and (2)? Typically, our theory choice in the philosophy of time is between the opaquely named A- and B-theories of time.[5] The A-theory of time is the view that there are some properties borne by entities in time that correspond to tensed predication. Thus, the sentence "the battle of Hastings is past" is true if and only if the event named (the battle of Hastings) bears the property of 'pastness'.[6] Similarly, the sentence "the creation of Mars outposts is future" is true if and only if the event "the creation of Mars outposts" bears the property of 'futurity'.

The theoretical insight motivating this thought is that if sentences are true then they are true *in virtue of* the world being some particular way. Thus, the sentence "the wall is white" requires two things of the world in order to be true. First, that there is a wall. Second, that

---

[3] For instance, McTaggart (1908) denies the reality of time; some sceptics (e.g. Unger 1971) deny that we can know anything; Berkeley (1999) denies that there is a mind-independent reality.

[4] Unless, of course, we are presented with reasons from historical enquiry to think that (1) or (2) are false. I assume throughout that such sentences are factual and that the key question for our enquiry is determining *how* such truths are true. For a more explicit statement of this viewpoint, see Bourne 2006, 1.

[5] Though there are some attempts to reconcile these: e.g. Tooley 1997, Corish 2005.

[6] For ease of discussion I'll leave out variations on the A-theory that claim that not all tensed predication corresponds to tensed properties. For instance, presentism is the view that only the present exists. and so some forms of presentism claim that only tensed predication ascribes a tensed property and use other ontological resources to make truth judgements about the past and future (see, e.g., Crisp 2007 and Bourne 2006).

there is some way of being that the wall instantiates—'whiteness'.[7] If we break with this schema, then it's hard to see how the subject predicate sentence "the wall is white" gets to be true.[8]

If we endorse this theoretical insight then sentences like (1), that ascribe a way of being (being past), to a particular event, seem to commit us to so-called 'tensed properties': in this instance, the property 'pastness'. This intuition, the intuition that truths require some part of reality to *make* them true, is often referred to in the contemporary literature as the 'truth-maker intuition' and may be expressed as:

TM: if some proposition <*p*> is true, then *p* is the case.[9]

So given the apparent truth of sentences like (1) and the need to satisfy TM, we might *prima facie* suppose that there are 'tensed' properties of 'pastness', 'presentness' and 'futurity'.

So much for (1), what about (2)? In the analysis of (2), we find the motivation for the B-theory. Unlike the A-theorist, the B-theorist denies that there are any properties that directly correspond to tensed predication. Instead, the B-theorist thinks that the reality of time consists in entities standing in the fixed and permanent relations 'earlier than' and 'later than'. Thus, the sentence "the battle of Hastings is earlier than the First World War" requires, in order to be true, that an 'earlier than' relation stand between the battle of Hastings and the First World War.

Again, the initial motivating thought here is that because sentence (2) explicitly predicates an 'earlier than' relationship to the events in

---

[7] For an introduction to the relationship between subject-predicate discourse and the metaphysics that underpins it, see Loux 1998, 92–129.

[8] One way of doing this might be to deny that the truth of a sentence has anything to do with the mind-independent world. Furthering this thought, perhaps 'truth' is simply a matter of convention, or societal agreement. I shan't deal with this objection here since it takes us too far from our discussion of the limits and constraints that we must place on a theory of time, but there are good reasons to think that this approach leaves something to be desired. See, for instance, Lowe 1998: 3–4.

[9] I follow the convention in using triangular brackets to indicate a proposition. I should also add that I'm not endorsing any particular version of Truthmaking as it is developed as a fully fledged theory (e.g. Armstrong 2004; Lowe 2005, chapter 12—see Bebee and Dodd 2005 for extensive discussion). All that I am doing here is using the intuition TM to explain quite how it is that philosophers think sentences like (1) and (2) constrain our theory choice.

question, so sentence (2), in conjunction with TM, seems to suggest the truth of the B-theory.

Of course, there might remain in the mind of the reader some concern as to why these sorts of considerations matter: why should a non-philosopher find these kinds of issues pressing? Isn't this debate a product of abstruse philosophical reasoning, allied to a restricted "ivory tower" view of the world that refuses to look beyond the narrow boundaries of philosophy?

I think that the temptation to pursue this line ought to be tempered with the realization that the debates sketched derive their motivation from our pre-philosophical intuitions. For instance, suppose we consider the sentence "the chair is blue." The natural supposition, as noted above, given the grammatical form of the sentence, is that there is a *thing* (a chair) and that the chair has some property (blueness): else, how does the sentence "the chair is blue" get to be true?[10] If such reasoning is remotely tempting, then, when we consider the sentence "the battle of Hastings *is past*," the natural temptation is to think that there is some *thing* (the battle of Hastings) and that it has some property (pastness): else, how does the sentence "the battle of Hastings is past" get to be true?

So, returning to sentences (1) and (2): what is motivating these philosophical theories are lines of reasoning that have pre-theoretical pull; these are intuitions that are supposed to be driven, not by flighty philosophical reasoning, but by natural consideration of the ways in which language works.

Accepting that sentences such as (1) and (2) place a constraint upon our theory choice, we are left with some problems that place a further constraint on our theory of time given the need to answer the following questions:

(3) Sentence (1) seems to require the existence of a tensed property; given TM and the fact that the B-theory *doesn't* include A-properties in its ontology, how can the B-theory be true?

[10] Of course, nominalists will object to the implication that there is a *property* involved here, but they do not typically deny that the form of the sentence offers *prima facie* motivation for the view that predicates denote properties.

(4) Sentence (2) seems to require the existence of a B-relation; given TM and the fact that the A-theory *doesn't* include B-relations in its ontology, how can the A-theory be true?[11]

The move typically made in response to these worries by both A- and B-theorists is to state the meaning of the problematic sentences in terms that are not so problematic. Allow me to explain.

The prevailing view in the philosophy of language is that to state the meaning of an expression, we need simply state the conditions under which it is true:

> To know the meaning of a sentence is to know what is the case if it is true. (Wittgenstein 2001, 4.024)
>
> To know the meaning of a sentence is to know in which possible cases it would be true and in which not. (Carnap 1947, 10)[12]

In order to make such *apparently* B-theoretic sentences as (2) true, the A-theorist then argues that the conditions under which said B-theoretic sentences are true can be expressed in *purely* A-theoretic terms. Thus, Craig 2000, 153:

> D_3': *e* is earlier than *e** ≡ e becomes present first and e* becomes present second
> *e* is later than *e** ≡ *e** becomes present first and e becomes present second[13]

In slightly less formal terms, for it to be true that "the battle of Hastings is earlier than the First World War" simply requires that the battle of

[11] Uniting A- and B-theories plausibly leads us into the contradiction first spelled out by McTaggart 1908. See Tallant 2005 and Oaklander 2004, 51–62 for introductory discussions. Though see also McCall 1994 and Tooley 1997 for attempts to unify the A- and B-theories. See Corish 2005 for an interesting objection to McTaggart's argument. Question: Doesn't talk of future, present, and past *imply* that the past is, for instance "earlier than" then present? And, if it does, why do we need this talk of B-relations?

[12] For an extensively worked out defence of this, see various papers in Davidson 1984.

[13] There might remain a concern that such talk of 'first' and 'second' presupposes some kind of 'earlier than' or 'later than' relation and, as such, that such tensed truth conditions implicitly presuppose some kind of B-theoretic metaphysics. In the face of such criticism it might be better for the tensed theorist to take the line suggested in Lowe 1998, 88–92, and give tensed token-reflexive truth-conditions for tensed sentences. However, an analysis of such will take us too far afield in this cursory introduction to the debate. I shall leave it to the reader to determine the viability of such strategies.

Hastings become present first and the First World war become present second. Because these conditions so stated make no reference to the B-theoretic relations 'earlier than' and 'later than', we have no need of such relations in order for sentences like (2) to be true.

Likewise, in order to make sentences such as (1) true, sentences that are apparently A-theoretic, the B-theorist posits tense*less* truth conditions.

Present tense sentences:

> For any token, *u*, of the form, 'x is present', *u* is true iff (if and only if) x is simultaneous with *u*.

Future tense sentences:

> For any token, *u*, of the form, 'x is past', *u* is true iff x is earlier than *u*.

Past tense sentences:

> For any token, *u*, of the form, 'x is future', *u* is true iff x is later than *u*.[14]

Thus, a sentence token of "the creation of Mars outposts is future" is true, according to the B-theorist, provided the creation of Mars outposts stands in a B-theoretic 'later than' relation to the time at which the sentence token is tokened.

The motivation behind these semantics manoeuvres is, once again, to satisfy the intuition expressed in TM and preserve the truth of sentences (1) and (2). Since (1) *appears* to report the existence of a 'tensed property', and the B-theory denies the existence of said properties, so the B-theorist needs something like the specified account of the meaning of (1) in order to preserve the truth of (1). Likewise, since (2) *appears* to report the existence of a B-relation and the A-theory (typically) denies the existence of said relations, so the A-theorist needs something like the specified account of the meaning of (2) in order to preserve the truth of (2).

Before closing this brief introduction to this portion of the philosophical terrain, let us recap. A key constraint upon our choice of either A- or B-theory of time is the need to preserve the truth of sentences

[14] Cf. Dyke 2002, 337, and see 2002, 331–336, for reasons to prefer the token reflexive analysis specified to the date analysis offered by Mellor 1998.

(1) and (2). This is a requirement of our theory in order to satisfy the truth-maker intuition expressed in TM. Both A- and B-theories then require us to consider some very particular semantics in order to preserve the truth of (1) and (2) within their respective ontologies.[15]

2. Other Constraints

But, of course, theory choice is not *solely* limited and constrained by the need to preserve the truth of these particular classes of sentences. The recent debates between A- and B-theorists indicate the need to take very seriously theoretical considerations from contemporary physics.[16] So it is clear that although there are semantic and truth-making considerations in the area of our discussion, the philosophy of time takes cognizance of scientific deliverances. This is just as well. Any philosophy that closes itself off from contemporary physics is surely a dogmatic philosophy and should probably best be ignored.[17]

But there is one *purported* constraint upon a theory of time that is less obvious: that the very possibility of scientific enquiry constrains our theory choice in the philosophy of time. According to Grunbaum, the first proponent of the view that the possibility of science so constrains our theory choice, the possibility of science and scientific practice reveals to us that the B-theory of time is true. Thus, the possibility of scientific enquiry limits our choice of theory to the B-theory and the B-theory alone. On the other hand, Dobbs, Ferre and Craig all argue, contra Grunbaum, that the possibility of scientific enquiry limits our theory choice to the A-theory of time.

The point of origin for this debate is Grunbaum's, *Modern Science and Zeno's Paradoxes*. Therein, Grunbaum suggests as follows:

> It seems to be of decisive significance that no cognizance is taken of nowness (in the sense associated with becoming) in any of the extant

---

[15] Notice, I do not say that either semantics is correct. I simply give these by way of example.

[16] See, for instance, Sider 2001, 42–52. For an excellent discussion of how metaphysics and physics relate to one another with particular consequence for the philosophy of time, see Hawley 2006.

[17] Another constraint upon a theory of time is that it is sufficient to explain the way in which we experience time—see Tallant 2007 for some worries about this in connection with the B-theory. I won't take consideration of this particular constraint further, here, since it takes us too far away from my intended target.

> theories of physics. If nowness were a fundamental property of physical events themselves, then it would be very strange indeed that it could go unrecognised in all extant physical theories *without detriment to their explanatory success.* (1968, 20)[18]

Take as a putative example a scientific formula, force = mass x acceleration ($F = m \times dt^{-2}$). In the application of such a law it matters not which time is 'present'; it simply matters which time (or times) we choose to apply the law to. Thus, if we wish to calculate the force of a particular body, *B*, at *t*, it doesn't matter a jot whether or not *t* is 'present'; all that matters is the mass and acceleration of B at *t*. Similar remarks can, Grunbaum thinks, be made about all aspects of scientific enquiry.[19] They are all such that no cognizance is taken of 'nowness'.

The rough thought then seems to be that because science is explanatorily complete without making recourse to any 'nowness' in the world, we have no reason to postulate such a 'now'. And as it stands, this reasoning is *prima facie* persuasive. Assume momentarily that Grunbaum is correct that science pays no heed to 'the now'. Science is an extremely complex discipline, one that provides us numbers of extremely accurate (if not true) claims about how the world is, how the world behaves, and why the world behaves in the way that it does. Provided we can make our scientific claims about how the world is and why the world behaves in the ways that it does *without* referencing a 'now', it would seem quite reasonable to agree with Grunbaum that the practices that are associated with scientific enquiry point to the truth of the B-theory.

If this line of reasoning is convincing, then the interdisciplinary consequences are fascinating: for each domain of enquiry that we have, if that domain of enquiry posits a fundamental 'now', then that domain of enquiry motivates an A-theory of time; otherwise, a B-theory. The philosophy of time then becomes a handmaiden to the language used in other disciplines; with theories to be chopped and

[18] Grunbaum's claim is that no cognizance is taken of 'nowness', by which he means some property of 'being now' that is key to the A-theory of time. Grunbaum does not deny that we will need to preserve tensed truths; but he claims that they are not to be accounted for in the terms specified by the A-theorist in section (1).

[19] I discuss below some of the other features of what 'scientific enquiry' consists in. The pre-existing debate is decidedly remiss in not specifying *precisely* what is required for 'scientific enquiry'. However, since a rough and ready notion is implicitly suggested and neither side seems to dispute said notion, I shan't go to any great lengths to specify quite what 'scientific enquiry' consists in.

changed in response to methodologies enshrined elsewhere. So, does Grunbaum's argument work?

3. Science as a pro-A-theoretic limit

Dobbs (1969) offers a four-fold argument against. The basis for his objections are the following claims: science *does* recognize the 'now'; science treats the past and future differently (and so is A-theoretic, thus entailing the existence of a 'now'); Grunbaum's theory is 'silly'; and that subjective mind-dependent 'now-ness' (discussed below) entails mind-independent 'now-ness'.[20]

But before we proceed we should get clear on the notion of 'now-ness' that Grunbaum thinks *is* legitimate. For although he denies that there is any such an A-theoretic property of 'now-ness' he, like all other philosophers, is constrained by the need to explain our experiences of time. And it seems to us, contra the B-theory, that we have experiences, *now*. There seems to be something very particular to the nature of (at least some of) our experiences such that they feel or seem to be now. To quote a B-theorist on this:

> Tenses, like the indexical 'now' and the past, present, and future, are foreign to calendar time, that is, B-time. They are part of the whoosh and the whiz I call *now-based time*, that is, A-time. (Falk 2003, 214–5)

Accordingly, B-theorists must then provide us with some account of how it is we experience time as 'whooshy' and 'whizzy', and they do so, typically, by explaining that this 'now-based-time', is not a feature of reality but simply a feature of our minds. Thus, Grunbaum:

> What qualifies a physical event at a time *t* as belonging to the present or as now is not some physical attribute of the event or some relation it sustains to other purely physical events; instead what so qualifies the event is that at least one human or other *mind possessing* organism *M* experiences the event at that time *t*, such that at *t M* is *conceptually aware* of the following complex fact: that this having the experience of the event *coincides temporally* with an awareness of the fact that he has it at all. (Grunbaum 1968, 17)

[20] An important point to reiterate when considering Dobbs' response to Grunbaum is that Dobbs focuses upon how science is carried out, rather than what science finds in the world. So Dobbs is *not* concerned to show that, for instance, the general theory of relativity gives us some preferred plane of foliation such that it may be identified as an objective 'now'. Instead, his strategy is to argue that science cannot be carried out in the absence of an objective now.

So imagine, then, that you hear a noise at some time; call it *t*. That noise only qualifies as 'now' because you are 'judgementally aware' of experiencing that noise and because the noise in question occurs (approximately) simultaneously with your judgement. Equally, imagine that the noise occurs at *t*, but that no one hears it. Indeed, imagine that no one *at all* exists at *t* (perhaps *t* occurs after all mind possessing organisms have ceased to exist). Where this to be the case then, according to Grunbaum, the noise would occur at *t* all right, but it would not occur *now*.[21] I call this view of the 'now' a, 'mind-dependent now' since, unlike the A-theorist, Grunbaum believes the 'now' to be a matter of conceptual awareness on the part of a temporally cognizant agent.

Dobbs begins his attack on Grunbaum by explaining that there *is* a portion of science that does recognise the significance of a mind-*in*-dependent 'now': "there is a very large and important body of physical theory which deals with *prediction* in which the concept of 'nowness' is basic" (Dobbs 1969, 318).

Dobbs quotes the statistical mechanics of Weiner in support of his claim. According to Weiner and Dobbs the physical applicability of certain mathematical expressions turns on whether the events they describe are past or future. As Weiner has it: "physically applicable operators of engineering allow us to work with the past of our data but not with their future" (1949, 8).

As Dobbs has it:

> the 'now' time-point t=0 separating the Past (negative values of t) from the Future (positive values of t) is of fundamental significance. This is not merely a matter of diagrammatic convenience (as in the case of Minkowski diagrams), but of the physical applicability of mathematical expressions—specifically those containing symbols called 'operators'—and of their realisability in terms of actual hardware. (1969, 318)

I think we might water Dobbs' claim down a little and still retain its significance. In order to engage in the act of prediction, whether that is in statistical mechanics as Dobbs has it or in science at large, we must take a particular starting point as our 'now'. For instance, imagine that

[21] In this respect, then, Grunbaum cannot endorse the token reflexive account of what it is for a present-tensed sentence to be true that was specified above. Let this slide for the time being since our concern here is with the particular constraint placed on a theory of time by scientific enquiry and not which account of the semantics necessary for the B-theory of time is best.

I am standing holding a pencil above the table at time *t*. What predictions will we form when asked what will happen if I drop the pencil at t*? Most likely we will predict that the pencil *will* fall. But, of course, we will only predict that the pencil *will* fall if we, ourselves, precede *t**; otherwise we will judge that either the pencil *fell* or *has fallen* or something suitably past tensed. Thus, in order to engage in the act of prediction we must suppose that our 'now' is suitably located with regards to that about which we are predicting.

Clearly, if the very act of prediction requires of physics that it pays heed to the temporal location of the 'now' or the 'present' (in a mind-*in*dependent sense), then physics is not such that it takes no cognizance of 'the now'.

The second Dobbsian point of note is that the act of prediction discussed above not only seems to yield the conclusion that there is something fundamental about the 'now', but also the conclusion that there is a genuine distinction between the past and future. For instance, recall the act of prediction described above. It surely makes no sense for me to make a prediction about what *has* happened to the pencil. For although I may try to *work out* what has happened to the pencil (for instance, discover its history) I cannot *predict* its history. I can only predict that certain things *will* happen.

So science *seems* to recognise a distinction between the past and the future. The 'past' and the 'future' are, at least *prima facie*, A-theoretic determinations, and the B-theory does not include such reference to such determinations. So it looks as if science takes cognizance of the past and the future. And if we are permitting such A-theoretic notions, then we are seemingly obligated to recognise 'the present', too, since 'the present' can simply be defined as the point between the past and future.

In the third of the objections Dobbs raises, he argues that Grunbaum's claim (that reality lacks an objective distinction between the past and future and also the quality of 'nowness'), is silly. By 'silly' what we mean is a theory that is not, and cannot, be taken seriously by non-philosophers or, indeed, philosophers themselves outside their technical deliberations.[22] To say that something is 'philosophically silly' is to say that, although we may assent to it in the depths of our

[22] Cf. Crisp 2004, 18; Armstrong 1978, 440–1.

philosophical considerations, we don't genuinely believe it to be true, and thus that our philosophical acceptance of it is, at best, a sham.

Dobbs (1969, 323–4) argues that Grunbaum's account of temporality is silly in just this sense. According to Dobbs we all (physicists, philosophers and the rest) believe that certain events are present in a mind-independent sense and we all believe that there is genuine becoming—that is, a change in what is mind-independently present.

The final objection to Grunbaum concerns our inter-subjective agreement that a moment is *now*. For instance, suppose you and I exist at a time *t*. We will agree that *t* is 'now'. However, we also exist at times other than *t*, *t**, *t*** etc. Now if you and I exist at both *t* and *t**, then on what grounds can we suppose that *t*, rather than *t** is 'now'? For if we exist at both *t and t** then we will judge that *both* times *t* and *t** are present. How, then, do we decide that *t*, as opposed to *t** is present? As Dobbs (1969, 321–2) then goes on:

> unless this agreement is founded on some objective common counterpart to the mental 'now', which *causes* our individual brains to register 'now' with practical simultaneity, it is impossible to see why *I* should not be contrary, and hold that 'now it is twelve noon on Queen Anne's birthday in the year 1700'. In other words, unless the *mental becoming* of the experience of the specious present is occasioned by the *physical becoming* present of the neurophysical events underlying the consciousness, the fact of intersubjective practical simultaneity would be explicable only in terms of a miraculous pre-established (spiritualist) harmony.

To put this argument very simply: if science is possible then there is a mind-independent now, since otherwise there is no inter-subjective 'now'. And science *requires* that there be an inter-subjective 'now' in order to carry out its most basic operations.

4. Grunbaum Replies, But to What Effect?

To each of the problems raised, Grunbaum offers us a solution. Grunbaum's solution to the first two problems (the role of the 'now' in both statistical mechanics and prediction, and the temporal asymmetry involved in prediction and recollection) can be best explained through the following:

> Dobbs' objections..., which invoke...temporal asymmetries, rest on his failure to heed my explicit *caveat* not to confuse the tenseless 'past-future' asymmetry of earlier and later than t_0 with the tensed one associated with the 'now' of becoming. (Grunbaum 1969, 151)

In other words, Grunbaum agrees with Dobbs that there is an asymmetry between the past and future, and agrees with Dobbs that physics acknowledges the 'now', but he argues that this is simply a product of the asymmetry of the 'earlier than/later relation of the B-series and that the 'now' is *nothing more* than judgement dependent. Thus, Grunbaum thinks, physics does not pay heed to a 'now' that would constitute support for the A-theory. And that, surely, is the key point when considering how physics *qua* physics limits and constrains our choice of philosophical theory.

But this is only partially cogent as a reply to Dobbs.[23] Grunbaum may be right to object to Dobbs that the B-theory can give us an account of the possibility of science whereby we treat the 'now' as mind-dependent. But that does not show that science ignores an A-theoretic 'now'. Instead it shows that it is *possible* that science could, in fact, be constructed in such a manner as to not have a cognizance of A-theoretic 'nowness'. If we do not already agree with Grunbaum that the past-future asymmetry should be equated with the earlier-later asymmetry then we have no reason to agree that the 'now' in question is B-, rather than A-, theoretic.

The key question, then, is why should we heed Grunbaum's caveat and admit that we have committed the apparently confused act of equating the tensed 'now' of the A-theory with the tenseless earlier-later asymmetry? The only reason that I can see to do so is to assume that B-theoretic accounts of temporality are correct and, hence, that we should construct physics in such a way as to pay no heed to the 'now'.

But this is a quite different claim. If Grunbaum's arguments were to show that physics does not take cognizance of A-theoretic 'nowness' then it would constitute an argument to motivate the B-theoretic account. If, as I've just suggested, we must first accept B-theoretic accounts that themselves pay no heed to A-theoretic 'nowness', and then construct an account of science that pays no heed to this A-theoretic 'nowness', then it is unsurprising that we should come to an account of physics that does not have a cognizance of A-theoretic 'nowness'. For in first adopting the B-theory and the tenseless seman-

[23] Cf. Ferre's (1970, 278) remark: "The mental events of several percipients, if *ex hypothesi* caused by practically simultaneously by intrinsically tenseless physical events, should be expected to show inter-subjective agreement on matters of tense."

tics that goes along with it, we have already explicitly ruled out the possibility of such A-theoretic determinations as 'nowness'!

Dobbs' claim that Grunbaum's theory is 'silly' for not including a 'now' can also be dealt with, though again the result is only partially satisfying. Although there is no property of 'nowness' in Grunbaum's ontology, by giving an appropriate semantic account of the meaning of tensed sentences, it is clear that we can still have all the truths that we could want about the present. Thus the problem clearly admits of a solution.

In fact, though, this isn't the route that Grunbaum takes. Grunbaum notes that Dobbs is correct, that his is *not* a theory that takes seriously our pre-theoretical beliefs and intuitions that there is a property of 'nowness', but that this *is not a problem*. For, argues Grunbaum (1969, 151), his task is not to describe a metaphysics that accords with our pre-theoretical beliefs and intuitions but to describe a metaphysics that can make sense of the reality of time *if* it turns out that time is a static ordering of B-theoretic relations. In order to do this, we must understand that 'nowness' is mind-dependent.

So notice, once again, the Grunbaumian claim is no longer that science pays no cognizance to an A-theoretic 'now' but instead that we can explain away the cognizance that science pays to a 'now' *if* we accept the B-theory of time. Well and good. But that does not establish that the mere possibility of scientific enquiry shows that reality is B-theoretic: instead, it shows that if we accept a B-theoretic account of reality then we can construct a viable account of scientific enquiry. That is, of course, a far weaker claim and one that is compatible with the A-theory being true and it most certainly does not *show* that physics takes no cognizance of an A-theoretic now. At best it shows that a B-theoretic account of physics is coherent.

Dobbs' final objection to Grunbaum, that science requires an intersubjective 'now' in order to be possible, is more difficult to reject. To see why, consider Grunbaum's (1969, 152) reply:

> I had noted that a physical event *tenselessly* occurring at time *t* can readily produce effects in each of several percipients such that at practically the time *t each* percipient is first aware of its occurrence.

So, an agent M, at t, will be aware of event E, as will another agent, N. At t*, M and N will not be aware of E, but of E*, because, at t, M and N are simultaneous with E, and, at t*, M and N are simultaneous with E*.

Both Ferre (1970) and (more recently) Craig (2000) have suggested that this reasoning is faulty. Ferre (1970, 279) explains.

> A_1, exists (occurs tenselessly) at clock time t_1, and that it is also true in my life's career that a different "now"—awareness event, A_2, exists (occurs tenselessly) at clock time t_2. In the universe depicted by Grunbaum both events have equal claim to existence; neither clock time t_1 nor t_2 has any intrinsic claim to privileged status as more "really" now [...] than the other. Suppose, further, that it is a phenomenological fact that A_2 fills my subjective field of awareness, not A_1. Why?
>
> It will not do for Grunbaum to answer that I am uniquely experiencing A_2 simply because it is clock time t_2. Objectively speaking, on his view, it is (tenselessly) no less clock time t_1, and at time t_1 a "now"—awareness event A_1 no less genuinely exists (occurs) for me. Yet I find myself involuntarily discriminating against time t_1 and its associated event A_1 *with no objective ground whatever* to account for this.

Thus, it *appears* that the experiences that we each have of mental becoming necessitate a genuine account of physical becoming.[24] There is nothing in the Grunbaumian account to explain the apparently unique nature of each moment of my temporal experience. So there must be more to 'the present' than our judging that a particular time is present.

The paucity of Ferre's argument rests upon the assumption that the perceived uniqueness of an agent's mental experience at t_1 or t_2 represents a genuine feature of reality—that it really is a unique mental experience. The B-theoretic thought, though, is not that any *one* time 'is now' but that *every* time 'is now' and that we agree with one another *at* particular times that *those* times are 'now' because we make the judgment *at* that time and *about that time* that it is 'now'. There is sufficient ground for such discrimination since we are, at each time, thinking about and talking about the time at which we are located.

Thus, the me-at-t_1 experiences t_1 as 'now' because the me-at-t_1 is conceptually aware of the contents of t_1 at t_1. Likewise, the me-at-t_2 experiences t_2 as 'now' because the me-at-t_2 is conceptually aware of the contents of t_2 at t_2. To flesh this out further, we-at-t_1 agree that t_1 is present, because we-at-t_1 are only conceptually aware of t_1 and we-at-t_2 agree that t_2 is present, because we-at-t_2 are only conceptually aware of t_2. The very fact that t_1 and t_2 are distinct times at which distinct

[24] Cf. Craig 2000, chapter 7.

instantiations of people exist *is* a sufficient ground for the discrimination that people draw *at* the times at which the draw them.[25]

In less loaded terminology there is no uniquely objective *now*; but that doesn't preclude the possibility of scientific enquiry, since any thought, experience, or sentence token that is present tensed is *about* the time at which it is tokened. This is the lesson we learn from getting the right account of tensed sentences within a tenseless ontology. So when any two scientists set out to discuss or think about what is happening *now* and succeed in doing so, Ferre and Dobbs' arguments do not trouble them because such present tense tokens are about the time of their tokening. And since thoughts as well as spoken sentences can be considered to be present tensed tokens (see Mellor 1981, 37, and Dyke 2002, 340 for discussion), so any two thinkers can, and will, agree that the time of their thought is present *at* the time at which they token them.[26] That, or so it seems, is all that is required for the possibility of scientific enquiry.

But, *again*, this is not sufficient to establish Grunbaum's original conclusion. The original conclusion was that science pays no cognizance to a 'now' of the sort associated with the A-theory. The current conclusion is that science pays heed to a 'now', but that given the right account of the semantics of tensed tokens, and the right account of our experience of time, we can explain this within a B-theoretic ontology. In other words, if the tenseless semantics or the B-theory failed, so would Grunbaum's argument, and if they succeed, then so does Grunbaum's account. This argument is, thus, entirely parasitic on the pre-existing debate between the A- and B-theorist of time and so the possibility of scientific enquiry *does not* place an additional constraint on our theory choice in the philosophy of time.

---

[25] For an excellent overview of the literature on persistence and what it means for 'distinct instantiations of the same person' to exist, see Haslanger and Kurtz (eds.) 2006.

[26] One legitimate objection to my defence of Grunbaum here might be that this will only work because we've drifted from the sorts of semantic account of the 'now' that Grunbaum gives, and the sort of semantic account that has been given more recently—e.g. Dyke (2002—and see above). Perhaps that's right, but then we need simply endorse a more modern B-theoretic semantics in order to defeat the attack from the A-theorist.

5. Conclusions

Let's step-back a moment and review the dialectic in which we're engaged. Grunbaum's original claim was that because our scientific practices are such to pay no cognizance of an A-theoretic 'now', so the possibility of physical enquiry seems to indicate that reality is B-theoretic. Contra Grunbaum, Dobbs *et al.* have tried to show that such enquiry *does* pay heed to a 'now'. Grunbaum then replies by assuming that reality is B-theoretic *and then* showing us how we can construct an account of science that pays no cognizance of an A-theoretic 'now'.[27]

Grunbaum's argument then amounts to nothing more than the following:

> If reality is B-theoretic, then we can give an account of physics that is B-theoretic.

But that is hardly the most illuminating of claims since it tells us little more than that the B-theory is coherent. Likewise, though, the arguments of those arranged against Grunbaum show nothing more than that:

> If the B-theory fails, then the possibility of scientific enquiry shows that reality is A-theoretic.

So here's the negative conclusion: the possibility of scientific enquiry does not limit or constrain our choice of philosophical theory. What limits and constrains our choice of philosophical theory is (in part) giving the right semantic and phenomenological account of the 'now', and although it's clear that part of that account will have to constitute compatibility with the possibility of scientific enquiry, both A- and B-theory at least seem *prima facie* up to the task.[28]

But it also seems right (especially in this forum) to draw attention to some implications for interdisciplinary research into the nature of time. The truth of sentences like (1) and (2) seem to give us good

[27] Unless, of course, one grants the further premise that either the A- or B-theory is true. I would have grave doubts about such a premise. For reasons why, see Tooley 1997, McCall 1994, and Barbour 1999.

[28] Though there are extensive debates about whether or not both sides have semantics in place that are adequate to the task—see Oaklander 2004 and Smith 1993 for a flavour of these debates. There are also numerous concerns about the phemenology of the 'now'. See, e.g. Oaklander and Smith 1994.

reason to think that something in the ball-park of either the A- or B-theory of time is going to be true for the reasons discussed in section1. Physical theories, such as the special theory of relativity, might give us pause since they *seem* to deny the existence of an absolute present—thus favouring the B-theory (provided we, as B-theorists, deny that there exists an objective simultaneity relation).[29] Cognitive science might yield results that show our perception of time favours one view over the other.[30] But the scope for further interdisciplinary enquiry seems limited in at least the following way: to the extent that particular domains of inquiry require particular propositions be true (so, for instance, if science requires that there is a 'now', or if history requires that we take seriously the idea of tense-less temporal relations) all that is *really* implied, then, is that the A- and B-theory must get their semantics right. Both theories must have a semantics adequate to preserving the classes of truth that we take ourselves to have. But this, as indicated in section1, is already a constraint on a good theory of time, quite independent of these further interdisciplinary concerns.

The upshot, then, is that cases such as the Grunbaumian one discussed above, are not such that they place any new constraints on our theory of time, and any interdisciplinary study should be live to that fact. The potentially exciting results alluded to at the end of section 2 do not come to pass.

Are there conclusions that one can reach, on the basis of this discussion, concerning interdisciplinary study? I think so. One of the assumptions that Fraser (1966, xx) makes explicit at the outset of his venture to put forward a systematized view of time, one that includes inputs from all disciplines, is that "[w]hen specialists speak of time, they speak of various aspects of the same entity." If we grant the further assumption that the manifestations of time are many, then the forgoing should serve as an attempt to make sense of at least one of these manifestations.[31]

[29] See Saunders 2002, though also Craig 2001, for an extensive and detailed argument to the contrary.

[30] See McKinnon 2003.

[31] One question obviously not addressed within this paper is how the putative A-properties/B-relations are supposed to have a bearing on our experience of time and duration. For instance, as Flaherty (1999: 4–5) asks, "How is the perceived passage of time shaped by the individual's transition from one to another situation or realm of social reality?" Clearly, if the A/B debate is to have substantial relevant to other

References

Armstrong, David. 1978. *Universals and Scientific Realism*. Cambridge: Cambridge University Press.

———. 2004. *Truth and Truthmakers*. Cambridge: Cambridge University Press.

Barbour, Julian. 1999. *The End of Time*. London: Weidenfeld and Nicolson.

Beebee, Helen and Julian Dodd, eds. 2005. *Truthmakers* Oxford: Oxford University Press.

Berkeley, George. 1999. *A Treatise Concerning the Principles of Human Knowledge*. Oxford: Oxford University Press.

Bourne, Craig. 2006. A Theory of Presentism. *Canadian Journal of Philosophy* 36: 1–24.

Carnap, Rudolp. 1947. *Meaning and Necessity* London: Chicago Press.

Corish, Dennis. 2005. McTaggart's Argument. *Philosophy* 80: 77–99.

Craig, William Lane. 2000. *The Tensed Theory of Time*. Dordrecht: Kluwer Academic Publishers.

———. 2001. *Time and the Metaphysics of Relativity*. Dordrecht: Kluwer Academic Publishers.

Crisp, Thomas. 2004. On Presentism and Triviality. *Oxford Studies in Metaphysics* 1: 15–20.

———. 2007. Presentism and the Grounding Objection. *Nous* 41: 90–109.

Dainton, Barry. 2001. *Stream of Consciousness: Unity and Continuity in Conscious Experience*. London: Routledge.

Davidson, Donald. 1984. *Inquiries Into Truth and Interpretation*. Oxford: Clarendon Press.

Dobbs, H. A. C. 1969. The 'Present' in Physics. *British Journal for the Philosophy of Science* 19: 317–324.

Dyke, Heather. 2002. Tokens, dates and tenseless truth conditions. *Synthese* 131: 329–351.

Falk, Arthur. 2003. Time Plus the Whoosh and Whiz. Pp. 211–50 in *Time, Tense, and Reference*, ed. Aleksander Jokic and Quentin Smith. Cambridge, Mass: MIT Press.

Ferre, Frederick. 1970. Grunbaum *vs* Dobbs: The Need for Physical Transiency. *British Journal for the Philosophy of Science* 21: 278–280.

Flaherty, Michael. 1999. *A Watched Pot: How we Experience Time*. New York: New York University Press.

Fraser, Julius. 1966. Introduction. Pp. xvii–xxiv in *The Voices of Time*, ed. Julius Fraser, London: Penguin.

Grunbaum, Adolf. 1968. *Modern Science and Zeno's Paradoxes*. London: Allen and Unwin.

———. 1969. Are Physical Events Themselves Transiently Past, Present and Future? A Reply to H. A. C. Dobbs. *British Journal for the Philosophy of Science* 20: 145–152.

Haslanger, Sally and Rosemary Kurtz, Rosemary, eds. 2006. *Persistence: Contemporary Readings*. London: IT Press.

---

domains of enquiry, work must be done to establish how it is that A-properties or B-relations *can* contribute to this manner of the perception of time and duration as well as how we work to socially construct a concept of time given those experiences. Sadly, that lies beyond the scope of the current work. Those interested in recent philosophical work on the *experience* of time might like to consult Dainton 2001, Prosser 2007, Oaklander and White 2007, Tallant 2007, and parts of Zimmerman 2008.

Hawley, Katherine. 2006. Science as a Guide to Metaphysics?. *Synthese* 149: 451–470.
Loux, Michael. 1998. *Metaphysics: A Contemporary Introduction*. London: Routledge.
Lowe, E. Jonathan. 1998. *The Possibility of Metaphysics*. Oxford: Clarendon.
———. 2005. *The Four-Category Ontology*. Oxford: OUP.
McCall, Storrs. 1994. *A Model of the Universe*. Oxford: Clarendon.
McKinnon, Neil. 2003. Presentism and Consciousness. *Australasian Journal of Philosophy* 81: 305–323.
McTaggart, John Ellis. 1908. The Unreality of Time. *Mind* 17: 457–474.
Mellor, D. Hugh. 1981. *Real Time*. Cambridge: Cambridge University Press.
———. 1998. *Real Time II*. London: Routledge.
Oaklander, L. Nathan. 2004. *The Ontology of Time*. New York: Prometheus Books.
Oaklander, L. Nathan and Quentin Smith, eds. 1994. *The New Theory of Time*. Yale: Yale University Press.
Oaklander, L. Nathan and Alan White. 2007. B-time: reply to Tallant. *Analysis* 67, 2007: 332–400.
Prosser, Simon. 2007. Could We Experience the Passage of Time?. *Ratio* 20.1: 75–90.
Saunders, Simon. 2002. How Relativity Contradicts Presentism. Pp. 277–92 in *Time, Reality and Experience*, ed. C. Callender. Cambridge: Cambridge University Press.
Sider, Thedore. 2001. *Four-Dimensionalism*. Oxford: Clarendon Press.
Smith, Quentin. 1993. *Language and Time*. New York: Oxford University Press.
Tallant, Jonathan. 2005. McTaggart: That to which we are Compelled to Respond. *Philosophical Writings* 28: 3–12.
———. 2007. What is B-time?. *Analysis* 67: 147–56.
Tooley, Michael. 1997. *Time, Tense, and Causation*. Oxford: Clarendon Press.
Unger, Peter. 1971. A Defense of Skepticism. *The Philosophical Review* 80: 198–219.
Weiner, Norbert. 1949. *The Extrapolation, Interpolation, and Smoothing of Stationary Time Series with Engineering Applications*. New York: Wiley.
Wittgenstein, Ludvig. 2001. *Tractatus Logico-Philosophicus*. London: Routledge.
Zimmerman, Dean Z. 2008. The Privileged Present: Defending an "A-theory" of Time. Pp. 211–225 in *Contemporary Debates in Metaphysics*, ed. Theodore Sider, John Hawthorne, and Dean Zimmerman. Oxford: Blackwell.

CHAPTER FOUR

TEMPORAL ASYMMETRY AND RELATIVITY[1]

Friedel Weinert

> The arrow of time is not implied by the fundamental equations, it nevertheless characterizes most solutions.
>
> —Popper 1956, 538

Summary

The purpose of this paper is to explore the compatibility of the Special theory of relativity (STR) with a dynamic notion of physical time. Such an approach seeks to establish the reality of time against the standard account, according to which the STR implies the unreality of time and shows that the physical world is static. The thesis of this paper is that a dynamic view of time within relativistic space-time physics can be recovered if both topologic and entropic aspects of the kinematic relations between reference frames are taken into account. Both need to be invariant. The frame-independence of some temporal processes is important to avoid the objection that there are "as many times as there are reference frames in the STR."

The STR does not require a reference to observers, only to reference frames in relative motion to each other. There is no need to talk about the metaphysics of a moving 'Now'. The 'earlier-later' relation is sufficient but it does not lead to a static view of time. This relation is also essential for the relational view of time, since, according to Leibniz, "time is the *order* of successive events."

1. Problem Situations

The debate about whether the world is fundamentally static or dynamic is very old. If the physical world just *is*, then the passage of time must be a human illusion. If, however, the physical world is fundamentally in *flux*, then the flow of time depends on physical events outside of human observers. Traditionally such debates relied on conceptual and logical arguments, on the basis of which metaphysical positions were staked. With the rise of the Special theory of relativity (STR, 1905),

[1] The author gratefully acknowledges the support of the British Academy by way of an Overseas Conference Grant, which helped to finance my attendance at the conference.

Jo Alyson Parker, Paul A. Harris and Christian Steineck (Eds), Time: Limits and Constraints, pp. 109–134

a genuine scientific theory became entangled with metaphysical issues about time. The old debate shifted ground: it came to rely on arguments, which are derived from the Special theory. Scientists, too, were drawn into the debate. The advent of the Special theory created a new problem situation. Two opposite philosophical views—the *Philosophy of Being* (Eternalism) and the *Philosophy of Becoming* (either Presentism or Possibilism)—appeal to a physical theory for evidential support. Most physicists and many philosophers believe that the STR favours the Philosophy of Being.

Eternalism is the view that the physical world is a block universe. For Laplace's omniscient demon, past, present, and future are equally real. The demon sees all events, stretching from the present into the past and future as if they were frames of a film. The apparent passage of time is a human construction. Presentism is the view that there is a privileged *Now*, in which events alone enjoy reality. As the *Now* constantly shifts—what is *Now* will become past, and what was future will become *Now*—Presentism is dynamic. It incorporates temporal becoming. McTaggart has offered metaphysical arguments against Presentism. And the Special theory seems to be incompatible with this view because it explicitly prohibits a privileged *Now*. There is an intermediate position: Possibilism is the view that the future is indeterminate whilst both present and past events are determinate and therefore, in a certain sense, real (see Savitt 2000; 2001). Possibilism also offers a dynamic view of time. For it to be viable it must resist McTaggart's arguments against the reality of time and forego a privileged *Now*.

In this paper I will use the phrases *Philosophy of Being* (Block Universe) or *Philosophy of Becoming* (Possibilism), for the simple reason that these views can be characterized in terms of space-time events. McTaggart taught us an important lesson. In discussions of the philosophy of time, it is crucial to distinguish between *tensed* language (events move through past, present and future) and *tenseless* language (events are earlier or later, before or after). The employment of the 'earlier-later' relation to characterize physical time immediately establishes a connection both with the theory of relativity and the relationist view of time. For the relationist, time is constituted by the order of physical events in the universe. The relationist characterizes time in the tenseless vocabulary of the 'before-after' relation. The standard view amongst philosophers and physicists alike is to accept that the tenseless 'before-after' relation commits us to a static view of time.

The paper will argue that tenseless physical time, which arises from the STR, is compatible with a dynamic conception of time.

The paper will first set out the new problem situation: how the Special theory came to be associated with metaphysical views of time. It will briefly review some of the main arguments that have been used to shore up the philosophy of being or becoming, respectively. After this review, this paper will reconsider an aspect of the debate, which Einstein already had in mind. It establishes a connection between entropy and physical becoming. This connection involves a relational view of time, of which Leibniz and Mach are the main proponents. But Einstein's view of space-time is relational too. This triad—relationism, entropy, and relativity—then points us in the direction of a philosophy of physical becoming. The conclusion is that contrary to accepted wisdom, the STR is compatible with objective becoming.

2. From Physics to Philosophy

The emergence of the STR in 1905 was a key event in both physics and philosophy. Einstein regarded his theory as a crowning of classical physics, rather than a "revolutionary act" (Einstein 1921a, 246). But the theory had radical consequences for a philosophical understanding of the natural world. Foremost, it led to a union of space and time, which Minkowki expressed in his celebrated phrase of four-dimensional 'space-time'. This conception of space-time, whose fruitfulness Einstein eventually accepted, means that the physical world can no longer be split into a '3+1' view: a three-dimensional space container and a one-dimensional time river. According to the Newtonian view of space and time, any event in the universe can be uniquely specified by three space coordinates and one time coordinate. Let *E* be the first determination of the circumference of the earth by the Greek mathematician Eratosthenes. The event occurred, say, at a particular moment in 240 BC. Commonsense tells us that all observers of the event, past and present, will agree on the unique assignment of three spatial coordinates to specify the location and one temporal coordinate to specify the time of the event. Newton added that to be useful in mechanics the spatial and temporal coordinates needed to satisfy two further conditions. They had to be both universal and absolute. Space and time coordinates need to be *universal*: all observers throughout the whole universe need to agree that the event, *E*, happened at the specified space and time location. They also need to be *absolute*: the temporal and spatial axes must be defined irrespective of particular material events. To illustrate, Newton resorted to a metaphor. He thought of space as an empty cosmic container, whose imaginary walls provided the three spatial dimensions. He thought of time as a metaphorical river whose regular flow provided the time dimension. The cosmic container allowed the precise placement of any event with respect to the container walls. The metaphorical river allowed the arrangement of the event along a unique time axis. These conditions needed to be imposed, Newton argued, to formulate the laws of motion. For material processes, like the rotation of the Earth, may not be as regular as the laws of mechanics require. But in the reformulation of Newton's views, for example, neo-Newtonian space-time, the notion of absolute space becomes redundant since Newton's laws respect the Galilean relativity principle with the effect that 'sameness of place' at different times cannot be defined; but it is still possible to refer different spatial locations to the same time axis.

Minkowski's space-time is no longer a '3+1' view. It is four-dimensional. The occurrence of every event must be specified by the particular space-time coordinates of the space-time location at which the event occurs. In the general theory of relativity, an extension of Minkowski's representation, space-time is neither absolute nor universal. Einstein cannot conceive of space-time as an empty cosmic container through which a metaphorical river meanders at a constant rate. Space-time is not *absolute*: it is made up of the total mass and energy in the world. Space-time does not exist apart from mass-energy. Nor is space-time *universal*: observers have different views of events in space-time, depending on their state of motion through it. Every observer assigns different space-time coordinates to a particular event in space-time. For reasons of simplicity we can restrict our considerations to inertial observers who are either at rest or in constant motion with respect to each other. Each of these observers has a particular window on space-time. The reason is that each of these observers occupies a particular reference frame. We can define an inertial reference frame as a system, which consists of a grid of rigid measuring rods with regular synchronized clocks attached to them. We can think of space-time as made up of these grids. These reference frames are either stationary or move with constant, non-accelerating motion with respect to each other. The spatial and temporal coordinates of the reference frames are determined by their respective motions. These coordinates are now defined by the synchronized clocks and rigid measuring rods, which are attached to the reference frames. The physics of the Special theory does not require the presence of observers in addition to the clocks and rods. For ease of presentation we can imagine the reference frames as scaffolds, carrying observers along. But the observers do no more than read what the clocks and rods record. The motions of the respective reference frames influence the behaviour of these clocks and rods. They experience what is known as *time dilation* and *length contraction*. From the point of view of a stationary reference frame, the clocks slow down in fast-moving reference frames and the rods contract. In short, the clocks and rods provide yardsticks for the perspectival measurement of events in respective reference frames.

This characterization of reference frames has an unavoidable effect on our conception of time. It is the simultaneity of events, which deserves particular attention. In classical physics, two events happened simultaneously for all observers in the universe, if they coincided with a particular value of the time axis. Thus all observers agreed that if two events, E_1 and E_2, happened at the same universal clock time, they would be simultaneous. In relativistic space-time, there is no universal clock time. Every reference frame records its own clock time. Observers in different inertial reference frames therefore cannot agree when two events are simultaneous. To illustrate, let bolts of lightning strike the front and rear end of a fast-moving train as it rushes through a station. For observers on the platform of the station, the events are simultaneous. The observers on the platform inhabit a reference frame relatively at rest with respect to the train. Their clocks record identical times for the two events. For observers sitting mid-way on the train the flashes do not hit the opposite ends of the train simultaneously. These observers are attached to a moving reference frame. From their point of view, the train rushes towards the light signal from the front end and runs away from the light signal coming from the rear end. Their clocks record the respective events at different times. For the signals need time to travel to the midway positions of the observers on the train. If there is no universal notion of time, to which all observers can appeal, time seems to pass at different rates for each observer, depending on the speed of their reference frame. In four-dimensional Minkowksi space-time, the Newtonian sense of absolute simultaneity of events is lost. Events are only simultaneous with respect to the class of reference frames, on whose coordinates observers agree. There is only relative simultaneity. As Einstein declares:

> No absolute meaning can be assigned to the conception of the simultaneity of events that occur at points separated by a distance in space [...]. If no coordinate system (inertial system) is used as a basis of reference there is no sense in asserting that events at different points in space occur simultaneously. It is a consequence of this that space and time are welded together into a uniform four-dimensional continuum. (Einstein 1929, 1073)

The phrase 'time of an event' has no meaning, until it is related to a reference-frame in which the event occurs. Every reference-frame records its own particular time. (Einstein 1920, ch. 9, 26) And so the lamination of the four-dimensional continuum into three-dimensional space and one-dimensional time is the work of the observer, attached to a particular reference frame. The perspectival observer holds on to a '3+1' view, but physics is four-dimensional. According to Einstein, *physics becomes a sort of statics in a four-dimensional continuum.* Due to the validity of the Lorentzian relativity principle, neither 'sameness of place' nor 'sameness of times' can be defined but the space-time interval, *ds*, has an invariant value for all inertial observers.

This is Einstein's expression of the idea of the block universe (Einstein 1920, ch. 17, ch. 26; 1921b, 783; 1929, 1072; Einstein and Infeld 1938, 199–208). In physics, the distinction between past, present and future is suspended. There is only a timeless block of events, eternally present. This is how Einstein saw the connection between physics and philosophy at that time. If we accept the notion of the block universe, then logic compels us to agree to an idealist account of the origin of time. For if the physical world just *is*, then the impression of change, of becoming and the flow of time, must be located in the mind of the observer. Idealist philosophies of time from Saint Augustine to Kant drew this conclusion. Physicists like Hermann Weyl and Arthur Eddington explicitly embraced the idealist view of time, which follows from the notion of the block universe. (Weinert 2004, ch. 4). We shall see that Einstein became ultimately dissatisfied with the relinquishment of objective becoming. When confronted with Gödel's idealist inference from relative simultaneity to the unreality of time, he offered an entropic counterargument from thermodynamics. This argument is worth spelling out because it supports a dynamic reading of space-time. But the predominant view after the emergence of the Special relativity was the block universe.

3. The Philosophy of Being, Defended

McTaggart argued for the unreality of time from the derivation of what he took to be a contradiction between the *A*-series (the dynamic ordering of events by the degrees to which they are past, present or future) and the *B*-series (an ordering of events according to a static 'before-after' relation). Many physicists and philosophers have found the step from the theory of relativity to the metaphysics of being compelling. They arrived at a philosophy of being by arguing strictly from the logic of the structure of Minkowski space-time. In McTaggart's terms, the argument is about the *B*-series. McTaggart simply asserts that the *B*-series, lodged in the four-dimensional structure of Minkowski space-time, is static. If it is static, it leads to the block universe and a philosophy of being. But McTaggart does not consider whether the *B*-series could be dynamic. If it is dynamic, it leads to dynamic space-time and a philosophy of becoming.

Consider first how a philosophy of being may be considered a *conceptual* consequence of the STR, even though it is not a deductive consequence. Although many authors contributed to this argument the main steps were worked out by Rietdijk (1966) and Putnam (1967).

The general idea is that events, which exist *now*, must be regarded as real.[2] But according to the STR, the *Now* has become frame-dependent, since temporal coordinates depend on the behaviour of clocks, and are read by human observers. Consider a coordinate system $\{x_1, t_1\}$ of observer O_1, who is in relative motion with respect to an observer O_2. O_1 will regard all events as real, which lie on O_1's space-like simultaneity planes $\{x_1\}$. *Space-like* events are events which cannot be connected instantaneously by any subluminal or luminal signals. These are all the events simultaneous with O_1's *Now*. Any two events, which happen simultaneously for O_1 in two different locations, are space-like separated. They can only be connected by the exchange of signals, which take time to propagate. But what, according to the Special theory, is *now* for one observer will not be *now* for another. There will be a

[2] This section summarizes a more detailed discussion in Weinert 2004, 175–6.

second observer O_2, relatively at rest with respect to O_1. O_2 is attached to a reference system with the coordinates $\{x_2, t_2\}$. (See Figure 1) The second observer's reference frame can be made coincident with that of O_1, by a convenient choice. That means that both coordinate systems can be made to coincide at the origin. Both observers will momentarily be in each other's present. There will be many events, which are simultaneous and hence real for one observer, that lie in the past or future for a second observer. Take an event E_1, which is already *past* for one observer, O_1, but still in the future of the other observer, O_2. If the observers are space-like separated, then the time order will not be unique for them. But the event E_1 cannot be determinate or real for one and indeterminate for the other observer. Putnam calls this basic requirement the principle of No Privileged Observer. It must be determinate for both observers. For what is real for one observer must be real for another observer in his presence. This conclusion generalizes to all events in space-time. Hence *all* events must already be determinate and equally real for all observers.

According to the STR, there is always an observer like O_1 for whom each *future* event in O_2*'s* frame of reference is already a *past* event. If an event is already determinate for one observer, it must be determinate for all observers. Equally, if an event is already real for one observer, it must be real for all observers. It seems to follow that the physical world is a block universe and that the passage of time is a human construction.

According to the proponents of the philosophy of being, the assertion of a block universe should be construed as the most appropriate philosophical consequence of the theory of relativity. That is, the STR provides us with a number of physical constraints, which must be respected in any consideration of time. If these constraints are taken into consideration, the assertion of *being* has a better grounding than the assertion of *becoming*. A proponent of becoming must show that the proponents of the philosophy of being have not taken into consideration all the constraints, which the STR obeys. A consideration of all constraints embedded in the theory of relativity should show that the STR is compatible with objective becoming as a conceptual consequence.

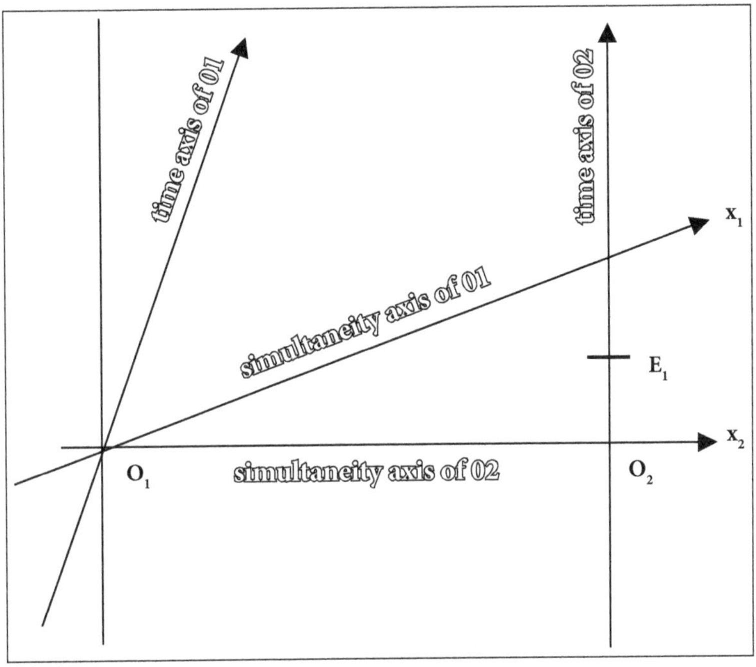

Figure 1: Determinate Reality and Relativity.
Two reference systems, in constant motion with respect to each other, each carrying an observer. Observer O_1 moves with constant velocity towards observers O_2. For O_1 event E_1 is already in the past. For O_2 the event E_1 is still in the future. As event E_1 is determinate from the point of view of O_1 it cannot be indeterminate from the point of view of O_2. For both it is equally determinate and hence equally real. Similar reasoning applies to an event E_2, which is simultaneous with O_1.

4. Relationism and the B-Series

Both the proponents of becoming and being will concede that appeal to human observers in the argument is more of a hindrance than a help. Human observers may indeed harbour illusions about the passage of time. The proponent of becoming must show that space-time itself is compatible with objective becoming, that it is not a block universe. Fortunately for the respective proponents in the debate, all observers in the STR can be replaced by the introduction of ideal clocks and rigid rods. All inertial reference frames, as indicated above, can be defined in terms of the behaviour of synchronized clocks and measuring rods. The clocks register the time in a particular reference frame, irrespective of whether it is read by a particular observer. With the removal of human observers, the need for McTaggart's *A*-series is eliminated. Removing the human observer removes the need for conceptual awareness and therefore the need for *tensed* language: the 'past-present-future' relation, which is the basis of calendar time. The language of physical occurrence of events at certain clock times is *tenseless* language, based on the 'before-after' relation. If physical events can be ordered along the 'before-after' relation, then there is, in a minimal sense, a succession of events. In the absence of human observers, who use tensed language, there will be a 'timeless' universe, since no observers will order events according to the 'past-present-future' relation. It is simply a universe devoid of calendar time. But this may not be a changeless universe, since in such a universe there will still be a succession of events, which happen at certain clock times. Consider the orbit of the moon around the earth. Even in the absence of conscious observers, the orbit of the moon establishes a 'period of time' (approximately 27.5 days in calendar time). These periods of time succeed each other, since even if each lunar cycle is physically identical in terms of its orbit, one cycle, C_n, is distinguished from another cycle, $C_{n\pm1}$, by the 'earlier-later' relation. Then 'timelessness' is not the same as 'changelessness'. It is at this point that the proponents of becoming find a natural alliance with the relational view of time; for the relational view shows that the 'earlier-later' series can be dynamic.

> Following Leibniz and Einstein, we have so far come to accept that there may be no meaning to time besides change. It has no absolute measure [...]. But this relational time is still a kind of time. It is grounded in change, but change then must be something real. (Smolin 1997, 355; cf. Clifton and Hogarth 1996, 359)

In his relational view, Leibniz held that time was constituted by the order of the succession of (actual and possible) material events. A Leibnizian relationist locates the order of the succession of events in the structure of classical space-time. The order of the succession of events is unique for all observers, because there is only one universal time metric, which applies throughout the universe. Leibniz, a contemporary of Newton, meant by order the classic 'before-after' relation. All events in classical space-time are uniquely ordered according to this relation because a universal simultaneity relation holds in classical space-time. As a relationist about time, Leibniz rejects the absoluteness in Newton's notion of time, for two reasons: (A) as Leibniz accepts the Galilean relativity principle it follows that 'sameness of place as a frame-independent notion' cannot be defined (Teller 1991; cf. Weinert 2006); (B) For Leibniz time is constituted by relations between actual and possible events, but it is still, as Aristotle said, "the same time everywhere." Leibniz therefore shares with Newton the view of the universality of time in the physical universe. The STR has also destroyed the notion of universal time: there are as many clock times as there are moving reference frames. As the notion of 'relative' simultaneity shows, the classic 'before-after' relation has become problematic with the STR. But the philosophy of becoming can still commit to the relational view. The relational view of time needs to be adapted to the constraints, which the Special theory has imposed. In particular, relationism needs to adopt relative simultaneity. Let us dub the alliance between relativity and relationism *space-time relationism*. The proponent of becoming will have to rely on a relational view of time as a succession in events in Minkowski space-time. This move opens the prospect that the *B*-series is not static. From the point of view of a relational view of Minkowski space-time, McTaggart's mistake lies in the identification of 'timelessness' with 'changelessness'. If the relational proponent of becoming can show that the *B*-series can be dynamic under the constraints of relativity, then the philosophy of becoming has gained a foothold in Minkowski space-time. Let us consider two arguments for a dynamic view of time in space-time relationism. First we consider a succession of events under *topologic* aspects and later turn to a consideration of *entropic* aspects.

5. The Philosophy of Becoming, Defended

Howard Stein (1991) proposes to define a relation *R*, whose job is to establish objective becoming in Minkowski space-time. That is, Stein proposes to define *R* using the geometric structure of Minkowski space-time. The term 'becoming' is stripped of the anthropocentric connection, which it still has in McTaggart's argument against the unreality of time.

> The concept of becoming acquires a meaning in physics: The present, which separates the future from the past, is the moment when that which was undetermined becomes determined, and "becoming" means the same as "becoming determined." (Reichenbach 1956, 269; cf. Dorato 2006)

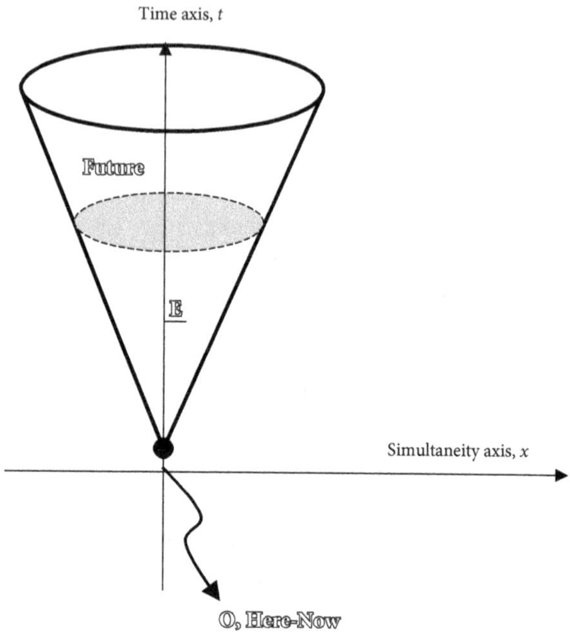

Figure 2a: Stationary reference frame and simultaneity plane to event *E*.

The geometric structure of Minkowski space-time offers the space-time events and the light cone structure as building blocks. Although light cones emanate from every space-time event at Here-Now, *O*, into the future and converge onto *O* from the past, the invariance of *c* for all observers means that light cones do not tilt. (See Figures 2a, b) As candidates for an objective becoming relation Stein considers both *null* and *time-like* vectors. The null vectors diverge at an angle of 45° from the origin at Here-Now, *O*, into the future; they converge at the same angle at *O* from the past. The light signals, which emanate from events in space-time, define *null-like* separated events. Time-like vectors always arrive at and leave *O* at an angle of inclination smaller than 45° (from *t*, *t'*). To speak of the past of an event *x* in Minkowski space-time is to identify a *past-pointing* vector from *x* to *y*, where *y* lies in or on the light cone, which converges onto *x*. The event *y* has become for the event *x* means: *y* lies in the past light cone of *x*, if and only if there exists a past-directed time-like curve from *x* to *y*. The event *z* has not yet become for *x* means: *z* lies in the future light cone of *x*, if and only if there exists a future-directed time-like curve from

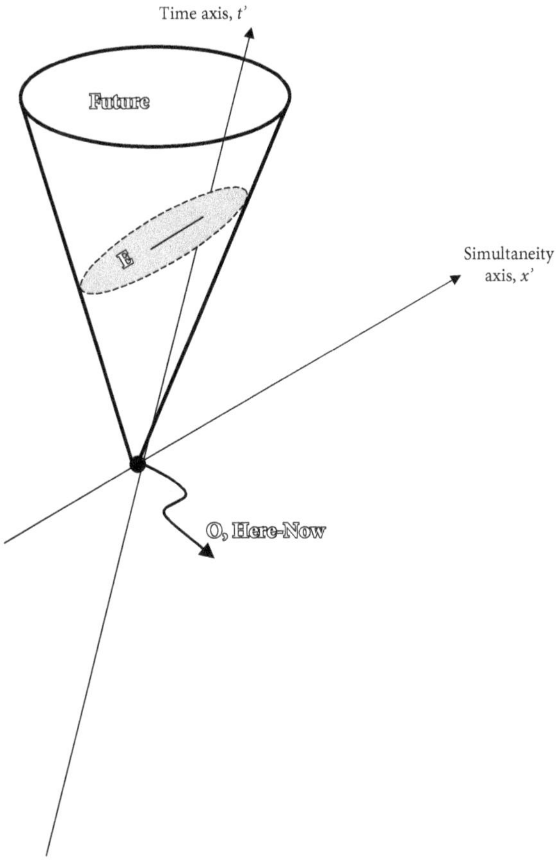

Figure 2b: Moving reference frame and simultaneity plane to *E*. Note that the light cone does not tilt for the moving observer.

x to *z* (see Stein 1991, 148–50; Clifton and Hogart 1996, 362; Savitt 2001, 16; Dorato 2006). Stein's relation *R* is a two-place relation, *yRx*, which establishes a temporal order between *y* and *x*: *yRx* says that from the viewpoint of space-time event, *x*, the space-time event *y* has occurred and is definite. In terms of Leibnizian relationism this means that *y* is a chronologically earlier event than *x*. Relation *R* satisfies two further requirements: *R* is transitive, *yRx* and *yRz;* and *R* is reflexive, *xRx*. The transitivity of *R* means that if there is a space-time event, *y*, which is already definite with respect to *x*, then *y* is also definite with respect to *z*, which lies in the future of *x*. Transitivity means that *yRx* & *xRz* entail that *yRz*. The reflexivity of *R* means that *x* has become with respect to itself, xRx. The fact that light cones converge and emanate from space-time events and that these light cones do not tilt, means that they have an invariant structure.

Note that this notion of objective, chronological becoming is not observer-dependent. The relation, R, defined on Minkowski space-time can be connected to the experience of a human observer by the following step:

> At a space-time point a there can be cognizance of—or information or influence propagated from—only such events as occur at points in the past of a. (Stein 1968, 16; cf. Stein 1991, 149)

Implicitly Stein appeals to two aspects of the relational view of time. In a primary sense, time is constituted by the order of the succession of events in space-time. The primary sense of time is physical time. Within these light cones, time-like connected events can be linked by slower signals, like sound. There is an asymmetry between an earlier and a later space-time event, such that signals can be sent from the earlier to the later event. The succession of time-like related events is a topological invariant, independent of our choice of reference frame. All observers moving relative to each other at constant speeds will agree on the *same* temporal order, but disagree on the duration and simultaneity of events. It depends in no way on human perception. McTaggart calls this succession the *B*-series, but it is not static because of the role of null- and time-like vectors. The relationist can also define McTaggart's *A*-series. According to Leibniz the *A*-series is a human abstraction from the observations of the order of the succession of physical events. The *A*-series speaks of the passage of events from past to present to future and of the moving *Now*. For the relationist this depends on human awareness. It is calendar time. But the Special theory only needs physical or clock time, which observers record.

For the space-time relationist the order of the succession of events is no longer absolute and universal. The order of succession must be defined by a Lorentzian metric, which is given by the material processes in the world. In the STR, these processes are constituted by the *time-like* and *null-like* connectibility of events. The order of succession of events does not extend to space-like connected events.

Note that the proponents of the two philosophies take different aspects of the Special theory into account. The proponents of a philosophy of being rely on space-like connected events and the relativity of simultaneity to conclude that the STR is only compatible with the block universe. For the space-time relationist there exists only a clear unambiguous order of succession for null-like and time-like connected events. This is a logically coherent position, but is it philosophically interesting? (See Callender 2000, §3). It is philosophically interesting if it produces an objective relation of becoming, which can be derived from Minkowski space-time.

Underlying Stein's argument is an important *empirical* aspect of time-like and null-like vectors, for example, the physical properties of the propagation of electro-magnetic signals in space-time (Cf. Dorato 2006, 570). The null-like and time-like vectors are geometric representations of signals, which connect events in space-time. These signals have physical properties, which bestow a chronological order on them. For a dynamic characterization of Minkowski space-time, it will be important to consider the underlying physical properties of signal propagation. The following section argues that entropic processes underpin the topologic order of succession of events, and result in an empirical grounding of dynamic Minkowski space-time. The finite propagation of 'causal' signals in Minkowski space-time implies that Minkowski space-time contains a dynamic element. They are causal signals because they happen between time-like and null-like related events, for example, events in space-time, which can be connected by the finite propagation of signals. The fundamental postulate of the Special theory—the finite propagation of light—confers a history on the world lines. It gives rise to objective becoming.

6. Relationism, Entropy and Objective Becoming

From time to time, Einstein seemed to endorse the unreality of time.

> For us believing physicists, the distinction between past, present, and future is only an illusion, even if a stubborn one. (Quoted in Hoffmann 1972, 257–8)

Yet he often speaks of *time-space* rather than *space-time*, and his whole theory of time-space is relational.[3] It points towards a philosophy of becoming.

[3] The question of space-time relationism in the context of the GTR has recently been the subject of a lively debate; see for instance Cao 2006; Stachel 2006, and further references there.

> I wished to show that space-time is not necessarily something to which one can ascribe a separate existence, independently of the actual objects of physical reality. (Einstein 1920, vi)

In particular, Einstein employed an entropic argument against Gödel's idealistic interpretation of the STR. Gödel had pointed out that the relativity of simultaneity offered "new and surprising insights into the nature of time." In particular, he held that through this theory one obtained 'unequivocal proof' for the *idealist* view of time. Gödel argued that the relativity of simultaneity and, by implication, assertions of temporal successions lose their objective meaning, "in so far as another observer with the same claim to correctness can assert that *A* and *B* are not simultaneous" (Gödel 1949, 557). If simultaneity is relative, the lapse of time will be affected: it cannot "consist of an infinity of layers of 'now' which come into existence successively, in an objectively determined way" (Gödel 1949, 558). Thus Gödel rejects, in agreement with proponents of being, the Newtonian character of time.

In his argument, Gödel puts emphasis on the *idealist* view of time, as a consequence of relative simultaneity, but he also embraces the block universe. For Gödel associates the idealist view of time with the denial of the objectivity of change. The *passage of time* is a subjective human experience.

When confronted with Gödel's argument, Einstein was not prepared to endorse it. In an explicit attempt to counter it, Einstein considers the question of the temporal direction of events in space-time. Imagine we send a signal, like a light pulse, from *B* to *A* through *P*. On thermodynamic grounds he asserts that the sending of a signal and light emissions generally are irreversible processes (Einstein 1909, 831). A time-like world line from *B* to *A* through *P* in a light cone takes the form of an arrow making *B* happen *before P* and *A after P*. (See Figure 3) This secures the "one-sided (asymmetrical) character of time [...], i.e., there exists no free choice for the direction of the arrow' (Einstein 1949, 687). This is true at least if points *A*, *B*, and *P* are sufficiently close in cosmological terms. The asymmetrical character of time is here based on an irreversible *before-after* relation between physical events without reference to an observer. It involves a process, according to which there is an event, *B*, at which the signal is emitted, and there is a later event, *A*, at which the signal is received. There is an entropy gradient between the state of events at *B* and *A*. The assessment of this differential entropy between the two locations does not depend

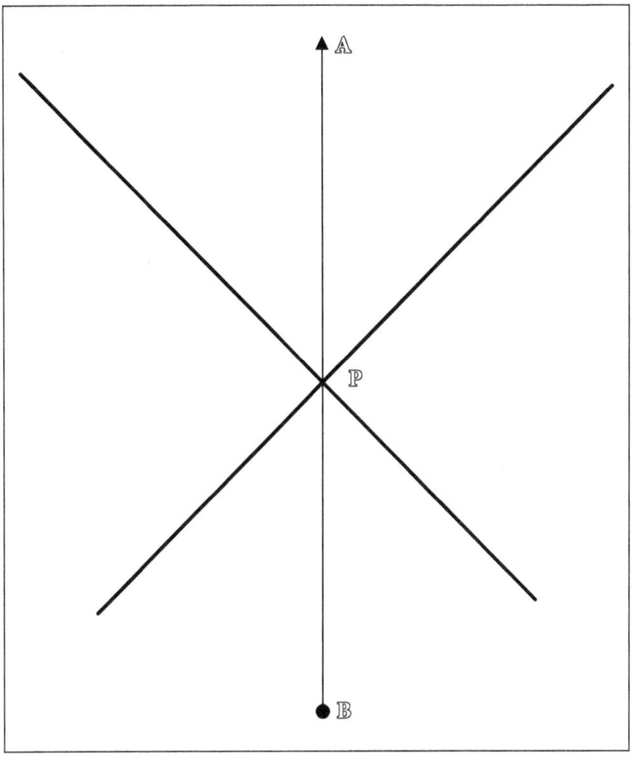

Figure 3: Einstein's consideration of the (local) direction of time in response to Gödel's idealistic interpretation of the special theory of relativity. A *time-like* world line exists between events A and B, which lies within, not outside, the light cone. A and B are linked by an irreversible signal.

on a particular reference frame. According to a fundamental result of the STR the entropy of a system is frame-independent (Einstein 1907, §15). This frame-independence is important for the argument. Stein had grounded the frame-independence of relation *R* in the topological invariance of the light cones. It is the notion of relative simultaneity, which Stein must avoid to establish his relation *R*. Relation *R* is defined in terms of the geometry of Minkowski space-time: it is an affine structure, whose topology is given by the *null-like* curves. Einstein appeals to the frame-independence of entropic processes in terms of thermodynamics. Without such frame-independent physical relations between space-time events, the space-time relationist would be in a dire situation. For the relationist, who wants to infer a dynamic view of time from the Minkowski space-time representation, cannot rely on the notion of relative simultaneity. The latter notion has been used by Gödel and others to infer a static view of time (block universe). But the space-time relationist will want to move from Einstein's thermodynamic considerations to a statistical notion of entropy, based on statistical mechanics. The latter shows how our asymmetric experience of temporal processes can be made compatible with the *t*-symmetry of mechanical laws and microscopic reversibility.

It has been objected that a statistical notion of entropy could not form the basis for a general theory of physical time (Popper 1956; Bridgman quoted in Grünbaum 1967, 187–8). This objection is based on Boltzmann's reconsideration of the 2nd law of thermodynamics in response to the reversibility objections. Boltzmann reinterpreted the 2nd law as a statistical law: $S = k \log W$, and regarded entropy as a measure of probability. W expresses the number of micro-conditions that are compatible with a given macro-condition of a physical system in a given thermodynamic state. (More precisely, W is related to the number of arrangements in μ-space, a six-dimensional space, which specifies the microscopic configuration (q_i, p_i) of an N-particle system.) The probability that an ensemble of physical systems will evolve from a state of low entropy to one of higher entropy is vastly greater than the transition from a high S to a low S: for example, the more ordered arrangements of molecules in a small area of μ-space are overwhelmingly likely to evolve into less ordered arrangements, occupying a larger area of μ-space. The statistical version of the 2nd law permits the reversibility of high entropy states of a physical system to lower entropy states—the reheating of a glass of cold water by a spontaneous rearrangement of the water molecules, without interference from the environment of the system. But the statistical version of the 2nd law is held to be valid for a majority of thermodynamic systems. The reinterpretation of the 2nd law made it compatible with the *t*-symmetry of the basic mechanical laws and thus reconciled the increase in entropy of an ensemble of macro-systems with the mechanical reversibility of individual micro-systems, as required by the dynamic laws of classical physics.

> If a system goes through a purely mechanical process, that is, a process during which its temperature, heat content and so forth, do not change, the process is reversible; therefore the entropy must remain constant. In other words, a change in mechanical parameters has no influence on the value of the entropy. Only when, in addition to mechanical changes, there are thermodynamic changes involved, such as heat generation

> through friction, will the entropy of the system change. (Reichenbach 1956, 50; Grünbaum 1967, §3)

But, so the objection continues, "it is absurd to link entropy to the existence of the arrow of time because of the existence of thermodynamic fluctuations" (Popper 1956, 382). Popper's purpose is to show that (a) classical processes, like divergent water waves from the centre of a pond, after a stone has been dropped into it, are much more likely than convergent water waves towards the pond's centre, which would require unlikely coherence conditions at the edges at the pond; (b) even so, in statistical mechanics all processes are still reversible, although this reversal is very unlikely if a majority of systems is taken into account. Popper therefore rejects the statistical theory of the arrow of time, since the arrow of time "does not seem to have a stochastic character"[4] (Popper 1965, 233; cf. Popper 1956, 538; Popper 1958, 403). However, the question of a dynamic reading of Minkowski space-time is not concerned with the cosmological arrow of time, but with the question of the anisotropy of time in Minkowski space-time. Popper's appeal to the "impossibility of coherence in initial conditions"—or the untypicality of initial convergent waves towards the centre of the pond—if generalized (Hill and Grünbaum 1957, 1296), can serve as an argument for the anisotropy of time. This generalization has been attempted by a 'law of conditional independence', according to which "causal influences coming from infinity in different directions are independent of one another" (Penrose/Percival 1962, 606; Davies 1974, §5.2).

A formal statement of this postulate is:

> If A and B are any two disjoint 4-regions [in the future light cones of Minkowski space-time], and C is any 4-region which divides the union of the pasts of A and B into two parts, one containing A and the other containing B, then A and B are conditionally independent given c. That is
>
> $p(a, b / c) = p(b / c)p(b / c)$ (all a,b).
>
> As usual, a, b and c are the histories of A, B and C respectively. (Penrose/Percival 1962, 611)

[4] However, Loschmidt's reversibility objection to Boltzmann's entropic view of time cannot easily be dismissed since the evolution from a non-equilibrium macrostate to equilibrium is mathematically as probable as the evolution from an equilibrium macrostate to a non-equilibrium macrostate; see Torretti 2007, 17.

Although the laws of mechanics permit reversible behaviour, electromagnetic waves and scattering particles diverge outwards from a space-time event, which constitutes their source. Consider two uncorrelated (statistically independent events) in space-time at t_1. They are made to interact for a brief interval, at t_2. The moment of interaction creates a conditional probabilistic dependence between the two systems, which forms the basis of temporal irreversibility between physical processes. The law of conditional independence is certainly a general principle that is satisfied in Minkowski space-time; it does not seem to rely on entropic considerations. The notion of conditional independence assumes the independence of causal influences coming from different directions. (Penrose/Percival 1962, 611) However, in Minkowski space-time such 'causal influences' are typically carried by electromagnetic signals, which, as Einstein's example illustrates, are subject to entropic processes. Note that events, not probabilities, are fundamental to four-dimensional Minkowski space-time. If we take into account the propagation of signalling processes, which link time-like events in Minkowski space-time, and the typicality of initial conditions, we find that entropic considerations cannot be avoided. The space-time relationist's concern is to show the compatibility of Minkowski space-time with a dynamic interpretation of physical time. Time is constituted by the order of physical events in space-time, and these events do not just include kinematic aspects. It was precisely the over-concentration on geometric relations, which encouraged the view that the STR was only compatible with the block universe, although Stein tried to counter it by a consideration of vectors between space-time events. Vectors have direction and this direction may be grounded in physical relations. Entropic processes therefore indicate an order of temporal processes irrespective of human awareness. They are compatible with the time-reversibility of the fundamental mechanical laws. Time is constituted by the order of invariant succession of events, related by electromagnetic signals. Locally, the entropy gradient points in the direction from *B* to *A*. All observers agree at least for time-like connected events. For the relationist this establishes, within local space-time regions, an order of the succession of events and thereby physical time.

Einstein was not alone in suspecting that entropic processes could serve as an indicator of the anisotropy of physical time. As Reichenbach observed "there is no such thing as mechanical entropy" (Reichenbach 1956, 50; cf. Penrose and Percival 1962, §1). By extension of this argument, a relational view of space-time must look beyond purely kinematic relations for such relations are subject to *t*-symmetric mechanical laws. What has to be taken into account is the history of world lines, the history of relations between space-time events, which includes dynamic and thermodynamic aspects (Cf. Huggett 2006). According to Reichenbach (1956, ch. 3, §§14–16), temporal asymmetry must be associated with the entropic behaviour of a majority of what he calls 'branch systems'. These subsystems become thermodynamically isolated for a brief period and then merge again with the wider system. The behaviour of *individual* systems may be time-symmetric. The anisotropy of time is indicated by the direction of entropy increase of the *majority* of branch systems in space ensembles.

> The direction of time is supplied by the direction of entropy, because the latter direction is made manifest in the statistical behaviour of a large number of separate systems, generated individually in the general drive to more and more probable states. (Reichenbach 1956, 135; cf. Grünbaum 1955, §3; Grünbaum 1967, §3; Davies 1974, §3.4; Albert 2000, §3.3)

For the space-time relationist, Reichenbach's hypothesis of the branch structure is an elaboration of the Leibnizian view that time is the order of successive (actual and possible) events, under the constraints set by the structure of Minkowski space-time. The relationist does not require a unidirectional succession of events in the whole universe. Rather, it follows from Reichenbach's hypothesis that "a time direction can be defined only for sections of the total entropy curve" (Reichenbach 1956, 127).

According to these considerations, physical time need not founder on *t*-symmetric laws. Regular irreversible processes in nature, even of a statistical kind, are sufficient to establish physical time. In his thought experiment with the one-directional signal propagation, Einstein combines both classical divergence phenomena and entropic behaviour. Such processes are not solely determined by the time-reversible equations of motion. They also depend on boundary conditions and thermodynamic aspects, which make them statistically irreversible.

The proponents of the philosophy of being may not be impressed with this argument, for their concern is to exploit the relativity of simultaneity for space-like connected events. No clear order of succession can be established for these events. Putnam's 'No Privileged Observer Postulate'—what has become real for one observer cannot remain unreal for another observer—seems to impose the block universe. But ontologically and metaphysically, space-like connected observers and time-like connected observers are not on a par. By definition no causal signals can be exchanged between space-like connected observers; such an exchange is the prerogative of time-like connected observers. Time-like connected observers are close enough in space and apart enough in time to communicate. Space-like connected observers are too far apart in space and too close in time to communicate. Whilst the existence of observers is not essential to establish relative simultaneity between events in separate reference frame, what is essential is the comparison of clocks. But the point about space-like connected observers is that their coordinate systems can only be made to coincide by conventions. This has a rather disturbing consequence for the proponent of the philosophy of being. When there can be no causal contact between events in space-time, as David Bohm observes,

> it makes no difference whether we say they are before or after each other. Their relative time order has a purely conventional character, in the sense that one can ascribe any such order that is convenient, as long as one applies his conventions in a consistent manner.

But the possibility of physical contact makes a significant difference to this situation:

> where such causal contact is possible, the order of events is unambiguous, so that the Lorentz transformation will never lead to confusion as to what is a cause and what is an effect. (Bohm 1965, 159–60; cf. Savitt 2001, 15–6)

If we can draw no metaphysically significant conclusion about time from the conventional order of the temporal relations between space-like connected events, then the important temporal relations are to be found amongst the time-like connected events. Between time-like connected events, we have found candidates for the order of the succession of events: topologic relations based on entropic relations. In the traditional debate between proponents of the philosophy of being *versus* the philosophy of becoming the question of a global cosmological arrow of time does not arise, because the STR does not deal with cosmological issues, which require other considerations. The debate concerns the possibility of the anisotropy of time in the setting of Minkowski space-time. The space-time relationist argues that the 'passage' of time must be grounded in physical relations, like the entropy gradient in branch systems, which become indicators of the anisotropy of time.

7. Conclusion

The theory of relativity may not have the physical resources to confirm or refute empirically the philosophy of being or the philosophy of becoming, respectively. This theory may however be more compatible with one view, at the expense of the other, if all the relevant constraints are taken into consideration. The standard view is that the STR is only compatible with a philosophy of being (Eddington 1920; Weyl 1952; Davies 1974; Lockwood 2005). The philosophy of being exploits the notion of relative simultaneity and the conventional coincidence of Here-Now between two space-like separated observers to conclude, in the words of Weyl, that the physical world "just is." The passage of time becomes mind-dependent, whilst the four-dimensional physical world exists as a block universe. The philosophy of becoming accepts the relativity of simultaneity but exploits the existence of a chronological order amongst time-like related events and the invariant structure of light cones to argue for a philosophy of becoming. In particular the propagation of 'causal' signals along entropic gradients in time-like connected events is an important part of a full appreciation of Minkowski space-time. This view emphasizes that space-like connected events are not as essential as time-like connected events. The paper has argued that the 'earlier-later' distinction is not static on the relationist view of time. The order of succession of events within Minkowski space-time—both under topologic and entropic aspects of signal propagation—confers a history on world lines and bestows chronological order on time-like connected events. Contrary to accepted wisdom the STR is compatible with objective becoming.

References

Albert, David Z. 2000. *Time and Chance*, Cambridge, MA, and London: Harvard University Press.

Bohm, David. 1965. *The Special Theory of Relativity*, London and New York: Routledge.

Callender, Craig. 2000. Shedding Light on Time, *Philosophy of Science* 67 (Proceedings): S587–S599.

Cao, Tian Yu. 2006. Structural Realism and Quantum Gravity. Pp. 40–52 in *The Structural Foundations of Quantum Gravity*, eds. D. Rickles, St. French, J. Saatsi. Oxford University Press.

Clifton, Rob and Mark Hogart. 1996. The Definability of Objective Becoming in Minkowski Spacetime. *Synthese* 103: 355–87.

Davies, Paul. 1974. *The Physics of Time Asymmetry*, London: Surrey University Press.

Dorato, Mauro. 2000. Becoming and the Arrow of Causation. *Philosophy of Science* 67 (Proceedings): S523–S534.
———. 2006. Absolute becoming, relational becoming and the arrow of time: Some non-conventional remarks on the relationship between physics and metaphysics. *Studies in History and Philosophy of Modern Physics* 37: 559–576.
Eddington, Arthur S. 1920. *Space, Time and Gravitation*. Cambridge: Cambridge University Press.
Einstein, Albert. 1907. Über das Relativitätsprinzip und die aus demselben gezogenen Folgerungen. *Jahrbuch der Radioaktivität und Elektronik* 4: 411–62.
———. 1909. Über die Entwicklung unserer Anschauungen über das Wesen und die Konstitution der Strahlung. *Physikalische Zeitschrift* 10: 821–26.
———. 1920/[15]1954. *Relativity. The Special and the General Theory*. London: Methuen.
———. 1921a. On the Theory of Relativity. Pp. 246–49 in *Ideas and Opinions*, A. Einstein. London: Alvin Redman Limited, 1954.
———. 1921b. A Brief Outline of the Development of the Theory of Relativity. *Nature* 106: 782–4.
———. 1929. Space-Time. Pp. 1070–73 in *Encyclopaedia Britannica*, vol. 20. 14th edition.
———. 1949. Reply to Criticisms. Vol 2: 665–88 in *Albert Einstein: Philosopher-Scientist*, 2 volumes, ed. P. A. Schilpp. La Salle, Il: Open Court.
Einstein, Albert and Leopold Infeld. 1938. *The Evolution of Physics*. Cambridge: Cambridge University Press.
Gödel, Kurt. 1949. A Remark about the Relationship between Relativity Theory and Idealistic Philosophy. Vol. 2: 557–62 in *Albert Einstein: Philosopher-Scientist*, 2 volumes, ed. P. A. Schilpp. La Salle, Il: Open Court.
Grünbaum, Adolf. 1955. Time and Entropy. *American Scientist* 43: 550–72.
———. 1967. The Anisotropy of Time. Pp. 149–86 in *The Nature of Time*, ed. T. Gold and D. L. Schumacher. Ithaca, NY: Cornell University Press.
Hill, E. L. and A. Grünbaum. 1957. Irreversible Processes in Physical Theory. *Nature* 1979: 1296–97.
Hoffmann, Banish. 1972. *Albert Einstein*. New York: Viking Press.
Huggett, N. 2006. The Regularity Account of Relational Spacetime. *Mind* 115: 41–73.
Lockwood, Michael. *The Labyrinth of Time*. Oxford: Oxford University Press.
Penrose, O. and I. C. Percival. 1962. The Direction of Time. *Proc. Phys. Soc.* 79: 605–16.
Popper, Karl. 1956. The Arrow of Time. *Nature*, 177: 538; *Nature*, 178: 382.
———. 1957. The Arrow of Time. *Nature* 179: 1297.
———. 1965. Time's Arrow and Entropy. *Nature* 207: 233–34.
Putnam, Hilary. 1967. Time and Physical Geometry. *Journal of Philosophy* 64: 240–47.
Reichenbach, Hans. 1956. *The Direction of Time*. Berkeley: University of California Press.
Rietdijk, Cornellis W. 1966. A Rigorous Proof of Determinism Derived from the Special Theory of Relativity. *Philosophy of Science* 33: 341–4.
Savitt, Stephen. 2000. There is no Time like the Present (in Minkowski Spacetime). *Philosophy of Science* 67 (Proceedings): S563–S574.
Savitt, Stephen. 2001. Being and Becoming in Modern Physics. *The Stanford Encyclopedia of Philosophy (Spring 2004 Edition)*, ed. Edward N. Zalta (ed.), URL = <http://plato.stanford.edu/archives/spr2004/entries/einstein-philscience/>.
Schilpp, Paul A., ed. 1949. *Albert Einstein: Philosopher-Scientist*, 2 vol. La Salle, Il: Open Court.
Smolin, Lee. 1998. *The Life of the Cosmos*. London: Phoenix.
Stachel, John. 2006. Structure, Individuality, and Quantum Gravity. Pp. 53–82 in *The Structural Foundations of Quantum Gravity*, eds. D. Rickles, St. French, J. Saatsi. Oxford University Press.

Stein, Howard. 1968. On Einstein-Minkowski Space-Time. *Journal of Philosophy* 65: 5–23.
———. 1991. On Relativity Theory and the Openness of the Future. *Philosophy of Science* 58: 147–67.
Teller, Paul. 1991. Substance, Relations and the Arguments about the Nature of Space-Time. *The Philosophical Review* C/2: 363–97.
Torretti, Robert. 2007. The problem of time's arrow historico-critically re-examined. *Studies in History and Philosophy of Modern Physics*, doi:10.1016/j.shpsb.2006.11.005.
Weinert, Friedel. 2004. *The Scientist as Philosopher*, Heidelberg, Berlin, and New York: Springer.
———. 2006. "Relationism and Relativity" in VIII. International Leibniz Congress, *Einheit in der Vielfalt*, eds. H. Breger, J. Herbst, and S. Erdner. Hannover 2006.
Weyl, Hermann. 1952. *Space, Time, Matter*, New York: Dover.

SECTION II

THE LIMITS OF DURATION

CHAPTER FIVE

TECHNICAL REPRODUCTION AND THE QUESTION OF MATERIAL DURATION

Heike Klippel

Summary

Reproduction and repetition are movements that deny the passing of time. They aim at a duration of past and present into the future, thereby trying to overcome time's limits and constraints. For Marx, reproduction coincides with production; it means eternal renewal of technical processes as well as of human beings. According to Benjamin, the ubiquity and timelessness of reproduction media try to evade the consequences of history. Nowadays the conditions of historical media show that they age indeed, and often no original matrix exists any more as a base from which to enter the reproductive circle again. The modernist promise of eternity has meanwhile been renewed through the notion of "loss free" digital reproduction, an idea that in itself is a repetitive too. This text will discuss the implicit assumptions and desires of these concepts of reproduction.

If there are strategies that resist the inevitable passing of time and, thus, its finitude and limitedness, repetition and reproduction are surely amongst them. Both aim to create everlasting presence, despite whatever restrictions may stand in the way. A characteristic feature of reproduction is that it exceeds the lifespan of an individual by producing the same again and again: in organic life, the same species is (re) produced; in technology, ideally, even 'identical' objects are (re)produced. Repetition, too, aims at overcoming the flow of time through creating the new as much as possible similar to the old, and it also has a tendency towards the identical. Daily repetitions transform the before in an endless after, thus working towards a duration of time suggesting endlessness. Repetitions that do not belong to the everyday often aim deliberately at the impossible task of defeating transience because they try to revive events and emotions that inevitably belong to the past. Repetition can be a chore as well as a wish-fulfilment—one can transform into the other, and both work on the border between life and death.[1]

[1] The most profound discussions of repetition in connection to life and death can be be found in Gilles Deleuze, *Présentation de Sacher-Masoch: Le froid et le cruel* (Paris: Éditions de Minuit, 1967) and Sigmund Freud, "Jenseits des Lustprinzips" [1920], *Gesammelte Werke XIII* (Frankfurt a.M.: Fischer, 1999), 1–69.

Jo Alyson Parker, Paul A. Harris and Christian Steineck (Eds), Time: Limits and Constraints, pp. 137–148

In opposition to repetition, reproduction seems to be much more concrete: it refers to objects or organisms, accepts their limited life span, and tries to produce them anew under present circumstances. Reproduction can be seen as a paradox since it needs death/finiteness as the basis of infinitude. In the following I will discuss this paradox not abstractly but in connection to a very concrete example, namely filmic reproduction. In order to do so, I will focus on Marx and Benjamin and their concepts of reproduction, which are contemporaneous with the development of industrial reproduction and film as an emerging mass medium of the twentieth century.

1. Reproduction

In Marx's economic theory, production and reproduction are inseparably related: every production necessitates reproduction, and every reproduction results in production.[2]

Reproduction simply means that all that is consumed will need to be replaced again for as long as one does not want the process to end. Here, the future is based on a conservative principle, or, put the other way around, he who proceeds conservatively ensures the future. However, it is a characteristic of a conservative principle, which is not directed backward but forward, to abandon the identity of what is to be conserved in order to maintain balance. It is central to Marx's concept that it exclusively includes structures and that it principally transcends the individual, be it an object or a living individual. If a process is to continue (that is, if it is to proceed in time), it is absolutely necessary that the singular object is being consumed or destroyed within that process and replaced by something new but in principle similar. Reproduction does not mean multiplying or reanimating a singular object but providing equivalents that always fulfil the same function. Thus it is based on representative relationships. A fundamental aspect in Marx's concept of reproduction is the constant transformation of varying modes of representation to which the elements in their continuous substitution processes are subjected.

Such a process, however, is not circular, returning always again to its beginning, but it continues to develop, open to the future. As it

[2] Karl Marx, *Das Kapital. 1. Band, Buch 1: Der Produktionsprozeß des Kapitals* [1867], Werke, Bd. 23. (Berlin: Dietz 1983).

is structural reproduction, as opposed to identical reproduction, the openness of the representative relationships allows for regeneration in the sense of something actually being new: the replacement of something earlier by something completely different, for as long as it is capable of fulfilling the same function within the process as a whole. Structural reproduction is thus one of the conditions of progress. That is, progress is not possible when identity is to be maintained, as insisting on identity would freeze the movement; and, further, progress is not possible without reproduction. Seen from a perspective that goes beyond the individual, the act of lending (restitution) results in life, while possession (holding on to) is similar to theft and leads to death.

Seen from this angle, Walter Benjamin's concept of reproduction yields very different, if not contrary perspectives.[3] His interest does not lie with reproduction as a resubstitution of something that has been consumed but with the technical reproduction of aesthetic objects. Here, we have to decide between the reproduction of artworks and the media of technical reproducibility of the nineteenth and twentieth centuries: photography and film. While the reproductions of artworks relate back to originals, the multiplied images in film and photography do not. Their material consists in a negative and the represented object, which may be viewed, according to Benjamin, as some sort of original.

Technical reproduction, in general, makes the non-particular, the non-unique, the non-mysterious available; it makes that which is readable and understandable by everybody available to everybody at all times. While art creates a unique material form for the particular that it expresses or represents, reproduction only scans the surface of every singularity, be it a human face or an artwork; it reduces being to its appearance and history to the moment. This moment is then given its own form of being, as reproduction mortifies its object. It preserves death masks. The moment that has passed may now be repeated anytime, but it does not have any meaning; it does not last, in the qualitative sense of simultaneous subsistence and its consummation due to historical changes. Continuity (according to Bergson's concept of *durée*)[4] contains all possibilities for differentiation, while

[3] Walter Benjamin, "Das Kunstwerk im Zeitalter der technischen Reproduzierbarkeit" [3. Fassung 1936], *Gesammelte Schriften. Bd. I,2* (Frankfurt a.M.: Suhrkamp 1991), 471–508.

[4] Henri Bergson, *Matière et mémoire* (Paris: Félix Alcan 1914).

the reproduced object simply remains the same, does not contain any future and no potential for meaning. Technical reproduction lifts the object out of the flow of history—it thus stands in contrast to biological and economic reproduction, which integrates the reproduced objects into the flow of history.

Thus we have the problem of two concepts of reproduction that obviously do not go together. Though using the same term, Marx and Benjamin seem to speak of entirely different phenomena. For Benjamin, reproduction is not the substitution of something that has been used up; rather, reproduction is the substitution of something unreachable, something far away. Put this way, however, we can come closer to what they have in common. In each concept, we deal with substitution; objects are being replaced by other objects of the same value, but not identical to the former ones. This lack of authenticity is of no interest to Marx, as it is enough if the replacement is able to keep the whole system going "with sufficient quality." For Benjamin, however, the inauthentic is a problem, as reproduction purports to be the same and still is not able to present the essential quality of the object referred to. The whole of the artwork or the auratic object—in contrast to its surface—is not reproducible because it is not renewable. The impossibility of reproduction suggests that the aliveness of the singular is based on the fact that it encompasses past and present, and its future is uncertain. It is subjected to transience and cannot be saved—not even the most faithful print will preserve the Mona Lisa if the original is destroyed. Systematic processes, on the other hand, remain alive, as the irretrievable is not relevant to them; they encompass present, past, and future, and their transience is the necessary condition for their progress.

While the reproduction of an artwork does refer to an original, there is no external point of reference outside of the process of technical reproduction for film and photography. Benjamin suggests the filmed or photographed object as some kind of an original, but this is a problematic assumption: Even if the pyramids remained forever in their locations, one would not ever be able to check, whether—in the moment of taking the picture—they appeared in exactly the light that is reproduced in the print. The light could have been manipulated in the process of taking the picture or during the process of developing the picture, or it could have been perceived differently by the human eye. The exact details of the aesthetic appearance of a photograph do not have a definitive equivalent in the external world. This is even less the case with regard to a feature film: it does not have an original reference in the world external to the film, and shooting has never been intended to assume an independent status. The film presents everything in the "here and now" although it does not locate this "here and now" as such. Its subjects are no longer unique, nor do they have a history, but in being autonomous and without reference they are "timeless." They are thus similar to "ghosts" because with them, too, time does not leave any traces, and they cannot be fixed in any location. In contrast to traditional artworks, the "new" artworks—and here Benjamin is thinking especially of film and photography—do not have an aura, as, from the very beginning, they are not conceived to be unique, but to provide mass-production.

Thus, technical reproduction coincides with the process of reproduction: that is, film- and photo-prints may be reproduced identically again and again by feeding the production process with new materials. Wouldn't that mean that we now have endlessly renewable artworks, though non-auratic ones, which still offer aesthetic pleasures that meet the demands of the modern industrial society? Instead of being charged with history like traditional artworks, instead of aging and yielding to destruction at some point, they are timeless and may be called into presence at any moment. In showing a thing of the past that actually has irretrievably vanished, they are, to a certain degree, inanimate, and prone to a loss of meaning, but as artefacts they are sure to be multiplied eternally.[5]

2. Film

When we now turn our attention to film, we have to realize, after more than a century of film history, that aura, or, to put it more simply, uniqueness cannot be shaken off that easily; it creeps back in again precisely via the process of reproduction. Especially technically produced images that do not refer to an "original" object, like film and

[5] Also Kracauer discusses the mortifying meaninglessness of photography and film." He compares them with the memory image that is condensed meaning, whereas the photograph only contains the "remains" of life and history. Cf. Siegried Kracauer, "Die Photographie" [1927], *Das Ornament der Masse* (Frankfurt a. M.: Suhrkamp 1977), 21–39.

photography, suffer from the process of reproduction because technical reproduction lacks renewal as well as identity. Reproduction as representation, as thought by Marx, is of course possible; but this does not necessarily include secure complete similarity as, ideally, one should be able to expect it from a process of technical reproduction. Taking film as an example, I will now elaborate this point.

Within film production, identical prints that look absolutely alike are not possible; rather, we have interchangeable copies, which means that they fulfil their purpose to the same extent. This interchangeability is defined by standards relating to narrative consistency and completeness (how many feet missing in an archival film print are tolerable?) as well as by technical characteristics of image and sound. The quality of each single print changes in the course of its existence, and, after dozens of projections, it cannot be used anymore in a first-run theatre; after being projected some hundred times it cannot be used anymore at all. Thus film prints have their own life, participating in history: the history of film production (they mirror the standards of film production at the moment of their creation), the history of their use and storage, and the history of the many ways they have travelled. The changes of location do not result in the omnipresence of the eternal same but imprint their traces onto the object. However, not only the historical status of individual film prints reduces their similarities, but the material position within the chain of reproduction is equally important. Most screening copies are taken from inter-negatives, not from the "original" negative; thus, copies are made from copies, which results in different image qualities.

But it is not only the process of multiplication that is a source of uncertainty; it is especially the fact that films only consist of reproducible material because there is often no definitive film version available. Corresponding to different requirements, films may be re-cut at any time, and still one could claim to deal with "the same film." In particular with regard to historical films, the evaluation of the existing material relating to a given "film" can be markedly problematic. The lack of an original means that there is no original film to which one could refer; therefore, different prints have to be compared to each other. If the production- and the reproduction-process would proceed ideally and result in mostly identical samples, this would not be a problem, but the reality is different. Only seldom are the different surviving positive and negative materials congruent, and often they differ very much. To look for a sample more "original" than the others would be a mistaken approach.

Other differences in several versions can be explained on the basis of the production technology itself. At the time of the genesis of Benjamin's third version of the artwork essay, to which I here refer, in 1936, film production had just reached the point where it became common use to let only one basic film version go into circulation; until then, multiple distribution copies were the rule. This practice was first due to the inadequate quality of the negative film stock available at the time, which did not allow the use of inter-negatives for the mass-production of screening copies for international distribution. Screening copies could only be generated from a master negative, which could not be used infinitely and which wore out in the process of creating ever more film copies; this happened even before enough prints were created, and therefore several master negatives of the same film were needed. Furthermore, it was too expensive for international distribution to export multitudes of screening copies; thus, a negative was sent instead, from which prints were created on the spot. If only one negative had existed, the danger that it could have been damaged or lost while it was shipped elsewhere would have been too great.

In order for film-makers to create several first-generation master negatives, several versions of a given film were produced by their recording every shot with at least two cameras and in several takes. Thus a master negative could be cut from the material of the first camera and another be based on the second. These two versions only slightly differed due to barely noticeable differences in perspective. From the second takes, one could generate two more negatives; these were mostly used for export. Sometimes even third takes were used in producing master negatives. These second- and third-take-negatives could very well show great differences from the first two master negatives. After the negative material had been improved in such a way that copies of sufficient quality could be created from inter-negatives, sound film provided the next hurdle. Before dubbing existed, different film versions were produced for export in other languages and often involving different actors and directors. Only in the mid-thirties were these technical difficulties overcome, and there were no longer any urgent technical reasons to produce multiple versions of one film. But even then, there were always numerous decisions relating to marketing, social policy, and technology that resulted in differing versions;

additionally, conditions of reproduction technology, as well as wear and tear and storage, could bring about differing versions. Moreover, censorship resulted in different film cuts, depending on the screening location. Therefore, a film consisted of a number of interchangeable materials rather than one artefact that was available in multiple identical copies.

These peculiarities of film production have several consequences in relation to the concept of an "original," as well as the concept of reproduction. First, the instability of the filmic text results in fictions of identity. The most obvious of these is created by spectators, who frequently are not interested in differing versions. Spectators assume that the film they have seen is an original because technical reproducibility is supposed to ensure the film's existence in identical and not different screening copies. As no original exists, all reproductions are considered authentic. Though they are subject to the course of history, they are often perceived to be timeless entities of eternal youth. This fiction leads to misguided conclusions; an example is the assumption that the bad sound and image quality of older films has its cause in supposedly inferior production conditions. However, often these problems are caused by wear and tear or by the fact that it is a copy of a later reproductive generation; with regard to sound, we are often faced with the difficulty that today's sound systems are not adequately suited to play back historical film sound. Thus it can happen that a film that originally entered the cinemas with a brilliant image and sound quality is misunderstood later to be an example of inferior production technology. If there is an awareness of the process of decay of technical materials, one can easily make the opposite mistake: a restored film copy of brilliant image quality is taken to be a faithful reproduction of the original state, even though its visual quality is "better" than the "original": that is, this visual brilliance could have only been produced using the technological means currently available. With regard to technically reproduced samples, it is assumed that they have always been that way: either "already not convincing back then" or "perfect back then"; the magic of the "always new and still always the same" proves its persistence.

Another inherent contradiction is provided by film restoration, which produces material fictive "originals." A new reproduction is generated from a group of existing versions according to different criteria that relate to technical integrity as well as integrity in content. This new version always has a dubious status and may be very close to, but just as well very far from, the historical appearance of the film at the moment of its creation. In this context, not only the assessment of the institution restoring the film plays a role, but also the reproduction technologies that have been reproduced in the meantime. Though film technology has remained relatively stable throughout the course of its long history, countless parameters have nonetheless changed: the nature and quality of the recording materials, its image formats, the chemical emulsions, and the lighting quality of the projector lamps. (Today's lamps, for example, would not produce the same projection even if we had a perfect copy of a seventy-year-old film available.) And the development of color technologies provides a particularly succinct example: during the early years of cinema, black-and-white positives were tinted and toned in monochrome dips, and with Technicolor, color was produced by printing the primary colors onto the black-and-white positive. The Eastman-Kodak-technique in use today, which works with a color negative, cannot faithfully reproduce the colors achieved by former technologies. One can approach them more or less closely, but it remains at best a more or less successful simulation. Thus, "film restoration" effectively is the production of a copy on an uncertain material basis and under partially inadequate technical conditions; and (this needs to be emphasized) the result is only *one* copy. Film restoration is not a mass phenomenon, and frequently it is underfinanced; therefore, it is not possible to produce more than one restored copy. This copy then travels throughout the world of film historical projections providing referential material for the film in question; it leads a monumental life as a "pseudo-original," which is totally inadequate with regard to technical reproduction. This would be it for technical reproduction, but not for renewal.

Looking at film as an example for technical reproducibility presents us with a rather unexpected result: technical reproduction of artworks, respectively aesthetic products, as has been conceptualized by Benjamin, is, at its core, representation; as such, it shows the characteristics of reproductive cycles as they have been described by Marx. With the "original," we lose not only "aura" but "identity" as well. Reproductions that comment upon and represent each other do not have to be exactly alike; it is sufficient for them to be convincingly similar and to fulfil the same purposes.

The problem of identity, however, is not yet solved. Photographs and films are aesthetic objects, and it is not without reason that they sometimes are considered "art"; therefore, we are dealing not only with a function within a process but often also with their concrete material quality, which simultaneously grants pleasure and meaning. The Technicolor-Red not only has its own particular beauty but also conveys associations and levels of meaning that are different from an Eastman-Kodak-Red. Technical reproduction is subject to the inaccuracies of the reproductive process and, at the same time, has to avoid these inaccuracies. It has to hold on to identity within the Un-identical. With regard to temporality, this means that it is supposed to allow for and enable eternal life. Films, though they may not refer to an original or have an aura, are firmly integrated into the history of their age, into social circumstances, economic structures, and the prevailing standards of production. They are witness to a particular historical moment, and only within that moment of their first appearance are they truly alive and up-to-date, so to speak. Right after, they begin to age: they change in their materiality; the circumstances and conditions of production under which they were created change, as well as the screening conditions and the audience. What they represented at the moment of their original publication is irretrievable. But reproduction still suggests their ever-present identity, and we are willing to accept its shortcomings while, in principle, we are expecting it to keep its promise. The demand for identical reproduction, as identical as possible, refuses to face loss and wants life from death or, rather, life within death: continuously reproducing zombies are expected to look like newborn babies. However, this demand is justified; it is for a good reason that we do not want to junk films as we junk cars by replacing them with newer and better models. Technical reproduction of aesthetic works has to solve a paradoxical task, and, despite all restrictions, it may be successful.

The fundamentally problematic nature of technical reproduction will always remain, and even the magnificent perspectives opened up by digital technologies will not change this situation. They, too, start off with the old promise to reproduce processes and identities at the same time. Loss-free reproduction, durability, general availability—all of these seem to be finally possible. Masses of photographs can be stored at varying levels of quality, to be retrieved at any time and to be reproduced without any problem. At the moment, video files are still very large, but they, too, are promised with the possibility of perfect preservation, a saving that allows for resurrection and availability at any given time. However, with the promise we find the difficulties reproduced as well: technological innovations require continuous copying, and it is more than doubtful that the same result can be achieved from repeated multiplication. From our everyday experience, we already know that files created using older hard- and software are lost if we did not back them up while the technologies and software were still compatible. When we are dealing with texts, differences in display and print may be negligible, but they can nonetheless cause disturbances, even on this simple level. With regard to all aesthetic objects, however, where visual appearance is of crucial importance, the familiar problems resurface: we deal with renewal that only creates a duplicate as similar as possible, but not the same.

By means of digital technology, one attempts to create a screening image satisfying the respective requirements on the basis of existing materials; in principle, one aims to create an image showing the characteristics of a particular period in film history. But if this particular version does not exist anymore in the form of film materials, but only as a digital file, it is by no means guaranteed that it will be preserved. As said before, parameters of hard- and software change continuously, resulting in a playback with different visual qualities that consequently render the material useless at some point. Then the reproduction cycle enters into a new generation; historical image quality has to be reconstructed using current technologies. This cycle finds a deliberate end when the interest in the respective film dies; it often ends inadvertently due to a lack of reproductive activities while the negative was still usable. A destroyed negative results in the loss of a film because film positives do not suffice as a basis for convincing reproductions; the cycle is no longer renewable at a sufficient level of quality.

What has been shown here for film is true for all media products, be they photographs, videos, or computer games, that require—for whatever reason—a certain aesthetic appearance. In these cases we do not deal with reproduction in its original sense; rather, reproduction is turned against itself: as soon as reproducibility no longer serves the preservation of processes, but of singular objects, its orientation towards the future is turned backwards. Representation turns into the simulation of something lost.

The media of technical reproducibility are thus threatened by loss, especially because of their alleged omnipresence. They lead a hybrid existence between commodities and cultural artefacts. Therefore, they are, on the one hand, submitted to the merciless law of economy and its asymmetries that, as Marx has shown, are open to imbalances and exposed to power interests. That which is unsuitable for the accumulation of capital is not being reproduced. On the other hand, reproducible aesthetic objects are considered worthy of preservation, not because they are auratic artworks, but because through them perceptiveness and understanding of tradition and history can be reproduced.

In conclusion I would like to adhere to Benjamin's ideas that technical reproduction, nowadays also comprising the digital, is believed to deliver objects of such similarity so as to make them appear identical and thus eternal. That the facticity is different obviously does not undermine this assumption very much, which makes it worth the discussion. I would also agree with Benjamin that technically reproduced artworks do not have an aura, and that attaching auratic qualities to them while at the same time expecting that finally loss-free reproduction is possible (a promise that is constantly renewed with any new reproduction technology) is very problematic. Benjamin (and also Kracauer) saw exactly in the non-auratic qualities of film and photography an emancipatory potential, which coincides with Marx's concept of the reproduction of processes. Nevertheless, the basic problem has no solution: identity is connected with singularity and is therefore terminal, but then infinitude has to give up the idea of the unique.

References

Benjamin, Walter, "Das Kunstwerk im Zeitalter der technischen Reproduzierbarkeit" [3. Fassung 1936]. In *Gesammelte Schriften. Bd. I,2*, 471–508. Frankfurt a.M.: Suhrkamp, 1991.

Bergson, Henri, *Matière et mémoire*. Paris: Félix Alcan, 1914.

Deleuze, Gilles, *Présentation de Sacher-Masoch: Le froid et le cruel*. Paris: Éditions de Minuit, 1967.

Freud, Sigmund, "Jenseits des Lustprinzips" [1920]. In *Gesammelte Werke XIII*, 1–69. Frankfurt a.M.: Fischer, 1999.

Kracauer, Siegfried, "Die Photographie" [1927]. In *Das Ornament der Masse*, 21–39. Frankfurt a.M.: Suhrkamp 1977.

Marx, Karl, *Das Kapital. 1. Band, Buch 1: Der Produktionsprozeß des Kapitals* [1867], Werke, Bd. 23. Berlin: Dietz 1983.

CHAPTER SIX

LIMITS ON THE DURATION OF COPYRIGHT: THEORIES AND PRACTICE

Tyler T. Ochoa

Summary

The question of how long a copyright should last has troubled scholars and policymakers ever since the first copyright statute was enacted. The controversy continues because there are two divergent views concerning the basic rationale underlying copyright law. Under the natural rights view, the author of a literary or artistic work has a natural right to profit from the fruits of his or her artistic labor. The logical extreme of the natural rights view is that copyright should be perpetual and is limited in duration only because of certain practical considerations. Under the utilitarian view, however, copyright exists primarily to encourage the creation and distribution of new literary and artistic works. With an exclusive right, a publisher can charge a higher-than-efficient price, earning excess profits that are used to compensate the author. Because the higher price is inefficient and contrary to the public interest in the long run, copyrights should last only as long as is necessary to accomplish their incentive function.

Historically, copyright terms have consistently increased over time, as common-law countries that initially adopted the utilitarian view have moved closer to the natural rights view in the interests of international harmonization. This increase is consistent with public choice theory, which posits that when the benefits of a law are concentrated but the costs of that law are diffuse, a small well-focused interest group will usually succeed in obtaining passage of the law, even if it does not benefit society as a whole.

How long should a copyright last? This question has troubled scholars and policymakers ever since the first copyright statute was enacted in England in 1710 and continues to do so today. The controversy continues because of fundamental differences of opinion concerning the basic philosophy and purposes of copyright law. As explained below, there are two principal schools of thought concerning the rationale for and purposes underlying copyright law: the utilitarian view and the natural rights view. While these alternative justifications for copyright often work in harmony and lead to similar public policy prescriptions, in the area of copyright duration they are in conflict and lead to diametrically opposed policy recommendations.

Jo Alyson Parker, Paul A. Harris and Christian Steineck (Eds), Time: Limits and Constraints, pp. 149–178

1. Theories of Copyright Protection

Copyright as a Natural Right

Under the natural rights view, the author or creator of a new literary or artistic work has a natural right to profit from the fruits of his or her artistic labor. The natural rights view finds support in the writings of John Locke, who famously posited that property results from the mixture of a person's labor with anything appropriated from the general state of nature (Locke 1988, 287–89). The application of Locke's theory to literary property was stated eloquently by William W. Ellsworth, speaking for the House Judiciary Committee in 1830:

> Upon the first principles of proprietorship in property, an author has an exclusive and perpetual right, in preference to any other, to the fruits of his labor.... If labor and effort in producing what before was not possessed or known, will give title, then the literary man has title, perfect and absolute, and should have his reward: he writes and he labors as assiduously as does the mechanic or husbandman. The scholar, who secludes himself, and wastes his life, and often his property, to enlighten the world, has the best right to the profits of those labors. (Gales and Seton 1831, 7: cxx).

This view also finds support in the Universal Declaration of Human Rights, which states that "Everyone has the right to the protections of the moral and material interests resulting from any scientific, literary or artistic production of which he [or she] is the author" (Art. 27(2)).

The logical extreme of the natural rights view is that the duration of copyright should be perpetual. Only a handful of countries, however, have ever enacted a perpetual copyright law, and most of those eventually thought better of it and restricted the term of copyright.[1] There are a number of reasons why copyright is limited in time, even in countries that generally accept the natural rights view. First, legal recognition of property rights is based in part on the economic efficiency of exclusive ownership: common ownership of tangible property leads to the so-called "tragedy of the commons" in which the asset is depleted by overuse. Unlike tangible property, however, intangible works of authorship are "nonrivalrous" in nature and can be possessed by many individuals simultaneously without restricting the possession of oth-

[1] *See, e.g.*, Law of May 27, 1927, Arts. 15(1) & 36 (Portugal). This law was repealed and replaced with a life-plus-50-years term in 1985. Law No. 45/85 of Sept. 17, 1985, Art. 31 (Portugal).

ers and without diminishing the value of the asset (Yen 1999, 550–53; Lemley 2005, 1050–52). As a result, for intangible property, greater net economic efficiency may be achieved by moderating the natural rights view.[2] Second, after a period of time, it often becomes difficult to identify, locate, and negotiate with all of the heirs and assignees who have a share of the copyright, meaning that many otherwise productive uses of the work will not be realized because of transactions costs (Ricketson 1992, 766; Dietz 1978, 161).[3] Third, there is the possibility that an author's heirs will try to suppress works of which they do not approve or will license the works only with unreasonably restrictive conditions that will harm the public's use and enjoyment of the work (Ricketson 1992, 767–68). For all of these reasons, there is a public interest in allowing unrestricted public access to and use of an intangible work of authorship after a limited period.

The Utilitarian View of Copyright

Under the utilitarian view, copyright is an exception to freedom of expression that exists primarily for the benefit of the public, in order to encourage the creation and distribution of new literary and artistic works. Without copyright, copiers would always be able to undercut the initial publisher's price because they have not had to bear the fixed cost of producing the work. Publishers would therefore be unwilling to pay authors for the creation of new works, and only authors who had other sources of income could afford to write. By eliminating competition from free-riding copiers, copyright enables publishers to charge more than the efficient marginal price, giving them profits from which to compensate the author (Lemley 2005, 1054–55; Landes and Posner 1989). In the words of Thomas Babington Macaulay, speaking in the House of Commons in 1841:

> The principle of copyright is this. It is a tax on readers for the purpose of giving a bounty to writers. The tax is an exceedingly bad one; it is a tax on one of the most innocent and salutary of human pleasures.... I admit, however, the necessity of giving a bounty to genius and learning.

---

[2] Other scholars reach the same conclusion using natural rights philosophy, criticizing the view that natural rights inevitably lead to perpetual copyright as "superficial" (Yen 1990, 554–557).

[3] Empirical support for this concern can be found in studies concerning so-called "orphan works," that is, those works for which a copyright owner cannot be located despite a diligent search. (Register of Copyrights 2006, 26–34; Vetulani 2008, 7–8).

> In order to give such a bounty, I willingly submit even to this severe and burdensome tax. (Macaulay 1853, 394)

But because this higher price is by definition inefficient, exclusivity should be granted only to the extent necessary to encourage the initial creation and distribution of the work, and the work should enter the public domain as soon as possible. Thus, in Macaulay's words:

> The advantages arising from a system of copyright are obvious. It is desirable that we should have a supply of good books; we cannot have such a supply unless men of letters are liberally remunerated: and the least objectionable way of remunerating them is by means of copyright.... [But] Copyright is monopoly, and produces all the effects which the general voice of mankind attributes to monopoly.... [T]he effect of monopoly generally is to make articles scarce, to make them dear, and to make them bad.... Thus, then, stands the case. It is good, that authors should be remunerated; and the least objectionable way of remunerating them is by a monopoly. Yet monopoly is an evil. For the sake of the good we must submit to this evil; but the evil ought not to last one day longer than is necessary for the purpose of securing the good. (Id., 390–92).

Of course, determining exactly what period of copyright is necessary to encourage the creation and distribution of the desired number of new works of authorship is a daunting task. At one extreme, Stephen Breyer (now an Associate Justice on the U.S. Supreme Court) once argued that the lead time necessary for a free-rider to duplicate a published book was sufficiently long that copyright might not even be necessary to encourage publishers to make an initial investment in producing the work (Breyer 1970, 299–302, 309–13; see also Tyerman 1971; Breyer 1972). Although this argument breaks down for works initially distributed in digital form (which are easily copied), the fact that publishers make investment decisions using very short time horizons nonetheless suggests that relatively short terms of copyright would be sufficient to encourage publishers to invest in distributing a copyrighted work (Breyer 1970, 325).

At the other extreme, William Landes and Richard Posner have questioned the basic assumption that works of authorship in the public domain can be freely copied without losing their value. They posit that "congestion externalities" exist, such that overexposure to a work will cause consumers to value it less, whereas wise management of a work over time will help it retain its value (Landes and Posner 2003, 484–88). Accordingly, while they agree that a relatively short fixed term (about 20 years) is a sufficient incentive for most works, they would allow copyrights to be renewed indefinitely for successive periods (Id., 517–18). Other scholars, however, dispute both the existence of congestion externalities and the belief that continued private management of a work benefits the public (Karjala 2006; Lemley 2004).

But even if short terms of copyright are sufficient to encourage publishers to distribute new works, they may not be sufficient to encourage enough authors to create new works. While it is undoubtedly true that some authors would create new works even if no copyright protection were provided, few would dispute that copyright encourages more people to become authors than would otherwise be the case. Ideally, we would like a term of copyright that is sufficient to enable an author to devote himself or herself to writing (or painting or composing) as a full-time profession (Guinan 1957, 74). This may include some provision for copyright to endure after the death of the author, as there is anecdotal evidence that authors are motivated by the need to provide for their heirs (See e.g. Brylawski and Goldman 1976, J116–17, J201). On the other hand, long terms of copyright may encourage a successful author to retire, instead of devoting himself or herself to further creative activity. In addition, economists agree that future benefits must be significantly discounted to take account of the time value of money, so that the prospect of earning additional royalties in the distant future may provide little motivation for the creation of new works in the present (Akerloff et al. 2002, 5–7, 23).

In the face of such disagreements in economic theory and lacking sufficient empirical data concerning the behavior of authors and publishers, we may be left with little more than gut feelings in deciding what the appropriate term of copyright should be. Accordingly, the behavior of legislators and interest groups takes on added importance in explaining how copyright terms are established in practice.

Public Choice Theory

Public choice theory is a branch of economics that applies game theory and decision theory to government action. One of the tenets of public choice theory is that because individuals tend to act rationally in their own self-interest, they will act to try to influence the legislature only if the perceived benefit to be gained is greater than the cost to the individual. The result is that a small interest group that has a lot of money at stake in a particular bill will expend more resources and will generally be more effective in lobbying the legislature than will the general public, for whom the cost of the bill on an individual basis may be very small (Olson 1965; Bard and Kurlantzick 1999, 216–28; Lessig 2004, 216–18).

Public choice theory works very well in explaining how decisions concerning the duration of copyright are made. Consider, for example, the Sonny Bono Copyright Term Extension Act of 1998 (CTEA), which added 20 years to the terms of all existing and future copyrights in the United States.[4] It was estimated that the aggregate amount of royalties that would flow to copyright owners from extending copyright terms by 20 years was approximately $317 million (Rappaport 1998, 16). This works out to about $2.58 per individual voter.[5] But the benefits of that extension would accrue largely to a handful of powerful media corporations and a small number of heirs of very famous authors. Thus, it is rational for a large media company to spend millions of dollars lobbying for copyright term extension (because the benefit it would receive is several times greater), but it is also rational for the individual voter to remain ignorant of the issue. As a result, Congress disproportionately heard about the costs and benefits of the law from copyright owners, and the CTEA passed by a substantial margin, notwithstanding the fact that most academic commentators believed that it represented bad public policy (See e.g., Bard and Kurlantzick 1999; Karjala 1998; Lessig 2004, 218, 292–93).

Public choice theory helps explain why the scope of copyright protection and the duration of copyright terms have steadily increased during the past three centuries and are rarely, if ever, diminished. In the rest of this chapter, we will examine how the duration of copyright terms has increased over time and summarize the current state of the law regarding duration. Before doing so, however, it may be helpful to discuss some international differences in the general attitude toward the foregoing theories of copyright protection.

An International Perspective

As a general matter, common-law countries have proceeded from the view that copyright exists primarily to serve the public benefit, while civil law countries have historically placed a greater emphasis on the

[4] P.L. No. 105–298, Title I, 112 Stat. 2827.

[5] According to the U.S. Census Bureau, there were about 123 million registered voters in 1998 (the year the CTEA was enacted). (Bureau of the Census 2000, 3). If calculated on the basis of the estimated 198 million people of voting age, the amount per person works out to about $1.60 per person.

natural rights of the author. This is apparent in the very language used to describe the law: while in common-law countries the preferred term is "copyright," in most civil law countries the more accurate translation in English is "authors' rights."[6] It is also apparent in the greater emphasis that civil law countries place on the "moral rights" of the author. In most countries, in addition to the economic rights enjoyed by the author, which can be assigned in exchange for monetary reward, the author also enjoys an inalienable right to control certain aspects of the public presentation of his or her work. Thus, for example, Article 6*bis* of the Berne Convention requires member nations to recognize the right of an author "to claim authorship of the work and to object to any distortion, mutilation or other modification of, or other derogatory action in relation to, the said work, which would be prejudicial to his honor or reputation."[7]

By contrast, while natural rights theories played a role in the development of copyright in Anglo-American law, it is clear that the primary justification for copyright in these common-law countries was a utilitarian rationale. For example, the Patent and Copyright Clause of the U.S. Constitution provides that "Congress shall have Power…to Promote the Progress of Science and useful Arts, by securing for limited Times to Authors and Inventors the exclusive Right to their respective Writings and Discoveries."[8] The parallel construction of the Clause indicates that copyrights are granted to "authors" for their "writings" in order to promote the progress of "science" (broadly meaning "knowledge" in the language of the eighteenth century), while patents are granted to "inventors" for their "discoveries" in order to promote the progress of the "useful Arts" (Walterscheid 2002, 11–12, 115–33).[9] Thus, copyrights exist to promote knowledge by encouraging the creation and publication of new works.[10] Both patents and copyrights, however, may be granted only "for limited Times," a restriction imposed in order to prevent the abuse of monopoly power that had existed in England prior to the Statute of Monopolies (which limited the duration of patents) and the Statute of Anne (which limited the duration of copyrights) (Ochoa and Rose 2002). In so doing, the Clause not only

[6] In French, *droit d'auteur*; in German, *Urheberrecht*; in Spanish, *derecho de autor*.
[7] Berne Convention for the Protection of Literary and Artistic Works, Sept. 9, 1886, revised at Paris, July 24, 1971, art. 6*bis*, S. Treaty Doc. 99-27, 1161 U.N.T.S. 30 [hereinafter Berne Convention].
[8] U.S. Const., Art. I, § 8, cl. 8.
[9] On the meaning of the word "Progress," see Pollack 2001, 794–809.
[10] *See* Twentieth Century Music Corp. v. Aiken, 422 U.S. 151, 156 (1975).

indicates that copyright should last no longer than necessary, but it also endorses the view that the progress of knowledge is best served by the creation of a "public domain," a body of works whose copyrights have expired and which can be freely copied by anyone (Walterschied 2002, 265–77; Ochoa 2003; Litman 1990).[11]

Despite these differences between nations concerning the principal rationale for copyright protection, it is the case that something approaching an international consensus has emerged concerning the basic term of copyright in literary and artistic works. The following history of the development of copyright terms demonstrates how, over time, the United States and other common-law countries have gradually moved away from their primarily utilitarian perspective and closer to the natural rights theories of continental European countries, in the interests of international harmonization.

2. History of Copyright Duration

Through the Eighteenth Century: From Privileges to Copyright

England

After the invention of moveable type in the fifteenth Century, European monarchs quickly realized that the printing press could be used as an instrument to spread sedition and heresy. Their response was to assert legal control over the new technology: no one could operate a printing press without royal permission. Such permission often took the form of letters patent, a document granting to a particular printer an exclusive privilege to print a particular book or class of books for a specified period of time. Publishers were also required to submit any manuscripts that they wished to print to government censors for their approval (Patterson 1968, 20–27, 78–90; Kaplan 1967, 2–3; Rose 1993, 9–11, 23–24).

In 1557, Queen Mary granted a charter to the Stationers' Company, a guild of London booksellers and printers, which provided that no one could operate a printing press in England unless they were a member of the Company or unless they received a printing patent from the monarch. Because the Stationers effectively had a monopoly, they could prevent competition among themselves by agreeing to a

[11] *See also* Sony Corp. of America, Inc. v. Universal City Studios, Inc., 484 U.S. 417, 429 (1984).

system of registration. Once secured by registration, the right to print a book continued forever and could be bequeathed or sold, but only to other members of the guild (Patterson 1968, 28–32, 42–56, 63–64; Kaplan 1967, 3–5).

Under a series of decrees and statutes, no book could be printed in England unless it had first been registered with the Stationers (Patterson 1968, 46–47, 115–39). In 1695, the last licensing act expired, throwing the book trade into disarray. The Stationers petitioned Parliament for relief, seeking a statute under which authors would have the exclusive right to print their works—an exclusive right that could, of course, be transferred to a publisher. As introduced, the proposed legislation did not limit the duration of these copyrights (Patterson 1968, 138–42; Rose 1993, 42–43). When the Statute of Anne was enacted in 1710, however, the term of copyright in new works was limited to 14 years, with the possibility of renewal for a second 14-year term if the author was still living at the end of the first. For books that were already in print, the act provided a single 21-year term.[12] These terms were based upon those in the 1623 Statute of Monopolies, which had limited patents for new inventions to a single term of 14 years; that period, in turn, was derived from the traditional seven-year period for most apprenticeships (Ochoa and Rose 2002, 677–81).

At first, the London booksellers simply ignored the term limit provision. But as the terms of copyright began to expire, Scottish booksellers began publishing competing reprints. This touched off a great debate in England concerning the nature of literary property. Established publishers argued that an author had a natural right of property in his works that passed to the booksellers when the manuscript was purchased. Under this view, the Statute of Anne merely provided supplemental remedies to an underlying common-law right that was perpetual. Competing booksellers argued that there was no common-law right after a work had been published, and alternatively that any such right had been extinguished by the Statute of Anne (Ochoa and Rose 2002, 681–83; Patterson 1968, 158–68; Rose 1993, 52–55, 69–78).

The common-law right was upheld by the Court of King's Bench in *Millar v. Taylor* in 1769 (Patterson 1968, 168–72),[13] but the defendant's view prevailed in the Scottish Court of Sessions in *Hinton v. Donaldson* in 1773 (Boswell 1975; Rose 1993, 83–85). Finally, in 1774,

[12] 8 Anne c. 19, §1 (1710) (Eng.).
[13] Millar v. Taylor, 4 Burr. 2303, 98 Eng. Rep. 201 (K.B. 1769).

in the landmark decision of *Donaldson v. Beckett*, the House of Lords, acting as the Supreme Court of Great Britain, rejected the claim of perpetual common-law copyright and established that copyright was limited in term under the Statute of Anne (Ochoa and Rose 2002, 683–84; Patterson 1968, 172–79; Rose 1993, 92–104).[14]

France

In pre-Revolutionary France, all books had to be approved by official censors, and the author or publisher had to obtain a royal privilege before a book could be published. Such privileges were exclusive and were usually granted for a period of six years, but they could be renewed indefinitely (Dawson 1992, 3, 7–10, 22–27; Davies 1994, 73–77).

In 1777, a series of royal decrees changed the nature of these privileges (Dawson 1992, 7–8, 18–19). The decree on the duration of privileges provided a minimum duration for all privileges of the longer of ten years or the life of the author (1777 Decree, arts. 3, 4).[15] The decree also prohibited the renewal of privileges and required a book to be augmented by at least a fourth to obtain a new privilege (Id., art. 2). Once a privilege had expired, anyone could obtain a "permission simple" to print or sell copies of the work (Id., art. 6). This decree, therefore, expressly recognized a public domain in books whose privileges had expired (Dawson 1992, 3).

After the French Revolution, a dispute arose concerning the exclusive privilege which had been granted to the *Comédie Française* to the public performance of all dramatic works (Ginsburg 1990, 1006). In 1791, the National Assembly voted to abolish the privilege and declared that the works of any author who had been deceased more than five years were public property (Id., 1006–07; 1791 Act, arts. 1–2).[16] In the same decree, the Assembly granted to authors the exclusive right to authorize the public performance of their works during their lifetimes, and extended that right to the author's heirs and assignees for five years after the author's death (1791 Act, arts. 3–5).

---

[14] Donaldson v. Beckett, 4 Burr. 2408, 98 Eng. Rep. 257 (H.L. 1774). For a fuller account of the debate, see Cobbett 1813, 17: 953–1003.

[15] *Arrêt du conseil portant règlement sur la durée des privilèges en librarie*, Aug. 30, 1777.

[16] Act of Jan. 13–19, 1791 (Fr.). An English translation is available in Sterling 2003, 1256–57.

In 1793, a new law was passed giving all authors, composers, and artists the exclusive right to sell and distribute their works, and extending the right to their heirs and assigns for a period of ten years after the author's death.[17] Although this law was based in part on the natural right theory, ninteenth-century commentators characterized the 1793 law as utilitarian and "a charitable grant from society" rather than a full recognition of the perpetual right of an author's heirs to the fruits of his labor (Ginsburg 1990, 1009–12).

The United States

In 1783, in response to several authors' petitions, a committee of the Continental Congress reported that it was "persuaded that nothing is more properly a man's own than the fruit of his study, and that the protection and security of literary property would greatly tend to encourage genius [and] to promote useful discoveries" (Continental Congress 1783, 24: 326). Under the Articles of Confederation, the Continental Congress had no authority to issue copyrights; instead, it passed a resolution encouraging the States

> to secure to the authors or publishers of any new books not hitherto printed... the copyright of such books for a certain time not less than fourteen years from the first publication; and to secure to the said authors, if they shall survive the term first mentioned,... the copyright of such books for another term of time not less than fourteen years (Id., 326–27).

Three states had already enacted copyright statutes earlier that year, and within three years all of the remaining states except Delaware had followed suit. Seven of the states followed the Statute of Anne and the Continental Congress' resolution in providing two 14-year terms. The five remaining States granted copyrights for single terms of 14, 20, or 21 years' duration, with no right of renewal (Ochoa and Rose 2002, 687–88).

At the Constitutional Convention of 1787, both James Madison of Virginia and Charles Pinckney of South Carolina submitted proposals to give Congress the power to grant copyrights for a limited time. These proposals resulted in the Patent and Copyright Clause of the U.S. Constitution (Ochoa and Rose 2002, 688–90). As noted above,

[17] Act of July 19–24, 1793, Arts. 1, 2 (Fr.). An English translation is available in Sterling 2003, 1260.

the Clause takes a utilitarian view of patents and copyrights, providing that exclusive rights may be granted "to Promote the Progress of Science and useful Arts" but only "for limited Times."[18]

The Copyright Act of 1790 granted copyrights for a term of "fourteen years from the time of recording the title thereof," with a right of renewal "for the further term of fourteen years" if the author survived to the end of the first term.[19] The Act covered "any map, chart, book or books already printed within these United States," as well as "any map, chart, book or books already made and composed, but not printed or published, or that shall hereafter be made and composed." Except for the addition of maps and charts, this language was copied almost verbatim from the Statute of Anne.

Because United States law was based primarily on a utilitarian theory of copyright, it made little sense to offer copyright protection for any length of time unless the author or publisher affirmatively claimed that he or she wanted the benefit of copyright protection. Thus, United States law required that the author or publisher comply with various statutory formalities as a condition of copyright protection. For example, the 1790 Act required that the author or publisher register the copyright with the clerk of the district court before publication and publish a notice of the registration in a newspaper for four weeks within two months of the registration (Id. §3, 1 Stat. 125). An 1802 amendment required that the notice be printed in each published edition of the work.[20]

In 1834, in an American replay of *Donaldson v. Beckett*, the U.S. Supreme Court ruled in *Wheaton v. Peters* that, although the author of an unpublished work had a common-law right to control the first publication of that work, the author did *not* have a common-law right to control reproduction following the first publication of the work, and that strict compliance with statutory formalities was required in order to recover under the federal statute.[21] Thus, in the United States there developed a dual system of copyright protection. Before a work was published or registered, it was protected under state law by common-law copyright, which provided a right of first publication of potentially

[18] U.S. Const., Art. I, §8, cl. 8.

[19] Copyright Act of May 31, 1790, c. 15, §1, 1 Stat. 124.

[20] Act of Apr. 29, 1802, c. 36, §1, 2 Stat. 171.

[21] Wheaton v. Peters, 33 U.S. (8 Pet.) 591 (1834). For a comprehensive discussion, see Joyce 2005.

unlimited duration. After a work was published, however, one of two things happened. If all of the statutory formalities were satisfied, the work received a federal statutory copyright of the specified duration; but if the formalities were not observed, the work immediately entered the public domain (Nimmer and Nimmer 2005, §4.01[B], §4.03). As a result, most works were not protected by copyright for any period of time after first publication; only those few works for which the author or publisher had complied with the statutory formalities received copyright protection for the duration provided by law.

The Nineteenth Century

Throughout the nineteenth Century, the durations of national copyright or author's rights laws were extended numerous times. In 1810, France extended its author's rights law to last for the life of the author, then for the life of the author's widow, plus an additional 20 years.[22] In 1814, reciting that "it would afford further encouragement to literature, if the duration of such copyright were extended," England consolidated its two 14-year terms into a single term of 28 years, "and if the author be living at the expiration of that time, till his death."[23]

In the United States, lexicographer Noah Webster took up the cause of extending the existing term of copyright. Webster advocated the view "that an author has, by common law, or natural justice, the sole and *permanent* right to make profit by his own labor" (Webster 1843, 176). In 1830, a report prepared for the House Judiciary Committee by Webster's son-in-law, William W. Ellsworth, stated that "an author has an exclusive and perpetual right, in preference to any other, to the fruits of his labor" (Gales and Seton 1831, 7: cxix–cxx). Despite this endorsement of perpetual copyright as a natural right, the 1831 Act provided only an initial term of 28 years from first publication and a renewal term of 14 years, which could be claimed by the author's heirs if the author was deceased.[24]

In 1837, Prussia adopted a term of life of the author plus 30 years after the author's death.[25] In the same year, in England, Thomas Noon

---

[22] Act of Feb. 5, 1810, Arts. 39–40 (Fr.).

[23] Copyright Act of 1814, 54 Geo. III, c. 156, §9 (U.K.). The circumstances leading to the enactment of this extension are described in Lowndes 1840, 64–72.

[24] Copyright Act of Feb. 2, 1831, §§1–2, 4 Stat. 436. This term was extended to all subsisting copyrights. *Id.* §16, 4 Stat. 439.

[25] Ordinance of July 11, 1837, §§5–6 (Prussia). See Lowndes 1840, 125–26.

Talfourd introduced a bill that sought, among other things, to increase the period of copyright to life of the author plus 60 years. The bill was controversial, in large measure because of the length of term it proposed, and it was debated at length over the next five years. Talfourd's bill was opposed by Macaulay, whose opposition killed the original proposal in 1841 (Seville 1999, 6, 18–19, 31). Macaulay proposed instead that the existing term be lengthened to the longer of 42 years from first publication or life of the author (Id., 31–31, 226). Eventually a compromise was reached, and the Copyright Act of 1842 provided an alternative term of life of the author plus seven years, or 42 years from first publication.[26]

In 1866, after a committee report that advocated perpetual rights, France extended the duration of its author's rights laws to life of the author plus 50 years after the author's death,[27] a term that the committee considered "a major concession to the public interest" (Davies 1994, 88–89). In 1870, the North German Confederation adopted a new copyright law which utilized the Prussian term of life of the author plus 30 years.[28] This law was extended to the unified German Empire in 1871.

In 1866, the United Kingdom, France, Germany, and seven other countries adopted the Berne Convention for the Protection of Literary and Artistic Works, under which member nations agreed to provide copyright protection to the citizens and residents of other member nations (Ricketson and Ginsburg 2005, 82).[29] At the time, however, the delegates were unable to reach agreement on a uniform duration of copyright, although they did agree on a non-binding resolution that recommended a minimum term of life of the author plus 30 years (Id. 2005, 536–38). Instead, the Convention provided for "comparison of terms" (also known as "the rule of the shorter term"), under which each country would protect works from other Berne countries for the same period of time granted to domestic works, but only so long as the work was still protected by copyright in its country of origin (Berne

[26] Copyright Act of 1842, 5 & 6 Vict. c. 45, §3 (U.K.).

[27] Act of July 14–19, 1866, Art. 1 (Fr.).

[28] Gesetz betreffend das Urheberrecht an Schriftwerken, Abbildungen, musikalischen Kompositionen und dramatischen Werken, June 11, 1870, §§ 8, 9 (N. Ger. Conf.). For a summary and English translation, see Jerrold 1881, 43–65.

[29] Berne Convention for the Protection of Artistic and Literary Works, Sept. 8, 1866, art. 2.

1866, art. 2). Once the copyright expired in the country of origin, it expired in all other Berne nations as well.

For the United States and other nations that remained outside the Berne Union, there was no legal obligation to provide any copyright protection to the works of foreign authors; and like most countries, the United States allowed the published works of foreign authors to be copied with impunity in the absence of any treaty obligation (Ochoa 2008, 167–71). Moreover, even after the United States began to extend copyright protection to some foreign authors beginning in 1891, it granted such protection only if the foreign author complied with the formalities required by United States law (Id., 172–173).[30] Thus, for most of the nineteeth century, the term of protection provided to foreign authors in the United States was zero.

The Twentieth Century

At the 1908 Berlin Conference to revise the Berne Convention, the German delegation proposed that the term of protection be that granted to domestic authors in the country in which protection was sought, without regard to the term in the country of origin. Although this would have effectively lengthened the term of copyright, the proposal met with considerable resistance without a uniform term of protection (Ricketson and Ginsburg 2005, 538–39). Ultimately, the 1908 Berlin Revision recommended a term of life of the author plus 50 years, but otherwise retained the "rule of the shorter term."[31] Although the Convention did not make the term mandatory, the United Kingdom nonetheless adopted a life-plus-fifty-years term in its 1911 Copyright Act.[32]

In 1905, in the United States, the Librarian of Congress convened a conference for the purpose of discussing a general revision of the copyright laws. At the conference, authors and publishers both expressed the view that the term of copyright ought to be as long as possible, and they suggested adoption of the French term of life of the author plus 50 years. The principal reasons advanced were that copyright was

---

[30] Act of March 3, 1891, c. 565, 26 Stat. 1106.

[31] Berne Convention for the Protection of Literary and Artistic Works, Sept. 8, 1886, revised at Berlin, Nov. 15, 1908, art. 7.

[32] Copyright Act of 1911, 2 Geo. V c. 46, §3 (U.K.). The Act provided, however, that after 25 years from the death of the author, anyone could reprint the work by giving notice and paying a royalty to the copyright owner of ten percent of the published price for all copies sold (Id.).

a natural right of the author; that authors ought not to outlive their copyrights; that it would provide income to the author's children and grandchildren; and that it ought not to be shorter than the term prevailing in many European countries (Brylawski and Goldman 1975, C3, C7, C11, C75, C78; Ochoa 2001, 33–39). At Congressional hearings in 1906, the star witness was Mark Twain, who believed that copyright ought to be perpetual (Brylawski and Goldman 1975, J116–20; Ochoa 2001, 36). Twain also remarked, however, that he had been able to negotiate a much higher price for his works at the time of renewal (Brylawski and Goldman, K20, K61–66, K88, K163, S14; Ochoa 2001, 37–38). As a result, in the 1909 Act Congress retained an initial term of 28 years, but it extended the renewal term to 28 years, for a maximum duration of 56 years from the date of first publication.[33]

Elsewhere, the march toward longer terms continued. In 1934, Germany adopted a term of life of the author plus 50 years, and Austria followed suit in 1936.[34] Finally, in the 1948 Brussels Revision of the Berne Convention, a minimum term of life of the author plus 50 years was made mandatory, while the "rule of the shorter term" was retained for those nations that had longer terms.[35] The basic term of life plus 50 years was carried forward in the 1967 Stockholm Revision and the 1971 Paris Revision of the Berne Convention.[36]

In the United States, the Copyright Act of 1976 adopted the Berne Convention term of life of the author plus 50 years for all works created on or after January 1, 1978 (except for "works made for hire," which were given the shorter of 75 years from first publication or 100 years from creation).[37] Those works that had been published or registered before 1978 and were still under copyright retained an initial term of 28 years but had their renewal terms extended to 47 years, for

[33] Copyright Act of 1909, §§23–24, 35 Stat. 1075, 1080–81.

[34] Act of Dec. 13, 1934, Art. 1 (Ger.); Act of Apr. 9, 1936, Art. 60 (Aus.).

[35] Berne Convention for the Protection of Literary and Artistic Works, Sept. 9, 1886, revised at Brussels, June 26, 1948, art. 7(1) (basic term); *id.* art. 7(2) (comparison of terms). The minimum basic term was not applied to cinematographic and photographic works. Id., art. 7(3).

[36] Berne Convention for the Protection of Literary and Artistic Works, Sept. 9, 1886, revised at Stockholm, July 14, 1967, art. 7(1) (basic term); id., art. 7(8) (comparison of terms); Berne Convention, *supra* note 7, art. 7(1) (basic term); id., art. 7(8) (comparision of terms).

[37] Former 17 U.S.C. §302 (1976), 90 Stat. 2451, 2572.

a maximum duration of 75 years from first publication.[38] As a transitional measure, those works created before 1978 that had not been published or registered (and which therefore were still protected by state common-law copyright) were accorded the same term given to new works, subject to a statutory minimum term of either 25 or 50 years.[39] The adoption of the life-plus-50-years term was principally motivated by the prospect of eventual U.S. adherence to the Berne Convention,[40] which finally occurred on March 1, 1989.[41]

In 1993, the Council of the European Community decided to harmonize copyright terms. At the time, only Germany and Spain had terms longer than life-plus-50 years; but rather than requiring Germany and Spain to reduce their copyright terms, the Council required all of the other European nations to increase their copyright terms to life plus-70-years.[42] In the same Directive, however, the Council mandated the use of the rule of the shorter term, so that works from the United States and other non-European countries would not receive the benefit of the longer term (Id., art. 7(1)). That, in turn, led American copyright holders to pressure Congress to adopt the life-plus-70 years term, which it did in the Sonny Bono Copyright Term Extension Act of 1998 (CTEA).[43] Shortly after its enactment, the CTEA was challenged in court on the grounds that it violated both the Patent and Copyright Clause and the First Amendment of the U.S. Constitution. On January 15, 2003, however, the United States Supreme Court held that the CTEA did not violate the U.S. Constitution,[44] meaning that the terms of protection set forth in the CTEA currently govern the duration of copyright in the United States.

---

[38] Former 17 U.S.C. §304 (1976), 90 Stat. at 2573–74.

[39] Former 17 U.S.C. §303 (1976), 90 Stat. at 2573. All unpublished works received a statutory minimum of 25 years until December 31, 2002; if the work was published during that period, the statutory minimum was extended through December 31, 2027 (Id).

[40] See H.R. Rep. No. 94–1476, at 135 (1976).

[41] See Berne Convention Implementation Act, Pub. L. No. 100–568, 102 Stat. 2853.

[42] Council Directive 93/98/EEC of October 29, 1993, Harmonizing the Term of Protection of Copyright and Certain Related Rights, art. 1(1), 1993 O.J. (L 290) 9.

[43] P.L. No. 105–298, Title I, 112 Stat. 2827.

[44] See Eldred v. Ashcroft, 537 U.S. 186 (2003).

2. Current Laws on Copyright Duration

Copyright and Neighboring Rights

It is important to note that most countries of the world distinguish between copyright in literary and artistic works and so-called "neighboring rights" of performers, producers of sound recordings ("phonograms" in international parlance), and broadcasting organizations. Various reasons have been given for the distinction. For sound recordings, "the objection was made that these were productions of an 'industrial character' and not capable of constituting literary or artistic creations" (Ricketson and Ginsburg 2005, 1205). Both broadcasts and sound recordings are typically produced through the collaboration of a large number of individuals, making it difficult to identify the "author" or "authors" of such works. Of course, similar objections were also raised with respect to both photographs and motion pictures, both of which are now protected under the Berne Convention, so the resulting division of labor between copyright and neighboring rights seems to be rather arbitrary, a matter of historical accident more than sound principle (Id., 1205–1209). Nonetheless, it is a fact that the term of protection accorded to such "neighboring rights" has lagged behind the term of protection granted to "literary and artistic works" under the Berne Convention; indeed, even though photographs and motion pictures succeeded in being brought under Berne, the term of protection provided to such works has not always kept pace with the basic term of protection.

International Agreements Concerning Copyright

Article 7 of the Berne Convention provides that "[t]he term of protection granted by this Convention shall be the life of the author and fifty years after his death" (Berne 1971, art. 7(1)). For works of joint authorship, the term is measured from the death of the last surviving author (Id., art. 7*bis)*. However, for motion pictures (or "cinematographic works" in international parlance), countries "may provide that the term of protection shall expire fifty years after the work has been made available to the public with the consent of the author," or if the work is not made available to the public, within 50 years after its making (Id., art. 7(2)). Anonymous and pseudonymous works are also protected for "fifty years after the work has been lawfully made available to the public," unless the author discloses his or her identity during that time (Id., art. 7(3)). Countries are permitted to provide shorter terms of protection to photographic works and works of applied art, but such terms must last at least 25 years from the time the work was created (Id., art. 7(4)). All the foregoing terms of protection run to the end of the calendar year in which they would otherwise expire (Id., art. 7(5)).

As of October 15, 2008, there were 164 members of the Berne Union (WIPO 2008a). Those countries that were already members of the Rome Act of the Convention and that had shorter terms on the date they signed either the Stockholm text or the Paris text were permitted to retain such shorter terms (Berne 1971, art. 7(7)). All countries are permitted to grant longer terms of protection (Id., art. 7(6)). "In any case, the term shall be governed by the legislation of the country where protection is claimed; however, unless the legislation of that country otherwise provides, the term shall not exceed the term fixed in the country of origin of the work" (Id., art. 7(7)).

Article 6*bis* of the Berne Convention requires that the author's moral rights of attribution and integrity "shall, after his death, be maintained, at least until the expiry of the economic rights." However, those countries that did not protect such rights after the death of the author at the time of their accession are permitted to retain terms that cease upon the author's death (Id., art. 6*bis*(2)).

The Agreement on Trade-Related Aspects of Intellectual Property Rights (TRIPS) makes all of the substantive provisions of the Berne Convention (except Article 6*bis*) enforceable between nations through the dispute resolution mechanism of the World Trade Organization (TRIPS, arts. 9, 64).[45] As of July 23, 2008, there were 153 members of the World Trade Organization, all of whom must abide by the TRIPS Agreement (WTO 2008). TRIPS does not change the basic term of protection provided by Berne except in one instance: Article 12 of TRIPS requires that photographs and works of applied art be given at least 50 years of protection from the year of authorized publication,

---

[45] Agreement on Trade-Related Aspects of Intellectual Property Rights, Apr. 15, 1994, Marrakesh Agreement Establishing the World Trade Organization, Annex 1C, Legal Instruments—Results of the Uruguay Round, 1869 U.N.T.S. 299, 33 I.L.M. 1197 (1994) [hereinafter TRIPS Agreement].

or if no such publication occurs, 50 years from the date the work was created (TRIPS, art. 12).

Article 9 of the WIPO Copyright Treaty requires that member nations give photographic works the same duration of protection as that provided to other literary and artistic works: namely, life of the author plus 50 years.[46] As of March 5, 2009, 70 countries (including the United States) were parties to the WIPO Copyright Treaty (WIPO 2009c).

In addition to these multilateral treaties, the United States (a net exporter of copyrighted works) has entered into a number of bilateral Free Trade Agreements that require some of its trading partners (such as Australia) to adopt a basic term of copyright protection of life of the author plus 70 years.[47] Other countries (such as Canada), however, have resisted pressure from the United States and the European Union to increase copyright terms beyond those provided in the Berne Convention.[48]

International Agreements Concerning Neighboring Rights

The Rome Convention for the Protection of Performers, Producers of Phonograms and Broadcasting Organizations was signed on October 26, 1961.[49] It provides that performers shall have the right to prevent fixations, reproductions, and broadcasts of their performances (Id., art 7); that producers of phonograms have the right to prohibit reproductions of their phonograms and to be compensated for any broadcasts

---

[46] WIPO Copyright Treaty, Dec. 20, 1996, art. 9, 36 I.L.M. 65.

[47] See, e.g., United States-Australia Free Trade Agreement, art. 17.4(4) (May 18, 2004). The Agreement also provides a minimum term for works made for hire of 70 years after first publication, or if the work is not published within 50 years, 70 years after creation (Id.). There are similar provisions in the Free Trade Agreements between the United States and Bahrain (Art. 14.4(4)), Chile (Art. 17.5(4)), Columbia (Art. 16.5(5)), Korea (Art. 18.4(4)), Morocco (Art. 15.5(5)), Panama (Art. 15.5(4)), Peru (Art. 16.5(5)), and Singapore (Art. 16.4(4)). In the Free Trade Agreement with Oman, the alternative term is 95 years after first publication, or if the work is not published within 25 years, 120 years from creation (Art. 15.4(4)). For the full texts of these agreements, *see* USTR 2009.

[48] See, e.g., North American Free Trade Agreement, Art. 1705 (life-plus-50-years); Copyright Act art. 6, R.S.C. 1985, ch. C-42, §6; R.S.C. 1993, ch. 44, §6 (Can.). Although NAFTA's minimum term remains at life-plus-50 years, in 2003 Mexico adopted a basic term of life-plus-100 years. *See* La Ley Federal del Derecho de Autor, Art. 29 (Mex.).

[49] International Convention for the Protection of Performers, Producers of Phonograms and Broadcasting Organizations, Oct. 26, 1961, 496 U.N.T.S. 44.

of their phonograms (Id., arts. 10, 12); and that broadcasting organizations have the right to prevent fixations, reproductions or rebroadcasts of their broadcasts (Id., art. 13). Under Article 14 of the Rome Convention, such protection shall last at least 20 years from the year in which the performance took place, the fixation was made, or the broadcast took place (Id., art. 14). As of February 13, 2009, 88 nations (not including the United States) were parties to the Rome Convention (WIPO 2009a). The United States, however, has acceded to the 1971 Geneva Phonograms Convention, which also provides a right against unauthorized reproduction for a minimum term of 20 years from the date of fixation or first publication.[50]

The TRIPS Agreement provides rights similar to the Rome Convention (except for the compensation for unauthorized broadcasts of phonograms), but it provides a longer term of protection for performers and producers of phonograms of at least 50 years from the year in which the fixation was made or the performance took place (TRIPS, art. 14(5)). The minimum term of protection for broadcasting organizations remains 20 years from the date the broadcast took place. (Id.).

The WIPO Performances and Phonograms Treaty (WPPT) provides similar protection for both performers and producers of phonograms (omitting broadcasting organizations), and provides to performers a minimum term of 50 years from the year in which the fixation was made, and to producers a minimum alternative term of 50 years from the year in which the phonogram was first published or 50 years from the year in which the phonogram was fixed.[51] As of December 18, 2008, 68 nations (including the United States) were parties to the WPPT (WIPO 2008b).

In addition to these multilateral treaties, the United States has entered into a number of bilateral Free Trade Agreements that require some of its trading partners (such as Australia) to adopt a basic term of protection for performers of life-plus-70 years, and for phonograms

[50] Geneva Convention for the Protection of Producers of Phonograms Against Unauthorized Duplication of Their Phonograms, Oct. 29, 1971, art. 4, 25 U.S.T. 309, 866 U.N.T.S. 71. As of February 25, 2009, there were 77 members of the Geneva Phonograms Convention (WIPO 2009b).

[51] WIPO Performances and Phonograms Treaty, Dec. 20, 1996, 36 I.L.M. 76, arts. 5–10 (performers); *id.* arts. 11–14 (producers); *id.* art. 17 (terms of protection).

of 70 years after first publication, or if the phonogram is not published within 50 years, 70 years after creation.[52]

The European Union

The European Union's Directive on the Term of Protection of Copyright and Certain Related Rights applies to the 27 members of the European Union and the three other members of the European Economic Area.[53] Under the Directive, "[t]he rights of an author of a literary or artistic work...shall run for the life of the author and for 70 years after his death (Directive, art. 1(1))." For works of joint authorship, the term is calculated from the death of the last surviving author (Id., art. 1(2)). For anonymous or pseudonymous works and collective works, the term is the longer 70 years from the year the work is lawfully made available to the public or 70 years from the year of creation, unless the identity of the individual author or authors is disclosed (Id., arts. 1(3), 1(4), 1(6)).

Original photographs are given the same term of protection as other literary and artistic works (Id., art. 6). For cinematographic works, the term is 70 years after the death of the last of the following to survive: the principal director, the author of the screenplay, the author of the dialogue and the composer of music written specifically for the film (Id., art. 2).

The neighboring rights of performers, producers of phonograms, and broadcasting organizations last for 50 years after the date of the performance, the fixation, or the broadcast, respectively; except that if a performance or phonogram is lawfully published or communicated

---

[52] See, e.g., United States-Australia Free Trade Agreement, art. 17.4(4). There are similar provisions in the Free Trade Agreements between the United States and Bahrain (Art. 14.4(4)), Chile (Art. 17.6(7)), Columbia (Art. 16.6(7)), Korea (Art. 18.4(4)), Morocco (Art. 15.5(5)), Panama (Art. 15.5(4)), Peru (Art. 16.6(7)), and Singapore (Art. 16.4(4)). In the Free Trade Agreement with Oman, the alternative term is 95 years after first publication, or if the work is not published within 25 years, 120 years from creation (Art. 15.4(4)). (USTR 2009).

[53] Directive 2006/116/EC of December 12, 2006, on the Term of Protection of Copyright and Certain Related Rights, art. 1(1), 2006 O.J. (L 374) 12; European Union, Europa: European Countries, Member States of the EU, at http://europa.eu/abc/european_countries/eu_members/index_en.htm (accessed Sept. 1, 2009); Agreement on the European Economic Area, May 2, 1992, 1994 O.J. (L 1) 3, annex XVII, ¶ 9f, at http://www.efta.int/content/legal-texts/eea/annexes/annex17.pdf (accessed Sept. 1, 2009).

to the public during that time, then the right lasts for 50 years after the earlier of such publication or communication (Id., art. 3).

The Directive also requires that when a previously unpublished work is published or communicated to the public for the first time, the publisher must be granted an exclusive right for 25 years from the date of such publication or communication (Id., art. 4). In addition, the Directive permits members to protect "critical and scientific editions of works which have come into the public domain." If such protection is granted, it may last no longer than 30 years from the year in which the publication was first lawfully published (Id., art. 5).

As mentioned previously, the Directive mandates the use of the "rule of the shorter term" with regard to works originating in and written by nationals of countries that are not covered by the Directive (Id., art. 7). All terms run to the end of the calendar year in which they would otherwise expire (Id., art. 8). If a member state provided a longer term of protection as of July 1, 1995, the Directive does not require the member state to shorten the term of protection (Id., art. 10(1)).

Finally, it should be noted that the Directive does not apply to the moral rights of an author, leaving a member state free to apply a longer (or shorter) term for such moral rights (Id., art. 9). In France, for example, the law expressly states that an author's moral rights are perpetual.[54]

In April 2009, the European Parliament approved a proposed amendment to the Directive that would extend the rights of performers and producers of phonograms to 70 years from the first authorized publication or communication to the public (European Parliament 2009). Although this amendment has not yet been approved by the Council, its probable adoption suggests that the march toward everlonger terms continues unabated.

United States

In the United States, all works published before 1923 were in the public domain before the Sonny Bono Copyright Term Extension Act of 1998 was enacted. The CTEA did not attempt to revive any expired copyrights, so all such works remain in the public domain (17 U.S.C. §304(b)). For works first published between 1923 and 1963, the term

[54] Law No. 92-597 of July 1, 1992 (Fr.), as amended, art. L. 121-1.

of copyright under the 1909 Act was 28 years, which could be renewed once;[55] the renewal term was extended twice and is now 67 years, for a maximum duration of 95 years from first publication (17 U.S.C. §304(b)). Copyright Office records, however, show that less than 15 percent of the works registered during this time period were renewed; the remaining 85 percent are in the public domain if they are works by American authors or works first published in the United States (Ringer 1960, 222).[56] In 1992, copyright renewal was made automatic, so all works first published between 1964 and 1977 have a duration of 95 years from the date of first publication (comprising a 28-year initial term and a 67-year renewal term) (17 U.S.C. §§304(a), 304(b)).

For works created in 1978 or later, the basic term is life of the author plus 70 years (Id., §302(a)). For so-called "joint works" (i.e., works of joint authorship), the term if life of the last surviving author plus 70 years (Id., §302(b)). For works made for hire, the term is 95 years from the date of first publication, or 120 years from the date of creation, whichever is shorter (Id., §302(c)).

Works created before 1978, but not published or registered before 1978, get the same term provided to new works: life of the author plus 70 years, or the alternative fixed term for works made for hire. These works, however, were subject to a statutory minimum term of 25 years from January 1, 1978, which has now expired (Id., §303(a)). As a result, any works created before 1978 that remained unpublished as of December 31, 2002, are now in the public domain if the author of the work died more than 70 years before the current year began (Reese 2007, 591; Gard 2006, 690). Those works that were created before 1978, but which were first published between 1978 and 2002, received the same term as new works, subject to a statutory minimum term, which has been extended to December 31, 2047 (17 U.S.C. §303(a)).

A special situation applies to copyright in sound recordings. Sound recordings were not added to the federal Copyright Act until February 15, 1972. Any sound recordings fixed on or after that date are entitled to the same term of protection as that granted to other works of

[55] Former 17 U.S.C. §24 (1909; repealed 1976).

[56] Works of foreign origin that were in the public domain in the U.S. for failure to comply with formalities such as notice or renewal, but that were not yet in the public domain in their country of origin, had their copyrights restored effective January 1, 1996. *See* 17 U.S.C. §104A(a)(1)(A); *id.* § 104A(h)(2) (defining "date of restoration"); *id.* §104A(h)(6) (defining "restored work").

authorship. Any sound recordings fixed before that date, however, are governed by state law rather than by federal law; and federal law provides that any state-law protection shall not be preempted by federal law until February 15, 2067 (95 years from the date sound recordings first became eligible for federal copyright protection) (Id., §301(c)). This leaves the term of copyright in such pre-1972 sound recordings up to the individual states.

Only one state has a statute regarding the duration of copyright in such sound recordings: California provides that all such sound recordings will be protected until February 15, 2047.[57] In one other state there is a court decision concerning the duration of copyright in such sound recordings. In 2005, the New York Court of Appeals (the highest court in the state of New York) held that sound recordings had a perpetual common-law copyright under New York law; that such sound recordings were not placed in the public domain when phonorecords of the sound recording were reproduced and distributed in New York; and that such sound recordings did not enter the public domain in New York when their copyrights expired in their country of origin.[58] Consequently, all sound recordings will remain under copyright in the state of New York until state law is preempted by federal law on February 15, 2067.

3. Conclusion

A rational choice concerning the duration of copyright cannot be made until some kind of consensus has been reached on the rationale for copyright protection. While common-law systems started from a utilitarian perspective, international harmonization has moved them much closer toward the natural right view of copyright that prevails in most civil law countries. Even in such countries, however, the countervailing considerations of increased access to literary and artistic works and the difficulty of identifying and locating copyright owners have led to some temporal limit on the duration of copyright.

The history of copyright duration demonstrates that while copyright terms have been lengthened a number of times, they are almost

---

[57] Cal. Civ. Code §980(a)(2).

[58] *See* Capitol Records, Inc. v. Naxos of America, Inc., 830 N.E.2d 250 (N.Y. 2005).

never shortened. This is consistent with the principles of public choice theory discussed above, which hold that when the benefits of a law are concentrated but the costs of that law are diffuse, a small well-focused interest group will usually succeed in obtaining passage of the law, even if it does not benefit society as a whole. However, because of the controversy concerning the CTEA, public awareness of the costs of extending the duration of copyright has been raised (Karjala 2007; Lessig 2004). Thus, while it can be expected that copyright owners will attempt to obtain further extensions of the copyright term in the future, it also should be expected that they will meet with more significant opposition than they faced in 1998. Whether opponents of copyright term extension will have the political power to succeed in defeating future attempts at copyright term extension, however, remains to be seen.

References

Akerloff, George A., et al. 2002. Brief *Amici Curiae* in Support of Petitioners, in *Eldred v. Ashcroft*, 537 U.S. 186 (2003) (No. 01–618). http://cyber.law.harvard.edu/openlaw/ eldredvashcroft/supct/amici/economists.pdf (accessed Sept. 1, 2009).

Bard, Robert L., and Lewis Kurlantzick. 1999. *Copyright Duration: Duration, Term Extension, The European Union and the Making of Copyright Policy*. San Francisco: Austin and Winfield.

Boswell, James. [1774] 1975. The Decision of the Court of Session Upon the Question of Literary Property in the Cause of Hinton Against Donaldson. In *The Literary Property Debate: Six Tracts, 1769–1774*, ed. Stephen Parks. New York: Garland.

Breyer, Stephen. 1970. The Uneasy Case for Copyright: A Study of Copyright in Books, Photocopies and Computer Programs. *Harvard Law Review* 84: 281.

———. 1972. Copyright: A Rejoinder. *U.C.L.A. Law Review* 20: 75.

Brylawski, Fulton E., and Abe Goldman, ed. 1976. *Legislative History of the 1909 Copyright Act*. South Hackensack: J. B. Rothman.

Bureau of the Census. 2000. Voting and Registration in the Election of November 1998. Washington, DC: Bureau of the Census. http://www.census.gov/prod/2000pubs/p20–523.pdf (accessed May 15, 2006).

Cobbett, William, ed. 1813. *The Parliamentary History of England*. London: T. C. Hansard.

Continental Congress. May 2, 1783. *Journals of the Continental Congress, 1774–1779*, 24: 326. Washington: US Government Printing Office.

Davies, Gillian. 1994. Copyright and the Public Interest. Vol. 14 of *IIC Studies in Industrial Property and Copyright Law*. Munich: Max Planck Institute.

Dawson, Robert L. 1992. *The French Booktrade and the 'Permission Simple' of 1777: Copyright and Public Domain*. Oxford: Voltaire Foundation at the Taylor Institution.

Dietz, Adolf. 1978. *Copyright Law in the European Community*. Alphen aan den Rijn: Sijthoff & Noordhoff.

European Parliament. 2009. Texts Adopted, Thursday, 23 April 2009, Strasbourg—Term of Protection of Copyright and Related Rights. http://www.europarl.europa.eu/sides/ getDoc.do?type=TA&reference=P6-TA-2009–0282&language=EN&ring=A6-2009-0070 (accessed Sept. 1, 2009).

Gales and Seton, comp. 1831. Report of House Judiciary Committee on Copyright, Dec. 17, 1830. In *Register of Debates in Congress* 7: cxix–cxx. Washington: Gales and Seton.

Gard, Elizabeth Townsend. 2006. January 1, 2003: The Birth of the Unpublished Public Domain and Its International Implications. *Cardozo Arts & Entertainment Law Journal* 24: 687.

Ginsburg, Jane. 1990. A Tale of Two Copyrights: Literary Property in Revolutionary France and America. *Tulane Law Review* 64: 991.

Guinan, James Jr. [1957] 1961. Duration of copyright. Study No. 30 of *Copyright Law Revision: Studies prepared for the Subcommittee on Patents, Trademarks, and Copyrights of the Committee on the Judiciary, United States Senate, 86th Congress, Second Session*, 10: 53. Washington, D.C.: U.S. Government Printing Office.

Jerrold, Sidney. 1881. *A Handbook of English and Foreign Copyright in Literary and Dramatic Works*. London: Chatto and Windus.

Joyce, Craig. 2005. "A Curious Chapter in the History of Judicature": Wheaton v. Peters and the Rest of the Story (of Copyright in the New Republic). *Houston Law Review* 42: 325.

Kaplan, Benjamin. 1967. *An Unhurried View of Copyright*. New York: Columbia University Press.

Karjala, Dennis S. 1998. Statement of Copyright and Intellectual Property Law Professors in Opposition to H.R. 604, H.R. 2589, and S. 505, "The Copyright Term Extension Act," Jan. 28, 1998. http://homepages.law.asu.edu/~dkarjala/OpposingCopyrightExtension/ legmats/1998Statement.html (accessed Sept. 1, 2009).

———. 2006. Congestion Externalities and Extended Copyright Protection. *Georgetown Law Journal*, 94: 1065.

———. 2007. Opposing Copyright Extension. http://homepages.law.asu.edu/~dkarjala/OpposingCopyrightExtension (accessed Sept. 1, 2009).

Landes, William W. and Richard A. Posner. 1989. An Economic Analysis of Copyright Law. *Journal of Legal Studies* 18: 325.

———. 2003. Indefinitely Renewable Copyright. *University of Chicago Law Review* 70: 471.

Lemley, Mark A. 2004. Ex Ante versus Ex Post Justifications for Intellectual Property. *University of Chicago Law Review* 71: 129.

———. 2005. Property, Intellectual Property, and Free-Riding. *Texas Law Review* 83: 1031.

Lessig, Lawrence. 2004. *Free Culture: How Big Media Uses Technology and the Law to Lock Down Culture and Control Creativity*. New York: The Penguin Press.

Litman, Jessica. 1990. The Public Domain. *Emory Law Journal*, 39: 965.

Locke, John. [1698] 1988. Of Property. Pp. 287–89 in *Locke: Two Treatises of Government*, ed. Peter Laslett. *Cambridge Texts in the History of Political Thought*. Cambridge: Cambridge University Press.

Lowndes, John James. 1840. *An Historical Sketch of the Law of Copyright*. London: Saunders and Benning.

Macaulay, Thomas Babington. 1853. Remarks on the Copyright Bill in the House of Commons (Feb. 5, 1841). Reprinted in Macaulay, Thomas Babington, *Speeches*, 1: 387. London: H. Vizetelly.

Nimmer, Melville B. and David Nimmer. 2005. *Nimmer on Copyright*. New York: Matthew Bender.

Ochoa, Tyler T. 2001. Patent and Copyright Term Extension and the Constitution: A Historical Perspective. *Journal of the Copyright Society of the USA* 49: 19.

———. 2003. Origins and Meanings of the Public Domain. *University of Dayton Law Review* 28: 215.

———. 2008. Copyright Protection for Works of Foreign Origin. In *The Internationalization of Law and Legal Education: Comparative Perspectives on Law and Justice*, ed. Jan Klabbers and Mortimer Sellers. New York: Springer.

Ochoa, Tyler T., and Mark Rose. 2002. The Anti-Monopoly Origins of the Patent and Copyright Clause. *Journal of the Copyright Society of the USA* 49: 675.
Olson, Mancur Jr. 1965. *The Logic of Collective Action: Public Goods and the Theory of Groups*. Cambridge, MA: Harvard University Press.
Patterson, Lyman Ray. 1968. *Copyright in Historical Perspective*. Nashville: Vanderbilt University Press.
Pollack, Malla. 2001. What is Congress Supposed to Promote?: Defining "Progress" in Article I, Section 8, Clause 8 of the United States Constitution, or Introducing the Progress Clause. *Nebraska Law Review*, 80: 754.
Rappaport, Edward. 1998. *Copyright Term Extension: Estimating the Economic Values*. Congressional Research Service Report No. 98–144 E.
Reese, R. Anthony. 2007. Public But Private: Copyright's New Unpublished Public Domain. *Texas Law Review* 85: 585.
Register of Copyrights. Report on Orphan Works, Jan. 2006. United States Copyright Office. http://www.copyright.gov/orphan/orphan-report.pdf (accessed Sept. 1, 2009).
Ricketson, Sam, and Jane C. Ginsburg, ed. 2005. *International Copyright and Neighboring Rights: The Berne Convention and Beyond*. Oxford: Oxford University Press.
Ricketson, Sam. 1992. The Copyright Term. *International Review of Industrial Property and Copyright Law (IIC)* 23: 753.
Ringer, Barbara A. [1960] 1961. Renewal of Copyright. Study No. 31 of *Copyright Law Revision: Studies prepared for the Subcommittee on Patents, Trademarks, and Copyrights of the Committee on the Judiciary, United States Senate, 86th Congress, Second Session*, 10: 105. Washington, D.C.: U.S. Government Printing Office.
Rose, Mark. 1993. *Authors and Owners: The Invention of Copyright*. Cambridge, MA: Harvard University Press.
Seville, Catherine. 1999. *Literary Copyright Reform in Early Victorian England: The Framing of the 1842 Copyright Act*. Cambridge: Cambridge University Press.
Sterling, J. A. L., ed. 2003. *World Copyright Law*. London: Sweet and Maxwell.
Tyerman, Barry W. 1971. The Economic Rationale for Copyright Protection for Published Books: A Reply to Professor Breyer. *U.C.L.A. Law Review* 18: 1110.
United States Trade Representative. 2009. Free Trade Agreements. http://www.ustr.gov/trade-agreements/free-trade-agreements (accessed Sept. 1, 2009).
Vetulani, Agnieszka. 2008. The Problem of Orphan Works in the EU. European Commission. http://ec.europa.eu/information_society/activities/digital_libraries/doc/report_orphan_stagiaire_2/report_orphan_vetulani (corrected version) (2).pdf (accessed Sept. 1, 2009).
Walterscheid, Edward C. 2002. *The Nature of the Intellectual Property Clause: A Study in Historical Perspective*. Buffalo: William S. Hein & Co.
Webster, Noah. 1843. Origin of the Copy-Right Laws in the United States. In *A Collection of Papers on Political, Literary and Moral Subjects*. New York: Webster and Clark.
World Intellectual Property Organization (WIPO). 2008a. Convention for the Protection of Literary and Artistic Works—Status on October 15, 2008. http://www.wipo.int/export/ sites/www/treaties/en/documents/pdf/berne.pdf (accessed Sept. 1, 2009).
——. 2008b. WIPO Performances and Phonograms Treaty—Status on December 18, 2008. http://www.wipo.int/export/sites/www/treaties/en/documents/pdf/wppt.pdf (accessed Sept. 1, 2009).
——. 2009a. International Convention for the Protection of Performers, Producers of Phonograms and Broadcasting Organizations—Status on October 15, 2008, http://www.wipo.int/treaties/export/sites/www/treaties/en/documents/pdf/rome.pdf (accessed Sept. 1, 2009).

———. 2009b. Convention for the Protection of Producers of Phonograms Against Unauthorized Duplication of Their Phonograms: Phonograms Convention—Status on October 15, 2008. http://www.wipo.int/export/sites/www/treaties/en/documents/pdf/ phonograms.pdf (accessed Sept. 1, 2009).

———. 2009c. WIPO Copyright Treaty—Status on March 5, 2009. http://www.wipo.int/ export/sites/www/treaties/en/documents/pdf/wct.pdf (accessed Sept. 1, 2009).

World Trade Organization (WTO). 2008. Understanding the WTO: The Organization—Members and Observers. http://www.wto.org/english/thewto_e/whatis_e/tif_e/org6_e.htm (accessed Feb. 15, 2009).

Yen, Alfred C. 1990. Restoring the Natural Law: Copyright as Labor and Possession. *Ohio State Law Journal* 51: 517.

CHAPTER SEVEN

WAITING AS A TEMPORAL CONSTRAINT

Florian Klapproth

Summary

In the first part of this paper, I present four criteria that constitute the wait as a temporal constraint. After that, the link between waiting and expectation is elaborated, and it is argued that temporal expectation is an important determinant of the experience of waiting. In the second part of the paper, I present a series of experiments that addressed the questions of whether temporal expectations affect the experience of the duration of the wait and how the wait is affectively evaluated by the waiting person. One main result was that temporal expectations governed judgments of the wait's duration and judgments of the wait's emotional impact in a different manner: If expectation did not affect time estimation, it affected the evaluation of the wait. And vice versa—if expectation did not influence the evaluation of the wait, it governed time estimation. At the end of this chapter, I offer possible explanations for this apparent contradiction and outline the contribution of temporal expectations to constraints caused by waiting.

Introduction

1. *What Is Waiting?*

What is waiting? It's a question that is not easy to answer. On the one hand, waiting seems to be an activity, as a waiting person is someone who waits. On the other hand, however, a wait is mainly characterized by a lack of activity. According to Paul Fraisse (1985), a famous French psychologist and time researcher, waiting emerges if there is a temporal gap between the appearance of a need and its satisfaction. Waiting implies accepting the deferral of that need's satisfaction—that is, exercising oneself in patience. Usually adults more readily accept waiting than children do. The latter commonly react with impatience and frustration when having to wait.

Whenever we wait, time occupies our perception. A waiting person recognizes that time is scarce and fears the loss of time due to the non-utilizable waiting period. The experience of duration is strongest when one has to wait (Fraisse 1985). Time appears as something that is

Jo Alyson Parker, Paul A. Harris and Christian Steineck (Eds), Time: Limits and Constraints, pp. 179–198

hostile: the waiting person wishes to kill time in order to escape from the enforced passivity. "Time arises from reluctance," says Novalis (quoted in Blumenberg 1985, 7). Unfortunately, the more people are aware of being imprisoned by time, the more slowly time passes for them (for example, Macar et al. 1994; Mattes and Ulrich 1998).

Someone who waits prepares himself or herself for events that shall happen in the future. The only barrier that separates the waiting person from an event that he or she is waiting for is time. And, usually, the waiting person also *experiences* this time as a barrier. Waiting, therefore, results from being constrained by time.

In order to analyze the essence of what it means to wait, I suggest that at least four defining elements characterize the state of waiting.

(1) Waiting inhibits goal-directed action. Even children usually consider waiting as a stoppage of action, meaning that they see waiting as being costly (Gasparini 1995; Piaget 1981). Frequently, people conceive waiting as a loss of time (Leclerc et al. 1995).

(2) During waiting, people spend time in a more or less passive way. Sometimes the term waiting is also applied to periods of time in which sometimes productive activities might happen. Such "waiting" periods might comprise waiting for one's birthday next summer or waiting for one's lover who is expected to arrive at the following weekend. However, many authors agree that real waiting is mainly determined by its lack of productive activity. If someone who waits is still carrying out an activity, the only purpose of this activity might lie in drawing off the attention from the boring waiting situation. Acting during waiting differs from acting in other situations by its subordinated purpose. Much as we strive for utilizing waiting time in a meaningful way, activities during waiting are still predominantly passive activities (Paris 2001, 708).

(3) Moreover, as waiting time usually cannot be chosen, but is dependent on external variables, people who wait are also dependent on these external variables. We do not terminate the wait—the event for which we wait does. We can abandon the wait, but only if we also surrender the event that we aim for.

(4) And finally, waiting time is often uncertain. Nobody knows exactly when a wait will be finished. According to Maister (1985, 118), a good example of the role of uncertainty in the waiting experience is provided by the "appointment syndrome." Clients who arrive early for an appointment will wait contentedly until the scheduled time.

However, once the appointment time is passed, even a short wait grows increasingly annoying because waiting beyond that point has no knowable limit.

2. *Waiting and Expectation*

Because the act of waiting implies the anticipation of a future event, waiting is intimately connected with *expectation*. A person who waits also expects the occurrence of an event after the expiration of the waiting time. Both waiting for something and expecting something involve a look into the future. To wait for something goes along with expecting something to occur after a period of time that is preventing the expected event from occurring immediately.

The similarity between waiting and expectation is mirrored in some languages. In some Romance languages and in the German language, a single word covers the two meanings for waiting and expectation: *attente* (French), *attesa* (Italian), *esperar* (Spanish), and *erwarten* (German).

Although waiting and expectation are related concepts, they nevertheless differ in some respects. Waiting, conceived as passively dwelling between the advent of a need and the instant of its satisfaction, is not object-related per se. In contrast, expectation is strongly and primarily related to objects as it is the object that is expected to occur. Heidegger emphasizes the discrepancy between waiting and expectation in his *Conversation on a Country Path about Thinking* (Heidegger 1966) by a fictive dialogue between a scientist, a scholar, and a teacher. He states that expectation has an object to which it is related whereas waiting has no object. In waiting, we rest in the act of waiting. As soon as we represent what we are waiting for, says Heidegger, we are not waiting anymore: "In waiting we leave open what we are waiting for." (1966, 68) In terms of cognitive psychology, Heidegger's postulate means that during the process of expectation we devote attention mainly to the desired object, whereas during waiting we rather focus our attention on the passage of time.

Psychology considers expectations as beliefs about a future state of affairs and subjective estimates of the likelihood of future events (Olson et al. 1996). Expectations do not necessarily imply a temporal gap between the present and the future. They just represent a present state of the mind, which is concerned with an event predicted to happen in the future. This conception of expectation leaves open what an individual does in the meantime between now and the time the event is expected to occur.

In the case of deliberate, conscious expectations, people can relate subjective likelihoods to their expectations. For example, they can indicate how likely a certain event is to occur in the future. This means that expectations can differ between the level of subjective certainty regarding whether the event will occur or not and at which time the event will occur. With waiting, however, people are seldom uncertain about the occurrence of the event, but rather they are often uncertain about when the event will occur. Usually, they believe that the expected event will definitely happen. Otherwise, their willingness to wait would decrease. An example might illustrate this. If I were uncertain about being called in the waiting room of a doctor, I surely wouldn't spend much time in the waiting room; instead I would choose another doctor.

Whereas event-based uncertainty corresponds rather rarely with waiting, uncertainty regarding the time of occurrence of an event is often the case. Such temporal uncertainty (which in a strict sense means lack of knowledge) might be counterbalanced by temporal expectancy (which is a kind of belief), true to the motto: the less I know, the more I have to believe.

3. *Short-term Waiting*

Although long waits are generally more annoying than short ones (Katz et al. 1991, Osuna 1985, Taylor 1994), actual waiting time may not be the most important factor of subjective well-being in waiting situations. For example, even short waiting at traffic lights may be annoying when a driver is in a hurry. Further examples of negative impacts of short waits on individuals' feelings come from computer systems and telephone queues. Although the development of computer systems has led to faster processing speeds, users still have to wait for system responses, which might influence performance and enhance the development of negative attitudes towards computers (Jay 1981) or might lead to feelings of impatience (Zakay et al. 1999). Telephone waiting times are generally much shorter than those of physical waits such as standing in a physical queue. Nevertheless, factors that mitigate negative reactions to waiting in a physical queue have also been proven to be effective in telephone settings (for example, Munichor and Rafaeli 2007).

In fact, most experimental studies concerned with waiting have investigated rather short waits for the following reasons. (1) Short waits are less demanding than long waits for people participating in those experiments. (2) Compared to long waits, short waits can be simply realized in laboratory settings in which experimenters can control for confounding variables. (3) Some variables that are assumed to affect the waiting experience are easier to realize with short waits than with long waits. One of these variables is the expected duration of the wait.

Antonides et al. (2002), for example, filled telephone waiting times for a commercial service with either music or information about the expected waiting time. Whereas music only marginally decreased perceived waiting time, information about the expected waiting time significantly reduced the overestimation of waiting time. Another study examining telephone waits was recently conducted by Munichor and Rafaeli (2007). In that study participants called a university lab to sign up for experiments. The experimenters varied the content during the wait, that is, whether the callers heard music during the wait, apologies for the wait, or information about their position in the waiting line. They found that call evaluations were most positive when information about the position in the line was provided.

Ahmadi (1984) let her participants wait for the beginning of an experiment slightly more than four minutes. In one experimental condition, the participants were provided with the (false) information that they had to wait two minutes. In another condition, no information was given regarding the expected duration of the wait. Perceived duration was shorter in the former condition than in the latter one. Similar results were obtained in a study conducted by Hui and Tse (1996) whose participants waited to be registered for a computer course. One group received (correct) information about the expected length of the wait, whereas the other group got no information. The group with duration information judged the waiting time more precisely than the group without information. Moreover, evaluation of the wait was more positive when the participants were informed about the wait's duration rather than not informed.

Knowledge about the remaining wait time has been shown to affect evaluations of computer-system delays as well. Kuhmann et al. (1990) could show with waits caused by delayed computer responses that variable waiting times, that is, waiting times that are hardly predictable, led to more negative emotions than predictable waiting times did.

Dellaert and Kahn (1999) investigated how negative waiting experiences could affect consumers' evaluations of Internet web sites. They found that when provided with information about the expected duration of the wait, the participants evaluated both the wait and the web sites more positively than when not provided with this information.

Results of these studies suggest that temporal expectations regarding the duration of the wait appear to be an important factor in one's determining how long and how unpleasant waiting will be experienced. Temporal expectations can stem from multiple sources. One source derives from verbal declaration. For example, service providers may tell customers waiting on the telephone that they will be served shortly. A second source of temporal expectations is prior experience. Most people waiting for an event to occur have collected experiences with similar situations and build expectations about the actual wait's duration based on these former experiences. Imagine a situation in which drivers (or even pedestrians) are waiting for traffic lights to turn green. Although nobody tells them how long the red light will last, they expect it to turn green after some seconds of exposure.

Despite these everyday manifestations of experience-based temporal expectations, no study so far has examined how temporal expectations based on prior experiences affect individuals' reactions to short-term waiting.

4. *A Series of Experiments*

Though expectations are ubiquitous and certainly guide our behaviour, there is still only a small number of studies that have investigated the effects of expectations on temporal judgments. Usually, experimental investigations of the role temporal expectations play on duration judgments have not considered how people affectively evaluated the conformity or nonconformity of their expectations with their experienced duration. Previous studies, primarily based on surveys, indicated (1) that disconfirmations of temporal expectations led to lower evaluations than confirmations (Houston et al. 1998), (2) that short waits corresponded with rather positive evaluations (Antonides et al. 2002; Dellaert and Kahn 1999; Hui and Tse 1996), and (3) that evaluations were, in general, marginally negatively correlated with perceived duration estimations (Houston et al. 1998).

In the following, I will present a series of experiments that we (that is, I and a couple of students)[1] conducted between March 2006 and July 2007. The main concern of this study was (1) to examine whether temporal expectancies affect the experience of the duration of the wait, and (2) to assess how the waiting period would be affectively evaluated.

I developed an experimental paradigm in which participants were to work on an attention test (Brickenkamp 2002). The task for the participants was to mark the letter "d," combined with two dashes, as frequently as possible within an array of similar letters. The participants had to begin the activity after hearing a high-pitched tone and had to stop the activity after hearing a low-pitched tone. After that period of activity, a break followed during which the participants were to watch a bright square presented on a monitor. Then, another period of activity followed, which was followed by another break, and so forth (see Figure 1). This sequence of activity and waiting was repeated a couple of times. The participants were encouraged to mark as many target letters as possible within the whole sequence. Thus participants would perceive every break as a barrier impeding the participants to fulfil their task successfully. With these breaks embedded within periods of activity, I aimed at realizing waiting situations that resembled those in real life settings, for example, waiting for downloading web pages while booking a journey on the Internet or waiting on a telephone line while trying to get some information from customer service.

Both the duration of the period of activity and the break duration (except for the final break) were kept constant within an experimental sequence. As the breaks prior to the final break were all of the same duration, it was assumed that the participants would anticipate the duration of each following break. The final break at the end of the sequence was of particular importance because, at the end of the experiment, the participants were asked to judge its duration.

The final break—or in more technical terms, the target interval—was experimentally varied. It was longer than, shorter than, or as long as the previous break intervals.

[1] I am grateful to Katrin Benner, Kathrin Raubach, Sarah Walzer, and Johanna Wegscheider, who helped to run the experiments.

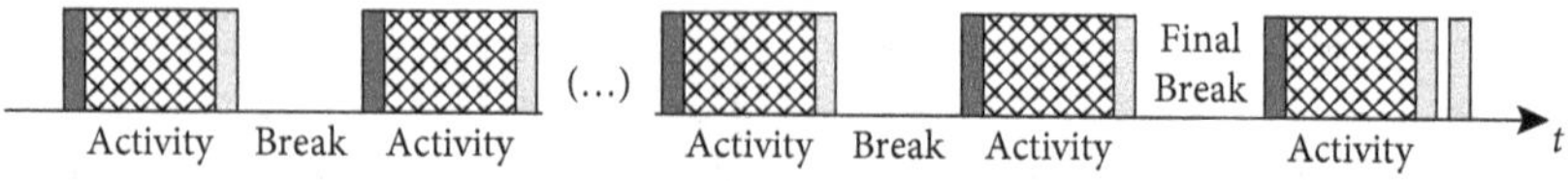

Figure 1: Idealized sequence of periods of activity and waiting realized in the experiments. The arrow symbolizes the passing time, dark-gray rectangles represent high-pitched tones, light-gray rectangles represent low-pitched tones. The patterned rectangles represent the time during which the participants were active. The breaks between periods of activity are indicated by the empty space between the patterned rectangles. The last break within the sequence was the final break, which was to be judged by the participants. The sequence stopped with the presentation of two low-pitched tones.

Dependent measures were (a) the retrospective judgment of the duration of the target interval and (b) the participants' feelings concerning their wait during the target interval. The participants judged the duration by means of a reproduction task. After the onset of a visual stimulus they pressed a button on the computer keyboard to terminate the stimulus when experiencing matching duration of that stimulus and the target interval. The affective evaluation of the target interval was assessed by a questionnaire comprising five bipolar pairs of adjectives that described the participants' mood during waiting. The adjectives, adopted from Antonides et al. (2002), were as follows: annoying/pleasant, boring/varied, unsatisfactory/satisfactory, irritating/not irritating, and unacceptable/acceptable. The task of the participants was to describe their feelings concerning the waiting time, using a seven-point scale to differentiate between the adjectives of each pair.

Experiment 1

Method

Twelve men and 48 women, between 20 and 44 years old (M = 27.4 years), participated in this experiment. The participants were tested individually in a moderately illuminated and quiet room. An IBM-compatible Pentium II computer controlled all experimental events and recorded the data. Responses were made on the computer keyboard.

The experiment started with a presentation of the instructions on the computer monitor. After that, the participants heard an 800-Hz sinus tone of 500 ms duration presented via headphones. This tone signalled the participants to start the activity. A 400-Hz sinus tone of 500 ms duration occurred after 15 sec to signal the end of the activity period. Simultaneously, a bright square (10 x 10 cm) was presented on the monitor, which was to be watched during the break. After 10 sec or 30 sec (dependent on the experimental condition), an 800-Hz sinus tone occurred again to signal the participants to start with a new period of activity.

After the tenth repetition of a period of activity (and therefore after nine repetitions of breaks), the target interval followed. The target interval was succeeded by a last period of activity, after which the participants estimated the duration of the target interval and rated its affective impact.

The values of the independent variables "actual duration of the target interval" and "expected duration of the target interval" (that is, the duration of the previous breaks) were 10 sec and 30 sec. Thus, four experimental conditions were realized. The participants were randomly and in equal number assigned to the four conditions. We predicted outcomes similar to those observed in the literature. That is, we assumed that the final wait should be judged more precisely when its expected duration matched the actual duration than when they differed. In the case of a discrepancy between expected and actual duration of the final wait, perceived duration should shift towards the expected duration. Moreover, long waits should be judged as being less pleasant than short waits, and disconfirmation of temporal expectations should lead to lower evaluations than confirmation of temporal expectations.

Results and Discussion

Figure 2 displays the results of Experiment 1. The left panel shows the results of the affective evaluation of the final wait; the right panel shows the results of the time-estimation task. With regard to the feelings of the participants during the final wait, there was a significant interaction between the expected and the actual duration of the target interval, $F(1, 56) = 4.06$, $p < .05$: the participants rated the wait more positively if the expected duration of the target interval corresponded to its actual duration. However, if the actual duration deviated from the expected duration, the wait was evaluated more negatively. Both the main effect of expectation, $F(1, 56) = 0.94$, $p = .34$, and the main effect of the target interval, $F(1, 56) = 0.50$, $p = .48$, were not significant.

Regarding time estimation, there was a trivial effect of the duration of the target interval: long target intervals were judged longer than

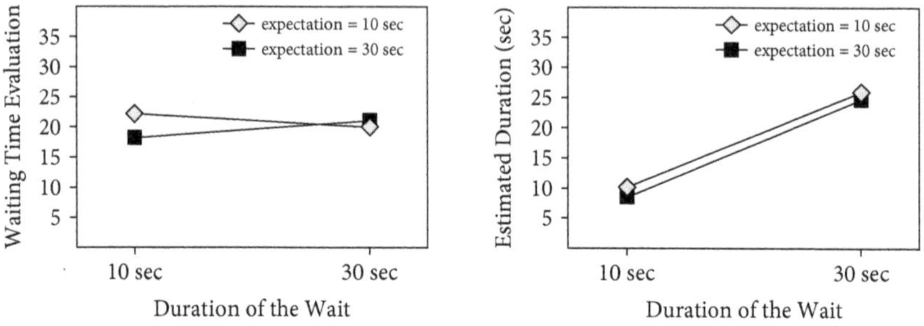

Figure 2: Results obtained from Experiment 1. Left panel: mean evaluation scores. Right panel: mean estimated duration in seconds. Note: The maximum score that could be achieved on the evaluation scale was 35, the minimum score 5.

short target intervals, $F(1, 56) = 54.70$, $p < .001$. In contrast to our hypothesis, there was no significant interaction between the expected duration of the target interval and its actual duration, $F(1, 56) = 0.01$, $p = .92$, nor was there a significant main effect of expectation, $F(1, 56) = 0.45$, $p = .50$.

Temporal expectation did not affect time estimation, but it affected the evaluation of the wait. Possibly, the waiting period was easy to encode and, hence, easy to remember so that perception of the time interval was hard to distort by expectation. However, the feelings regarding the wait may reflect the participants' disappointment when their temporal expectations were not met. Surprisingly, the duration of the wait had no impact on its evaluation.

Experiment 2

Method

We sought to generalize the results obtained from our first experiment by varying the durations of the stimuli. Therefore, shorter durations were chosen for both the expected and the actual duration of the final wait. These were 2 sec and 5 sec, respectively (instead of 10 sec and 30 sec as in Experiment 1). Except for the durations, the rest of the material as well as the procedure remained the same.

In Experiment 2, eight men and 40 women participated. Their age was between 21 and 44 years ($M = 27.7$ years).

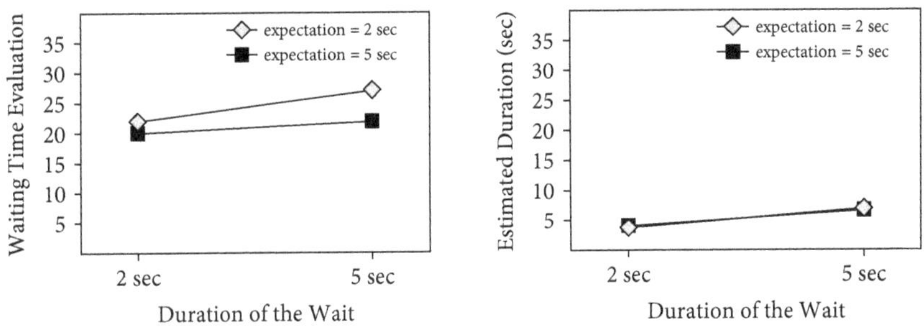

Figure 3: Results obtained from Experiment 2. Left panel: mean evaluation scores. Right panel: mean estimated duration in seconds. Note: The maximum score that could be achieved on the evaluation scale was 35, the minimum score 5.

Results and Discussion

In Figure 3 the results of Experiment 2 are shown. As in the previous experiment, there was the trivial finding that long waits were estimated as longer than short waits, $F(1, 44) = 9.77$, $p < .01$. Moreover, neither a main effect of expectation on time estimation, $F(1, 44) = 0.01$, $p = .94$, nor an interaction between expected and real target duration, $F(1, 44) = 0.06$, $p = .80$, was obtained.

However, in contrast to Experiment 1, analysis of variance yielded two significant main effects on waiting time evaluation: a main effect due to expectation, $F(1, 44) = 5.05$, $p = .03$, and a main effect due to waiting time duration, $F(1, 44) = 4.81$, $p = .03$. The interaction was not significant, $F(1, 44) = 1.09$, $p = .30$.

Summing up the results, both experiments yielded no effect of temporal expectation on time estimation. Yet, expectation affected the participants' feelings during the last waiting period of the sequence. If the expected duration matched the actual duration, the average value of the participants' pleasure was medium (about 21 points). However, if there was a discrepancy between expected and actual duration, participants' pleasure deviated from that medium value. Whereas in Experiment 1, waiting was rated lower if expected and actual duration differed than if they were the same, this effect was dependent on the actual duration of the target interval in Experiment 2. If the target interval was short and its expected duration rather long, the wait was rated less pleasurable than if the target interval was long and its expected duration rather short. It appears as if the participants were positively surprised when they could wait for a longer time than expected. Altogether, it can be stated that disconfirmation of temporal expectations seems to play a significant role in evaluating a waiting period.

Experiment 3

Two results of the previous experiments guided the planning for the third experiment.

On the one hand, we found that temporal expectations did not affect time estimations of the final waits. We assumed that this result originated from the fact that, after the presentation of the target interval, a period of activity followed in both experiments. It might be that the effect of expectation on time estimation was attenuated due to a delayed recall. To avoid this possible source of error, the participants of Experiment 3 conducted the time estimation task immediately after the final wait.

On the other hand, long waits were rated more positively than short waits in Experiment 2. Whereas in the literature, waiting almost always has a negative connotation, some authors claim that waiting can be seen positively as a temporal resource, particularly for making decisions (Gasparini 1995). Waiting can also be a temporal resource for recreation (Boucsein 1988). Possibly, the quick succession of periods of activity and waiting in Experiment 2 could have been demanding for the participants. Therefore, every break between intervals filled with activity could have been served as a breather. And if the participants used the breaks for recreation, they should have enjoyed long breaks more than short breaks.

We presumed that the context within which waiting is embedded interacts with temporal expectation. Therefore, in Experiment 3 the context of waiting was varied: that is, the valence of the activity was manipulated. In a preliminary investigation, we let a number of participants assess two versions of attention tests. One group of participants was presented with a modified version of the original test. In this version, the letters to be marked followed a rhythmic structure. The other group of participants received the original test form. We found that the participants evaluated the rhythmic version more positively than the original version. As this difference was statistically not significant, the original version of the test was made more aversive by the simultaneous presentation of white noise to the participants. With Experiment 3, three hypotheses were tested: (1) We assumed that waiting should be evaluated to be more acceptable if waiting periods were embedded within less pleasurable activities than if waiting periods were embedded within more pleasurable activities. (2) We hypothesized that more acceptable waiting periods would be estimated as being shorter in duration than less acceptable waiting periods. (3) Due to the change of the procedure described below, temporal expectations should shift time perceptions towards the expected duration. Hence, with positive contrast (that is, expected duration is shorter than actual duration), the waiting period should be judged shorter than with negative temporal contrast (that is, expected duration is longer than actual duration).

Method

In this experiment 26 men and 54 women participated, aged between 18 and 67 years, with a mean of 39.1 years. We realized only disconfirmations of expectations, that is, positive and negative temporal contrasts. This was due to the fact that in the previous experiments effects of expectation were strongest if there was a discrepancy between the expected and the actual duration of the wait.

Both the duration of the target interval and the duration of the period of activity were kept constant in all conditions (15 sec). The duration of the breaks presented prior to the target interval was 10 sec in the positive-contrast conditions and 20 sec in the negative-contrast conditions. Furthermore, we varied the type of activity. Half of the participants were presented with the attention test that had already been used within the previous experiments. Additionally, they heard white noise while working on this test. The other half of the participants were presented with a modified version of this attention test, in which the to-be-marked letters were arranged in a regular order. We assumed that their working on the original test, along with their being presented with white noise, should result in a strong need for overcoming the waiting period, whereas working on the modified test was assumed to be experienced as enjoyment.

Contrary to the previous experiments, the sequence of periods filled with activity and break periods ended with the target interval. Thus, the participants judged both the duration and the pleasantness of the target interval immediately after the presentation of the target interval.

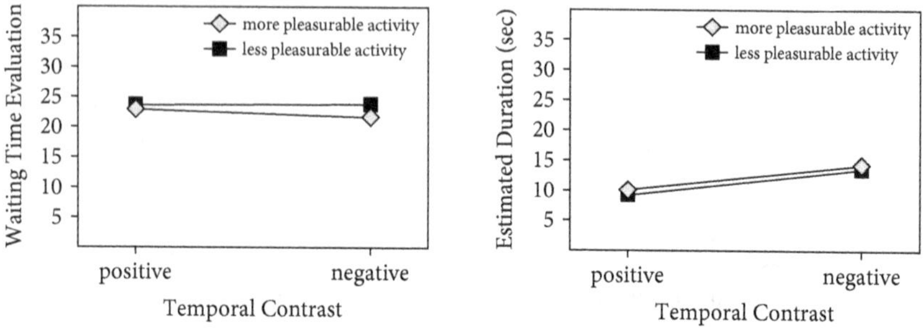

Figure 4: Results obtained from Experiment 3. Left panel: mean evaluation scores. Right panel: mean estimated duration in seconds. Note: The maximum score that could be achieved on the evaluation scale was 35, the minimum score 5. The actual duration of the target interval was 15 sec. Positive contrast means that actual waiting time is longer than expected waiting time; negative contrast means that actual waiting time is shorter than expected waiting time.

Results and Discussion

Figure 4 shows the results of Experiment 3. Regarding the wait evaluation, there was a marginal difference depending on the kind of activity. Waiting followed by less pleasurable activity was evaluated to be more positive than waiting followed by more pleasurable activity, $F(1, 76) = 1.39$, $p = .24$. However, this difference was not statistically significant. Furthermore, evaluation of the wait was not dependent on temporal contrast, which is indicated by a lack of significance in both the main effect of temporal contrast, $F(1, 76) = 0.61$, $p = .44$, and the interaction between temporal contrast and the kind of activity, $F(1, 76) = 0.004$, $p = .95$.

However, as is indicated in the right panel of Figure 4, time estimations were largely affected by temporal contrast. With positive temporal contrast (that is, waiting longer than expected), subjective waiting time was rather short, whereas with negative temporal contrast (that is, waiting less than expected), subjective waiting time was much longer. Accordingly, we obtained a significant main effect due to temporal contrast, $F(1, 76) = 9.46$, $p < .01$. This result differs entirely from the previous results obtained in Experiment 1 and Experiment 2. Neither the main effect of the kind of activity, $F(1, 76) = 0.08$, $p = .78$, nor the interaction between the kind of activity and temporal contrast, $F(1, 76) = 0.01$, $p = .91$, was significant.

A simple explanation for the outcome of Experiment 3 could be that the participants didn't judge the duration of the target interval but the average duration of the previous break intervals. If this assumption holds true, judging the target interval might have been enhanced if the markers that defined the previous break intervals (that is, the tones and the period of activity) were the same as the makers that defined the target interval. According to Penner (1976) who investigated duration discrimination with rather short intervals, changing the markers of an interval will result in poorer discrimination, compared to unchanged markers. However, if intervals are difficult to discriminate, participants are looking for information that might facilitate their judgment. They prefer to rely on relational rather than absolute judgments: that is, they use anchors in order to judge the target interval (cf. Thomas et al. 1991). In Experiment 3, the series of repeated breaks presented prior to the target interval could have served as such an anchor. Experiment 4 was designed to test the hypothesis that sameness of interval markers should result in better recall of the target interval, whereas differences between interval markers should lead to worse recall.

Experiment 4

Method

In Experiment 4, 46 men and 44 women participated. Their age was between 18 years and 64 years, with a mean age of 30.6 years. Six conditions were realized. We varied both what the direction of the temporal contrast was (positive, negative, and zero) and whether the markers of the target interval were the same as the markers of the previous intervals or whether they were in part different. As in Experiment 3, the duration of both the target interval and each of the periods of activity was set to 15 sec. The duration of the previous break intervals was 10 sec with the positive contrast, 15 sec with the zero contrast, and 20 sec with the negative contrast.

Assuming that changing the interval markers would result in a relational duration judgment, we hypothesized that temporal expectations should affect time estimation if the markers of the target interval differed from the markers of the previous intervals. However, if the

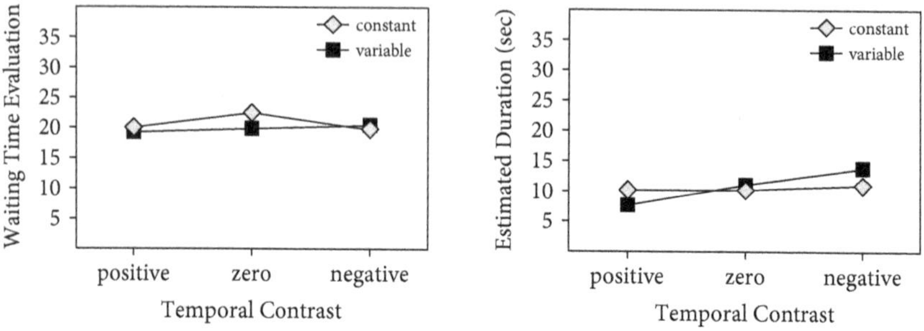

Figure 5: Results obtained from Experiment 4. Left panel: mean evaluation scores. Right panel: mean estimated duration in seconds. Note: The maximum score that could be achieved on the evaluation scale was 35, the minimum score 5. The actual duration of the target interval was 15 sec. Positive contrast means that actual waiting time is longer than expected waiting time, negative contrast means that actual waiting time is shorter than expected waiting time, and zero contrast means equivalence between actual and expected waiting time.

markers remained constant, the participants were supposed to judge the target interval rather absolutely, that is, without relating the target interval to the previous break intervals. Hence, in that condition time estimations should be unaffected by temporal expectations. Regarding the waiting time evaluation, exactly the opposite was assumed: the occurrence of an effect of temporal contrast was expected only if the interval markers were constant because only then a discrepancy between expected and perceived final waits would arise.

Results and Discussion

Figure 5 displays the results of Experiment 4. Concerning the evaluation data, observed differences between the conditions did not reach statistical significance. On a descriptive level, temporal contrast affected the wait evaluation only in the conditions with constant interval markers. Waiting was evaluated as being more positive in conditions with zero contrast and as being rather negative in conditions with negative or positive contrast.

However, with respect to the time estimation data, the results of Experiment 4 confirmed the hypotheses. As hypothesized, an effect of temporal contrast on time estimation was observed for conditions in which the interval markers changed at the end of the final wait, but not for conditions in which the interval markers were kept constant. The main effect of temporal contrast was statistically significant,

$F(2, 84) = 4.24$, $p = .027$, and the interaction between interval markers and temporal contrast was marginally significant, $F(2, 84) = 2.58$, $p = .082$. However, the main effect of interval marker failed to become significant, $F(1, 84) = 0.75$, $p = .79$. Moreover, we obtained a trend we also found in Experiment 3: the target interval in the negative-contrast condition was estimated as being longer than the target interval in the positive-contrast condition. Estimations of the target interval in the zero-contrast condition were in between both conditions.

With Experiment 4 the results of the previous experiment were both replicated and extended. People judged waiting time longer if they expected a long wait and shorter if they did the reverse. Therefore, temporal expectation was very predictive in forecasting how people would estimate the wait's duration. Furthermore, it could be shown that marker variability indeed interacted with expectation. The change of the markers at the final break obviously impeded remembering the duration of that final wait and, hence, let the participants revert to the duration of the previous breaks which had been encoded and memorized before.

General Discussion

With four experiments using the same experimental paradigm, I am able to draw some preliminary conclusions about the role temporal expectations play on waiting.

We observed that temporal expectations governed judgments of the wait's duration and judgments of the wait's pleasantness in a different manner. If expectation did not affect time estimation, it affected the evaluation of the wait. And vice versa: if expectation did not influence the evaluation of the wait, it governed time estimation. We can solve this apparent contradiction when we consider the psychological processes that might lie behind the obtained effects. When expectation controlled time estimation (particularly in Experiment 3 and in the variable-markers conditions of Experiment 4), the participants were apparently unable to perceive a difference between the expected and the real target interval. They judged the target interval as they expected it to be. It could furthermore be shown that the evaluation of the final wait depended on the existence of a perceived discrepancy between the expected and the real duration of the target interval (particularly in Experiment 1 and Experiment 2). In these experiments, the participants judged the duration of the final wait more or less accurately, at least without being biased by their previous expectation. In contrast to their counterparts in Experiment 3 and Experiment 4, these participants were indeed able to identify a difference between the expected and the actual duration of the target interval. It seems obvious that perceiving a discrepancy between expected and experienced wait durations should be a necessity for evaluating the wait differently dependent on its expected duration. Otherwise, if there is no experienced discrepancy, no difference with respect to expectancies would be made in evaluating the wait.

I propose that stability of interval markers would alleviate encoding and recall of the duration of the final wait, whereas the change of interval markers would impede it. If recall of the target interval can be done easily, no (or just minor) effects of temporal expectations would occur. If recall is pretty difficult, however, people would rather "remember" the average of those intervals they had encountered repeatedly prior to the target interval (cf. Helson's adaptation-level theory, Helson 1948). Thus, time estimation would strongly be influenced by the expectation people had built up before encountering the final wait.

In the light of the present results, how do temporal expectations contribute to constraints caused by waiting? I presented cases where expectation guided feelings concerning the wait. I also presented cases where expectation guided the subjective duration of the wait. Hence, both feelings regarding the wait and perceived duration of the wait were affected by temporal expectations. Temporal expectations may help decrease the constraints caused by waiting, as they allow the individuals to plan and schedule future activities. However, violation of expectations may lead to the experience of increased uncertainty regarding the wait's duration. But, as was already documented, this seems not to be the case if the waiting period can be enjoyable. Provided that the waiting period is not enjoyable, one can say that expectations play a two-fold role on the experience of waiting: they can either help the individual perceive waiting as more controllable and hence less restrictive; or they can—if violated—raise the individual's concerns about the wait and thus make the waiting person feel more constrained by time.

References

Ahmadi, Kate S. (1984). Effects of social influences and waiting on time judgment. *Perceptual and Motor Skills* 59: 771–776.

Antonides, Gerrit, Peter C. Verhoef, and Marcel van Aalst. 2002. Consumer perception and evaluation of waiting time: a field experiment. *Journal of Consumer Psychology* 12: 193–202.

Blumenberg, Hans. 1985. *Lebenszeit und Weltzeit*. Frankfurt/M.: Suhrkamp.

Boucsein, Wolfram. 1988. Wartezeiten am Rechner—Erholung oder Streß? *Zeitschrift für Arbeitswissenschaft* 42: 222–225.

Brickenkamp, Rolf. 2002. *D2-Aufmerksamkeits-Belastungs-Test* (9. Aufl.). Göttingen: Hogrefe.

Dellaert, Benedict G. C., and Barbara E. Kahn. 1999. How tolerable is delay? Consumers' evaluations of internet web sites after waiting. *Journal of Interactive Marketing* 13: 41–54.

Fraisse, Paul. 1985. *Psychologie der Zeit*. München: Ernst Reinhardt Verlag.

Gasparini, Giovanni. 1995. On waiting. *Time and Society* 4: 29–45.

Heidegger, Martin. 1966. *Discourse on Thinking*. New York: Harper and Row.

Helson, Harry. 1948. Adaptation-level as a basis for quantitative theory of frames of reference. *Psychological Review* 55: 297–313.

Houston, Mark B., Lance A. Bettencourt, and Sutha Wenger. 1998. The relationship between waiting in a service queue and evaluations of service quality: a field theory perspective. *Psychology & Marketing* 15: 735–53.

Hui, Michael K., and David K. Tse. 1996. What to tell consumers in waits of different lengths: an integrative model of service evaluation. *Journal of Marketing* 60: 81–90.

Jay, Timothy B. 1981. Computerphobia. What to do about it. *Educational Technology* 23: 47–48.

Katz, Karen L., Larson, Blaire M., and Larson, Richard C. 1991. Prescription for the waiting-in-line blues: entertain, enlighten, and engage. *Sloan Management Review* Winter: 44–53.

Kuhmann, Werner, Schaefer, Florian, and Boucsein, Wolfram. 1990. Effekte von Wartezeiten innerhalb einfacher Aufgaben: eine Analogie zu Wartezeiten in der Mensch-Computer-Interaktion. *Zeitschrift für experimentelle und angewandte Psychologie* 37: 242–265.

Leclerc, France, Bernd H. Schmitt, and Laurette Dubé. 1995. Waiting time and decision making: Is time like money? *Journal of Consumer Research* 22: 110–119.

Macar, Françoise, Simon Grondin, and Laurence Casini. 1994. Controlled attention sharing influences time estimation. *Memory & Cognition* 22: 673–686.

Maister, David H. 1985. The psychology of waiting lines. Pp. 113–23 in *The service encounter*, ed. John A. Czepiel, Michael R. Solomon, and Carol Suprenant. Lanham, MD: Lexington Books.

Mattes, Stefan, and Rolf Ulrich. 1998. Directed attention prolongs the perceived duration of a brief stimulus. *Perception & Psychophysics* 60: 1305–1317.

Munichor, Nira, and Rafaeli, Anat. 2007. Numbers or apologies? Customer reactions to telephone waiting time fillers. *Journal of Applied Psychology* 92: 511–518.

Olson, James M., Neal J. Roese, and Mark P. Zanna. 1996. Expectancies. Pp. 211–38 in *Social psychology: Handbook of basic principles*, ed. Edward T. Higgins and Arie. W. Kruglanski. New York: Guilford.

Osuna, Edgar E. 1985. The psychological cost of waiting. *Journal of Mathematical Psychology* 29: 82–105.

Paris, Rainer. 2001. Warten auf Amtsfluren. *Kölner Zeitschrift für Soziologie und Sozialpsychologie* 53: 705–733.

Penner, Merrilynn. J. 1976. The effect of marker variability on the discrimination of temporal intervals. *Perception & Psychophysics* 19: 466–469.

Piaget, Jean. 1981. *La développement de la notion de temps chez l'enfant*. Paris: PUF.

Taylor, Shirley. 1994. Waiting for service: the relationship between delays and evaluations of service. *Journal of Marketing* 58: 56–69.

Thomas, David R., Mood, Kelly, Morrison, Spencer, and Wiertelak, Eric. 1991. Peak shift revisited: A test of alternative interpretations. *Journal of Experimental Psychology: Animal Behavior Processes* 17: 130–140.

Zakay, Dan, Block, Richard A., and Tsal, Yehoshua. 1999. Prospective duration estimation and performance. In *Attention and performance XVII*, ed. Daniel Gopher and Asher Koriat, 557–580. Cambridge, MA: MIT Press.

SECTION III

CREATIVE CONSTRAINTS

CHAPTER EIGHT

DAVID MITCHELL'S *CLOUD ATLAS* OF NARRATIVE CONSTRAINTS AND ENVIRONMENTAL LIMITS

Jo Alyson Parker

> Time's Arrow became Time's Boomerang....
>
> —David Mitchell, *Cloud Atlas*

> Writing bends time.
>
> —David Mitchell, Interview for "The Agony Column"

> It may be observed in general that the future is purchased by the present. It is not possible to secure distant or permanent happiness but by the forbearance of some immediate gratification.
>
> —Samuel Johnson, *Rambler* 178

Summary

David Mitchell's 2004 novel *Cloud Atlas* pushes the limits of narrative construction through its use of narrative constraints: thematically linked embedded narratives that come to closure only during their second iteration, wherein they proceed in reverse order. In this paper, I explore how the self-imposed narrative constraints that Mitchell employs enable him to drive home a message about exceeding environmental limits. Through the temporally innovative narrative structure, which moves back from a post-apocalyptic "future" to a past seemingly still in flux, Mitchell encourages us to consider the far-reaching ramifications of our present actions.

In his 2004 novel *Cloud Atlas*, David Mitchell pushes the limits of narrative construction. Like the title to a fictitious musical composition in the text, the novel comprises "*Matryoshka Doll Variations*"[1]: it features six separate but thematically and imagistically linked novellas, the first of which is embedded in the second, the second of which is embedded in the third, and so forth, with each story breaking off at a crucial point as we move to the next. The sixth or chronologically last story serves as a fulcrum, the novel then boomeranging back through the

[1] David Mitchell, *Cloud Atlas* (New York: Random House, 2004), 52. Subsequent references to this edition are included in the text.

Jo Alyson Parker, Paul A. Harris and Christian Steineck (Eds), Time: Limits and Constraints, pp. 201–218

preceding stories, each of which then comes to closure. In the process, Mitchell takes us from the 1850's South Pacific voyage of Adam Ewing to a post-apocalyptic Hawaii seemingly several hundred years in the future wherein the last survivors of the "Fall" attempt, perhaps in vain, to keep the spark of "Civ'lize" glowing.

With its embedded narratives, suspensions of closure, boomerang trajectory, and five-hundred-year-plus range, *Cloud Atlas* plays audacious games with narrative temporality. Yet these games have a serious purpose. In the chronologically final story, Mitchell shows us the dire future that present action (or inaction) may trigger and thereby drives home a message about the global consequences of immediate gratification. The novel thus serves as a response to one of the great challenges facing us today—how to get individuals and nations to make the changes (even sacrifices) in the present to ensure a habitable future on earth.

1. The Narrative Constraint

Cloud Atlas is one cohesive novel built of six novellas, each written in a different genre and taking place in a different time and place. The novellas are as follows:

(1) A travel diary written by good-hearted, naïve estate-agent Adam Ewing, set in the South Pacific of the 1850s. Ewing saves Autua, a Moriori tribesman, from death at the hands of the Maori (a traditional enemy of the Moriori), and he is in turn saved by Autua from being killed for his money by the corrupt doctor Henry Goose. This story takes place not long after the Maori conquest and destruction of the Moriori of the Chatham Islands, and it shows the consequences of colonialism in the South Pacific.

(2) An epistolary narrative written by the unscrupulous, charming, and artistically gifted composer Robert Frobisher, set in Brussels in 1931. Frobisher takes on the job as amanuensis for the syphilitic composer Vyvyan Ayrs, hoping thereby to win fame and fortune, but he discovers that Ayrs intends to claim Forbisher's compositions as his own. In despair and bereft of resources, Frobisher takes his own life.

(3) A mystery/suspense page-turner featuring intrepid investigative reporter Luisa Rey in 1970s California.[2] Rey discovers a plot by the Seaboard Corporation to put a flawed nuclear reactor online. Beset by Seaboard henchmen intent on killing her, Rey (as well as her exposé) is saved by Joe Napier, a former Seaboard lackey who finds redemption but dies in the process.

(4) A satire of contemporary Britain, put forward as the memoirs of Timothy Cavendish, the venal owner of a vanity press. Through a series of unfortunate events, Cavendish is locked up in a nursing home where the horrific attendants prey on their fragile charges, and he finds new meaning in life by joining several of the more robust residents in a daring escape.

(5) The transcription of an interview with the "ascended fabricant" Sonmi-451, set in a near-future Korean "corpocracy." A member of an enslaved clone race who has achieved consciousness, Sonmi joins a doomed revolt against the system, but, before she is executed, she pens her *Declarations*, which will serve as the basis for a new religion, as we discover in the following novella.[3]

(6) A supposed oral account by Zach'ry, one of the few surviving humans, set in the Hawaiian Islands several hundred years in the future after a "Fall" has wiped out civilization. Zach'ry initially distrusts but then befriends Meronym, one of the "Prescients," a group that has attempted to preserve knowledge. When his family and village are destroyed by the vicious Konas, Zach'ry and Meronym help one another escape to Maui.

Despite the novellas' differences in genre and setting, a single theme resonates throughout *Cloud Atlas*: the clash between those who pursue immediate gratification at the expense of all else and those who are future-thinking and life-affirming. We might say that the six seemingly disjointed stories work together to advance one plot: the plot of

[2] The name "Luisa Rey" is a pointed allusion to Thornton Wilder's *The Bridge of San Luis Rey*, which similarly is structured around several stories that tie together. In fact, Mitchell even has Luisa plunge from a bridge, but unlike Wilder's ill-fated characters, she survives. In Mitchell's earlier novel *Ghostwritten*, which also comprises interrelated stories, a character names Luisa Rey also appears, but in this instance she is the writer rather than the protagonist of crime stories.

[3] Sonmi's full name is, of course, an allusion to Ray Bradbury's *Fahrenheit 451*, which similarly tells a story of a dystopic future.

a "good man" coming to the realization that, in the words of Adam Ewing, "a purely predatory world *shall* consume itself" (508) and vowing to work toward "shaping a world" that "divers races & creeds… can share" (508). The strange temporality of Mitchell's text enables him to enforce his meaning.

In fashioning *Cloud Atlas*, Mitchell set himself a particular narrative constraint: "how to make the various novellas fit inside each other and to come up with ways of making the preceding narrative appear as an 'artefact' of the succeeding narrative."[4] Each narrative except Zach'ry's is embedded in the subsequent one, with the composer reading the travel diary, the reporter reading the composer's letters, and so forth. As Mitchell points out, the structure fittingly melds with the theme of predatory behavior that the novel examines: "Each block of narrative is subsumed by the next, like a row of ever-bigger fish eating the one in front."[5] Frobisher even speaks of "gobbling… down" Ewing's diary (460). After the fulcrum narrative, we boomerang back through the other narratives in reverse order. If we extend Mitchell's conceit of the fish, we might say that, in the second half of the novel, the "ever-bigger fish" are ultimately swallowed by the ones they consumed, this structure serving as the narrative equivalent of one of M. C. Escher's paradoxical drawings. More significant for my purpose, the chronologically prior narrative ends up swallowing the chronologically later narrative in this second half.

Interestingly, this first constraint works in tandem with a second—the suspended ending. For example, the first part of Adam Ewing's diary cuts off in mid-sentence, and we are immediately plunged into Robert Frobisher's letters. Within the letters, Frobisher recounts his discovery of Ewing's diary, but, as he announces in exasperation, only the first half: "To my great annoyance, the pages cease, midsentence, some forty pages later, where the binding is worn through" (64). Frobisher's desire for closure, like the reader's, is temporally frustrated. When we return to Frobisher's story during the boomerang's return journey, however, Frobisher has managed to discover the second half

4 "Silver Dagger and Russian Dolls—David Mitchell, author of *Cloud Atlas*, in interview," by Shane Barry, *Three Monkeys Online* (March 2005) <http://threemonkeys-online.com/threemon_article_david_mitchell_cloud_atlas_interview.htm> (21 May 2006).

5 "Q&A: Book World Talks With David Mitchell," *washingtonpost.com*, 22 August 2004, http://www.washingtonpost.com/up/dyn/articles/A17231-2004 Aug19.html (21 May 2006).

of the journal, and the conclusion of his story leads to the resumption of Ewing's.

This constraint pertains to all the narratives except—necessarily—Zach'ry's: the embedded narratives cut off at the middle, and, only once we boomerang back through them do they come to closure. Those who have read Italo Calvino's *If on a Winter's Night a Traveler* may find the truncated-narrative concept familiar. Mitchell, in fact, has named Calvino's novel as an important source: "I wondered what a novel might look like if a mirror were placed at the end of a book like Calvino's so that the stories would be resolved in reverse."[6] Unlike the narratives in Calvino's novel, however, which remain unresolved, the narratives in *Cloud Atlas* do conclude. But, by opening up five separate narratives and then deferring their closure, Mitchell plays with the way in which we depend upon conclusiveness for meaning, and he pushes the limits of how far we can sustain the open-ended. Five narratives intervene between the beginning of Ewing's story and its ending; suspension builds suspense so that, by the time we reach this ending, which is also the ending of *Cloud Atlas*, it is a consummation devoutly to be wished. After all, as Frobisher points out, "A half-read book is a half-finished love affair" (64).

The novel's boomerang structure destabilizes the tradition of the embedded narrative. Customarily, embedded narratives are provided within the story itself. Taking one of the most well known instances of narrative embedding as a model (*The 1001 Nights*), we recall that Scheherazade engages in telling tales to the Sultan in order to defer her execution. She thus often breaks off the story at the most exciting moment or has one of her own characters launch into a tale, which, in turn, might feature a tale-within-a tale. Such a structure implicitly follows a temporal constraint, for each tale-within-a-tale is presupposed as antedating the tale that frames it, a past history that can only be recounted, not reworked.

If the novel followed the Scheherazadian tradition of embedding, it would begin with Zach'ry telling the story of listening to Sonmi's interview and then follow with Sonmi watching the film of Cavendish's life, Cavendish reading Luisa's story, Luisa reading Frobisher's letters, and Frobisher reading Ewing's journal. As the chronologically earliest narrative, Ewing's journal would be the most deeply embedded

[6] "Book World Talks With David Mitchell."

of the narratives. Instead, Zachry's narrative takes this role, the post-apocalyptic future that it recounts enclosed by a prelapsarian past inhabited by its own particular Adam. Frame narrative and embedded narrative change places depending on perspective so that the outside (frame) and inside (embedded narrative) are one and the same, the smallest matryoshka doll enclosing the large doll that would seem to enclose it. Although Ewing's narrative antedates all the other narratives, its enclosure of them troubles the temporally linear transmission implicit in traditional embedded stories, thus suggesting that history can be not only recounted but also reworked, at least within the constraints of fiction.

2. The Boomerang and the Eternal Recurrence

This notion of reworking or revising the past so as to impact the future lies at the heart of Mitchell's novel. During "the Night of the Lemon Prize Awards" (Mitchell's sly satire of Britain's Orange Prize), a sloshed Timothy Cavendish notes that "Time's Arrow became Time's Boomerang" (147). Unlike Martin Amis's backwards-moving *Time's Arrow*, which Mitchell may be playfully referencing here, *Cloud Atlas* moves both forward and backwards, and, in doing so, premises not only that past events can affect the future but also that future events can affect the past. (See Figure.) Or, more accurately—because we readers return to the chronologically earlier stories *after* confronting the post-apocalyptic future that Mitchell envisions, that the created (fictional) future affects the meaning of the created (fictional) pasts.

In a self-referential passage, Isaac Sachs, the doomed scientist who helps Luisa Rey expose the Seaboard cover-up, formulates a theory of time that draws upon the metaphor of the matyroshka doll:

> *One model of time: an infinite matryoshka doll of painted moments, each "shell" (the present) encased inside a nest of "shells" (previous presents) I call the actual past but which we* perceive *as the virtual past. The doll of "now" likewise encases a nest of presents yet to be which I call the actual future but which we* perceive *as the virtual future.* (393; Mitchell's italics)

The "now" of each particular novella "encases a nest of presents yet to be." If we think in terms of the sequence of our reading, as opposed to chronological sequence, the second part of, say, Adam Ewing's story is encased in Zach'ry's. Essentially, the logic of plot dictates that the

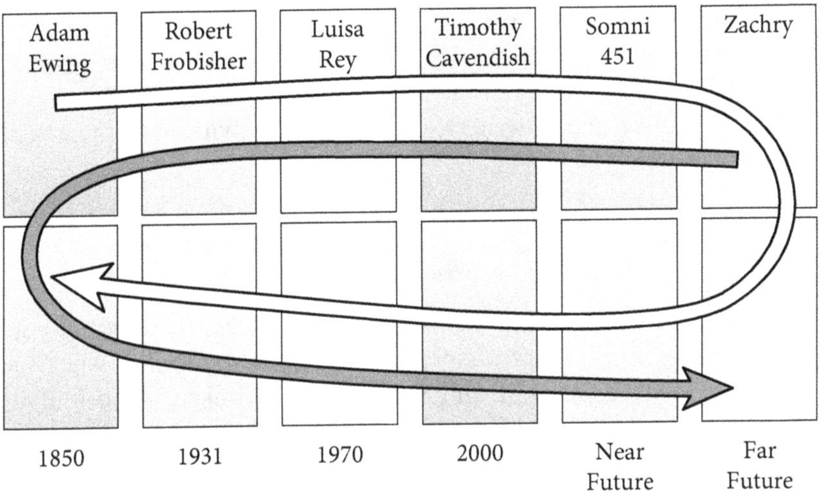

Figure: This figure shows the temporal structure of *Cloud Atlas*. As the top row demonstrates, the first part of the novel proceeds in chronological order; as the second row demonstrates, the second part proceeds in reverse chronological order. Generally, with regard to embedded narratives, we expect each successive embedding to reach further back into the past, as occurs in *The 1001 Nights*; the dark arrow demonstrates how such a structure might work. In *Cloud Atlas*, however, the chronologically earliest narrative encloses all the others, as the light arrow demonstrates.

future Zach'ry describes engenders the past in which Adam acts—to significant effect, as I discuss in the following section.

The distinction Isaac Sachs makes between actual and virtual pasts is the distinction between the past as it really happened and the past as construction:

> *The actual past is brittle, ever-dimming +ever more problematic to access + reconstruct: in contrast, the virtual past is malleable, ever-brightening + ever more difficult to circumvent/expose as fraudulent.* (392; Mitchell's italics)

Similarly, the virtual future is "*constructed by wishes, prophecies + day-dreams,*" and it "*may influence the actual future, as in a self-fulfilling prophecy*" (393). Through Sachs as mouthpiece, Mitchell makes clear that all we can have of the past is indeed a reconstruction or reconstructions of it and that all we can have of the future is a visualization that we create based on our present reality.

To a certain extent, creating virtual pasts and futures is the very work of fiction-writers such as Mitchell, who draw on wishes, prophecies, and daydreams to craft virtual worlds. Of course, others do so as well, but, unlike fiction-writers, they present the virtual as the actual. As Sacks notes,

> *The present presses the virtual past into its own service, to lend credence to its mythologies + legitimacy to the imposition of will. Power seeks + is the right to 'landscape' the virtual past.* (392; Mitchell's italics).

We see such "landscaping" when the Maori recast their brutal conquest of the Moriori—as a story of the Maori reclaiming what was rightfully theirs, as Adam Ewing describes: "A drunken Maori mulatto told me that the entire history of the Aboriginals had been dreamt up by the 'mad old Lutheran' & Mr. D'Arnoq preaches his Moriori gospel only to legitimize his swindling land claims against the Maori, the true owners of Chatham, who have been coming to & fro in their canoes since time immemorial!" (16). This passage, we might note, also recalls the landscaping performed by the other conquerors, the British, who renamed the Moriori's Rehoku as the Chatham Islands. As Mitchell implies, landscaping the virtual past becomes a means of enforcing ideological agendas.

It is not surprising that the words "will" and "power" occur in close proximity in Sach's discussion of the landscaping of the past, for the shade of Friedrich Nietzsche haunts *Cloud Atlas*. Ayrs plans "a final symphonic major work, to be named *Eternal Recurrence* in honor of his beloved Nietzsche" (84), and, along with the mythical *Matryoskha Doll Variations*, it might serve as commentary on the text itself. Certainly, there is a sense throughout of history repeating itself. The *Prophetess*, the ship upon which Ewing sails, reappears as the "Best-Preserved Schooner in the World" at the marina where Napier saves Luisa's life, and a "Great Ship o' the Prescients" (247) brings Meronym to Zach'ry's island. The vicious Maori of Ewing's Chatham Islands give way to the vicious Kona of Zach'ry's Hawaii. The fabricants of Sonmi's Nea So Copros, locked in endless service in Papa Song's "golden arches," are latter-day version of the sweatshop workers in Luisa Rey's California. In his train-trip across England, Timothy Cavendish remarks on the Science Parks "where those Biotech Space Age cuboids now sit cloning humans for shady Koreans" (168), the dreadful outcome of which appears in Sonmi's story. Henry Goose collects the teeth that cannibals expelled when dining on their victims, Cavendish calls out

"Soylent Green is made of people" as he attempts his first escape from the nursing home, and Sonmi discovers that the "Soap" that serves as the food supply for her and her fellow fabricants is made of recycled fabricants "who have reached the end of their working lives" (343).[7] In a convergence of Nietzschean ideas, what seems eternally to recur throughout the stories of *Cloud Atlas* is the will to power, summed up in Henry Goose's cynical aphorism "The weak are meat, the strong do eat" (503).

Yet, as bleak a vision of the future as he presents, Mitchell suggests that it is not only the will to power that recurs. Adam Ewing, Robert Frobisher, Luisa Rey, Timothy Cavendish, and Meronym all sport a comet-shaped birthmark in the hollow of the shoulder, the implication being that a single soul has transmigrated across the centuries. Indeed, some of the characters themselves believe so: "in his loonsome old age [Zach'ry] b'liefed Meryonym the Prescient was his presh b'loved Sonmi" (309), and Luisa finds the music of Frobisher's virtually unknown masterpiece, *Cloud Atlas Sextet*, "intimately familiar" (408) and ponders whether "the molecules of Robert Frobisher's hand" are "now swirling in my lungs, in my blood" (436). When evaluating the Luisa Rey manuscript, Cavendish sniffily dismisses the reincarnation motif as "[f]ar too hippie-druggy—new age" (357), Mitchell thus playfully undercutting the motif even as he reinforces it by having Cavendish subsequently comment on his own comet-shaped birthmark.[8] But the reader need not subscribe to this "new age" metaphysics in order to get Mitchell's message. In effect, what he suggests is that, in counterbalance to the eternal recurrence of predatory behavior, is the eternal recurrence of Ewing's "good man."

---

[7] In the 1973 film *Soylent Green*, the protagonist discovers that the people's bodies are recycled into a synthetic food substance. The film's final line is "Soylent Green is people!" *Soylent Green*, Dir. Richard Fleischer, Perf. Charlton Heston (MGM), 1973.

[8] A disconcerting aspect of *Cloud Atlas* is that Luisa Rey's story is presented as a fiction, penned by one Hilary V. Hush. Of course, all of the narratives in the text are fiction, but, with fiction in general, we tend to suspend our disbelief during the reading process. By pointedly presenting Luisa's story as a fiction within the fiction, Mitchell undermines our suspension of disbelief with regard to not only Luisa's story but the others as well, for, if fictional (within the fiction) Luisa Rey refers to Robert Frobisher, must not he be fictional (within the fiction) as well? Such metanarrative games may be within keeping of Mitchell's point about the potential of fiction to effect change even if it is simply fiction.

3. The Book of the Long Now

But what entails the "good man"? Mitchell suggests it has to do with one's relation to time, with what I would call an embrace of the "Long Now"—"thinking, understanding, and acting responsibly over long periods of time," in the words of Stewart Brand, one of the originators of the Clock of the Long Now project.[9] Spanning centuries, *Cloud Atlas* encapsulates a Long Now, providing a vision of the future linked to the past and present, which is intended to make clear the importance, the necessity, of acting responsibly.

Regular clocks figure significantly in several key instances in *Cloud Atlas*, their accuracy being an important issue. Adam Ewing, for example, is preoccupied with making sure he has the correct time; he notices that "a somnambulant Grandfather clock" is "at odds with my own pocket watch by a margin of hours," asserting that "one valued import from New Zealand is the accurate time" (9). Later, as he recounts a climbing adventure in his diary, he says he cannot tell when he reached the summit, "for I neglected to wind my pocket watch last night." Adam's concern with the correct time, even in the midst of a seeming tropical paradise, is perhaps not surprising. His story takes place during the middle of the nineteenth century, when the standardization of time became a necessary adjunct to industry and travel—indeed, a necessary adjunct to what was regarded as progress. Later, Timothy Cavendish will note that "clocks in disagreement are worse than no clock at all" (171).

In Zachry's tale, the implications of no clock at all are made clear. Whereas Adam resides in a world that increasingly depends on the clock, Zach'ry resides in one where the clock no longer has any real significance beyond its function as a sort of museum piece in the "school'ry," where an Abbess attempts to instill some modicum of learning into her charges. The clock is "the only workin' clock in the Valleys an' in hole Big I, Hole Ha-Why." For Zachry, it is "[t]the greatest of 'mazements," but its language is lost to him: "Abbess'd teached us Clock Tongue but I'd forget it, 'cept for *O'Clock* an' *Half Past.*" Zach'ry does remember, however, why Clock Tongue mattered:

---

[9] The term comes from Brand's book *The Clock of the Long Now: Time and Responsibility* (Basic Books, 1999) and the phrase from p. 4.

> I mem'ry Abbess sayin', *Civ'lize needs time, an' if we let this clock die, time'll die too, an' then how can we bring back the Civ'lize Days as it was b'fore the Fall?* (247; Mitchell's italics)

If we take the conclusion of Zach'ry's story as the conclusion of the novel, seemingly one cannot bring back the "Civ'lize Days," for, when Zach'ry and Meronym return to the Valley after the vicious Kona have been there, they discover that "the school'ry was razed, yay, jus' a black shell now, the last books an' the last clock" (298). The image takes on particular poignancy when we consider it in light of Adam's pride in having the accurate time during his own sojourn in the South Pacific.

The destruction of the clock serves as a synechdoche for the destruction of civilization. Prescience Isle, the last repository of knowledge, is probably doomed, for Meronym's last communication with it indicates that it is threatened by "a terrorsome sick what our Smart can't cure" (295). And, in the tale's final passage, both Meronym and Zach'ry have died, and Zach'ry's son, referring to the "orison" containing what is a hologram of Sonmi during her interview, comments that "the beautsome ghost girl…speaks in an Old-Un tongue what no one alive und'stands nor never will" (309). With the end of Zach'ry's story, history has apparently come to an end.

Again, however, Zach'ry's story is not the end of the book. Although not actually appearing in the novel, the clock that may be most relevant to *Cloud Atlas* is the Clock of the Long Now. This clock, the brainchild of the Long Now Foundation, is described in Brand's eponymously titled book. The Foundation envisions a clock that "is both mechanism and myth": "a large (think Stonehenge) mechanical clock" that "ticks once a year, bongs once a century, and the cuckoo comes out every millennium."[10] Such a clock would be designed to last 10,000 years, a span of years into the future equivalent to the 10,000-year-span into the past that marks the beginning of civilization.[11] As Brand points out, what such a clock would do is "embody deep time for people"[12]—and what seems to be important is not so much whether the Stonehenge-sized Clock is built somewhere in the high desert of the Southwest but

[10] Ibid., 3. The website for the Foundation of the Long Now is as follows: http://www.longnow.org/.

[11] Ibid., 4.

[12] Ibid., 3.

whether the very thought of it can compel us to think in terms of a real future, a future wherein our great-great-great-great-great-grandchildren might flourish. As Michael Chabon explains in an incisive discussion of the Clock, we have lost our "belief or interest in the Future, which has in turn produced a collective cultural failure to imagine that future, any Future, beyond the rim of a couple of centuries"; the Clock, he suggests, can brings us back to future-thinking:

> The point of the Clock of the Long Now is not to measure out the passage, into their unknown future, of the race of creatures that built it. The point of the Clock is to revive and restore the whole idea of the Future, to get us thinking about the Future again...to reintroduce the notion that we don't just bequeath the future.... We also, in the very broadest sense of the first person plural pronoun, inherit it.[13]

Although we never know how many centuries into the future Zach'ry's story takes place, through it, *Cloud Atlas* offers its own harrowing glimpse into the Long Now.

This glimpse compels us to consider what sort of future we bequeath and inherit. In his interview for *The Agony Column*, Mitchell claims, "Writing can bend time." Zach'ry notes that the Abbess would tell her charges that "pretendin' can bend bein'" (283). Through bending time in his fictional creation, Mitchell hopes to bend being—to bend our own thinking so that we might act in terms of the Long Now—"thinking, understanding, and acting responsibly over long periods of time."

From the get-go, the novel makes clear that a lack of future-thinking has devastating effects. Before the British and the Maori overran the Morioris' Rehoku, it indeed seemed a sort of earthly paradise. As Adam Ewing recounts,

> In their virgin state, the Moriori were foragers, picking up *paua* shellfish, diving for crayfish, plundering bird eggs, spearing seals, gathering kelp & digging for grubs & roots. (11)

Along with this idyllic hunter-gatherer existence, the Moriori eschewed bloodshed:

---

[13] Michael Chabon, "The Omega Glory," http://www.longnow.org/press/articles/Michael_Chabon_-_The_Omega_Glory.pdf, first published in *Details* (Jan. 2006).

> the Moriori's priestly caste dictated that whosoever split a man's blood killed his own *mana*—his honor, his worth, his standing & his soul. No Moriori would shelter, feed, converse with, or even *see* the persona non grata. (12; Mitchell's italics)

But a succession of snakes lead to a Fall: sealers make "the surf pink with seals' blood," virtually eliminating the species (12); bushfires used to clear the land "smolder beneath the peat for many seasons, surfacing in dry spells to sow renewed calamity" (13); the cats and rats brought by whalers destroy "the burrow-nesting birds whose eggs the Moriori so valued for sustenance" (13); Western diseases decimate the population; and, finally, the Maori, facilitated by the British, engage in whole-sale slaughter of their possibly racial brothers—Cain effectually wiping out Abel.

The ecological destruction, disease, and savage conquest that beset Rehoku are based on historical accounts.[14] In his dire vision of the future, Mitchell implies that what happened in the microcosm of Rehoku could happen on the macrocosm of the entire planet. The defective Hydra reactor in Luisa Rey's tale threatens to blow the California coast to "Kingdom Come" (100). In Sonmi's corpocracy, much of the terrain has become "deadlands so infected or radioactive that purebloods perish there like bacteria in bleach" (206). In Zach'ry's tale, Ha-Why is one of the last spots where, as Meronym says, they might find "*good earth to plant more Civ'lize*" (295); but, even here, people die by age fifty of "mukelung," and "freak birth" babies are born without eyes and noses. Too, the Kona slaughter and enslave those who would attempt to live in community. Zach'ry's world may be regarded as the logical outcome of what each chronologically earlier tale puts forward.

Indeed, what these chronologically earlier tales recount is that the very things that signify progress may also signify destruction. In his book *The Weather Makers*, Tim Flannery remarks,

[14] Mitchell's account of the destruction of Rehoku has its basis in fact. See Jared Diamond, "A Natural Experiment of History," chapter 2 of *Guns, Germs, and Steel* (New York: W. W. Norton, 1997). Mitchell cites his indebtedness to Diamond's book in the "Book World" interview."

> Climate change is difficult for people to evaluate dispassionately because it entails deep political and industrial implications, and because it arises from the core processes of our civilization's success.[15]

In *Collapse*, Jared Diamond similarly equates technological advances with destructive consequences:

> All of our existing problems are unintended negative consequences of our existing technology. The rapid advances in technology during the 20th century have been creating difficult new problems faster than they can solve old problems: that's why we're in the situation in which we now find ourselves.[16]

As Meronym tells Zach'ry, "*human hunger birthed the Civ'lize, but human hunger killed it too*" (273), civilization bringing about its own fall:

> *Old Uns' Smart mastered sicks, miles, seeds, an' made miracles ord'nary, but it din't master one thing, nay, a hunger in the hearts o' humans, yay, a hunger for more.* (272)

As Flannery, Diamond, and Mitchell/Meronym suggest, it is not progress per se that is at fault but progress that thinks in terms of the now rather than the later.

Each tale recounts the clash between those who desire immediate gratification and those who are future-thinking. Meronym explains to Zach'ry that it is not laws that determine civilization but foresight:

> *The savage sat'fies his needs now. He's hungry, he'll eat. He's angry, he'll knuckly. He's swellin', he'll shoot up a woman. His master is his will, an' if his will say-soes "Kill" he'll kill.... Now the Civ'lized got the same needs too, but he sees further. He'll eat half his food now, yay, but plant half so he won't go hungry 'morrow. He's angry, he'll stop'n' think why so he won't get angry next time. He's swellin', well he's got sisses an' daughters what need respectin' so he'll respect his bros' sisses an' daughters. His will is his slave, an' if his will say-soes, "Don't!" he won't, nay.* (303; Mitchell's italics)

In fact, Mitchell reinforces Meronym's message through the very structure of his novel, the suspended conclusions compelling us to delay our need for closure—unless, of course, we skip ahead, in which

[15] Tim Flannery, *The Weather Makers: How Man Is Changing the Climate and What It Means for Life on Earth* (New York: Atlantic Monthly Press, 2005), 4.

[16] Jared Diamond, *Collapse: How Societies Choose to Fall or Succeed* (New York: Viking), 505.

case we will miss the point.[17] Brand makes a similar point about immediate gratification:

> The worst of destructive selfishess is not *Me!* But *Me! Right now!* The generous opposite could be phrased as *All of us for all of time.*[18]

Civilization entails replacing the will to power with the will to preserve if there is to be a future at all.

There may be no future according to Zach'ry's tale. Even in this bleak narrative, however, a "flea o' hope," in Zach'ry's words, is put forward by Meronym, who forecasts a possible remaking of civilization:

> *Some savages what I knowed got a beautsome Civ'lized heart beatin' in their ribs. Maybe some Kona. Not 'nuff to say-so their hole tribe, but who knows one day? One day.* (303; Mitchell's italics)

More important, the boomerang structure entails that, rather than concluding conclusively on a bleak note, Mitchell holds out more than a flea of hope. As we move back through the earlier stories, we see flickers of hope at the ending of each. Sonmi, although facing execution, has written her Declarations, and, as she tells her interviewer, her "ideas have been reproduced a billionfold" (349). After his daring escape, Timothy Cavendish has discovered that, although "[m]iddle age is flown... it is attitude, not years, that condemns one to the ranks of the Undead, or else proffers salvation" (387). Luisa Rey's exposé of Seaboard has been printed, presumably leading to the dismantling of the HYDRA nuclear reactor—one potential source of deadlands eliminated, at least temporarily. Robert Frobisher prepares to kill himself, but he has left as his legacy the *Cloud Atlas Sextet*, and, as he writes his friend and lover,

> One writes music because the winter is eternal and because, if one didn't, the wolves and blizzards would be at one's throat all the sooner. (82)

And, finally, Adam Ewing resolves to work in service of what is essentially the Long Now:

> A life spent shaping the world I *want* Jackson [his son] to inherit, not one I *fear* Jackson shall inherit, this strikes me as a life worth the living. Upon my return to San Francisco, I shall pledge myself to the

---

[17] I am indebted to Sabine Gross for this insight.
[18] *Long Now*, 9; Brand's emphasis.

> Abolitionist cause, because I owe my life to a self-freed slave and because I must begin somewhere. (508; Mitchell's italics)[19]

Adam, I must point out, has the last word.

By giving Adam the last word, by boomeranging back from terrible future to malleable past/present, Mitchell suggests that we may be able to avoid the fate of Zach'ry and his world. Let us return to Isaac Sach's theory of time. In the virtual times created in the text, when we reach the second part of Adam Ewing's story, the chronological future that Zachry's tale illustrates is in the novel's past—that is, it is a prior reading experience, *preceding* the chronological past of Adam's resumed story. From the perspective of the reader, when we are reading Zachry's tale, Adam's resumed story is the novel's future. When Robert Frobisher discovers the second part of Adam's diary, he remarks, "Happy, dying Ewing, who never saw the unspeakable forms waiting round history's corner" (460). But we readers *did* see them, and the text remembers them, providing a clear warning. That vision of the future influences the past/present in which Adam resides. Indeed, if we accept Roland Barthes's argument that "the mainspring of narrative is precisely the confusion of consecution and consequence, what comes *after* being read in narrative as what is *caused by*," we cannot help but regard Adam's pledge to work for the public good as the logical outcome of all that has come before in *Cloud Atlas*.[20] And this pledge may ensure that the "virtual future" and its "unspeakable forms" may be averted.

At one point, a young neighbor asks Luisa Rey, "Can you change the future or not?" (401). Thinking to herself, "*Maybe the answer is not a function of metaphysics but one, simply, of power*," Luisa responds, "It's a great imponderable" (401). This question would seem to be Mitchell's as well. Although he cannot ultimately answer it, he implies that the Adam Ewings, the Luisa Reys, the Sonmis, and even the Timothy Cavendishes of this world may help shape it in a way that we would want our children to inherit. Adam envisions his father-in-law's scornful response to his newfound determination to shape the world:

---

[19] In his interview with *The Washington Post*, Mitchell echoes Adam's sentiments. When asked what distinguishes *Cloud Atlas* from his other novels, he replies, "It has more of a conscience," and then adds, "I think this is because I am now a dad. I need the world to last another century and a half, not just see me to happy old age." See "Book World Talks With David Mitchell."

[20] Roland Barthes, "Introduction to the Structural Analysis of Narratives," in *A Barthes Reader*, ed. Susan Sontag (New York: Hill and Wang, 1982), 266.

> Naïve, dreaming Adam. He who would do battle with the many-headed hydra of human nature must pay a world of pain & his family must pay it along with him! & only as you gasp your dying breath shall you understand your life amounted to no more than one drop in a limitless ocean! (508–09)

Adam's very last words (the very last words of the novel itself) stirringly recontextualize the metaphor: "Yet what is any ocean but a multitude of drops?" (509). A sufficient multitude of drops may effect change.[21]

References

Barthes, Roland. Barthes, "Introduction to the Structural Analysis of Narratives." In *A Barthes Reader*, ed. Susan Sontag. 251–95. New York: Hill and Wang, 1982.

Brand, Stewart. *The Clock of the Long Now: Time and Responsibility*. Basic Books, 1999.

Chabon, Michael. "The Omega Glory," http://www.longnow.org/press/articles/Michael_Chabon_-_The_Omega_Glory.pdf. First published in *Details* (Jan. 2006).

Diamond, Jared. *Collapse: How Societies Choose to Fail or Succeed*. New York: Viking, 2005.

———. "A Natural Experiment of History." Chap. 2 of *Guns, Germs, and Steel*. New York: W. W. Norton, 1997.

Flannery, Tim. *The Weather Makers: How Man Is Changing the Climate and What It Means for Life on Earth*. New York: Atlantic Monthly Press, 2005.

Mitchell, David. *Cloud Atlas*. New York: Random House, 2004.

———. *Ghostwritten*. New York: Vintage, 2001.

———. "Q&A: Book World Talks With David Mitchell." *washingtonpost.com*, 22 August 2004. http://www.washingtonpost.com/up/dyn/articles/A17231-2004 Aug19.html (21 May 2006).

———. "Silver Dagger and Russian Dolls—David Mitchell, author of *Cloud Atlas*, in interview." By Shane Barry. *Three Monkeys Online*, March 2005. <http://threemonkeysonline.com/threemon_article_david_mitchell_cloud_atlas_interview.htm> (21 May 2006).

Soylent Green. Dir. Richard Fleischer. Perf. Charlton Heston. MGM, 1973.

[21] I wish to thank my colleague Jason Mezey for stimulating conversations about *Cloud Atlas*. His unpublished manuscript "'A Multitude of Drops': David Mitchell's *Cloud Atlas* and the Ethics of Globalization" provides an insightful discussion of the novel's critique of globalization. I also wish to thank the students in my graduate "Narrative and Time" course, whose enthusiasm for and insights about *Cloud Atlas* spurred my own.

CHAPTER NINE

CLOSURE AND "COLORED PEOPLE'S TIME"

John Streamas

Summary

In social settings "Colored People's Time," or CPT, is an in-joke among people of color. I argue that it is also the time-sense of what Du Bois called "double consciousness," the twinned perspectives that people of color learn to adopt in relations with whites. CPT serves as a vehicle for visions of justice, and in narratives by people of color it serves as what Gary Saul Morson calls "sideshadowing," a parallel imagining of what might have happened. This function is most apparent in endings of stories. Whereas Western apocalypse narratives, for example, aspire to an end-time as a reward for individual virtue, for people of color narratives of freedom and justice aspire to realization of the dream of Martin Luther King, Jr. CPT is, then, the time-sense of the struggle, measuring both story-time and storytellers' time. The dream will come true when sideshadowed narratives merge with living history, making CPT no longer necessary.

A Samoan college student, arriving late for graduation, explains that he is on "Samoan time." A Native American poet writes that, because he is on "Indian time," he will be late for his own funeral.[1] These claims are more like jokes than apologies. Geneva Smitherman defines "African People's Time," however, in all seriousness:

> Reference to the African American concept of time. Being in tune with human events, nature, seasons, natural rhythms, not a slave to the artificial time of the man-made clock. Being "in time," in tune with emotions, feelings, the general flow of things, is more critical than being "on time." The challenge continues to be how to reconcile an "in time" philosophy with the "on time" demands of mainstream America.[2]

The more general term for "Samoan time," "Indian time," and "African People's Time" is "colored people's time" or CPT. In general social

---

[1] See the title poem in Luke Warm Water's book *On Indian Time* (Rapid City, SD: Trickster Arachnid, 2005), 86. The Samoan student made his remark at the graduation ceremony of Asian American/Pacific Islander students of Washington State University in May 2006.

[2] Geneva Smitherman, *Word from the Mother: Language and African Americans* (New York: Routledge, 2006), 21.

Jo Alyson Parker, Paul A. Harris and Christian Steineck (Eds), Time: Limits and Constraints, pp. 219–235

settings, CPT becomes, as the student and the poet demonstrate, a self-effacing joke on late arrivals or slow and languid performance. But it is also, as Smitherman importantly explains, a politics and a philosophy, a way of living socially "in time." Elsewhere, Smitherman defines the phrase "what time it is" as "[t]he real deal, the truth, what's actually occurring at the moment, reference to political or psychological 'time.'"[3] More than a century earlier, W. E. B. Du Bois wrote of the emergence of a "double consciousness" among African Americans, a knowledge that one is not only looking but is also being looked at. Double consciousness provides a frame for understanding and communicating that is more than a mere coping and survival strategy but that is also a site in which to imagine and even work toward liberation from racism.[4] Yet CPT existed long before Du Bois advanced his theory, and so double consciousness creates a frame for its own emergence in a "double time." That is, CPT may well be, for all racially marginalized peoples, the time-dimension of that double consciousness.

I propose to examine the relationship between CPT and narrative closure, to determine whether the former, often simplified as a circular sense of time or as a sardonic rationalization for lateness, frustrates or even thwarts the latter. Because Western European narrative theory is mostly cloistered in the formalisms of story-grammars and thus practices a colorblind leveling that denies the possibility of a strategic double consciousness to texts coming from the racial margins, most critics have not examined narratives as racialized, which means that few have examined narratives for double consciousness and fewer still for CPT.[5] My work is therefore mostly speculative and provisional,

[3] Smitherman, 47.

[4] Actually, Du Bois, writing in 1903 in *The Souls of Black Folk*, defines double consciousness as a "sense of always looking at one's self through the eyes of others, of measuring one's soul by the tape of a world that looks on in amused contempt and pity" (1903; Millwood, NY: Kraus-Thomson, 1973), 3. If his definition seems radically gloomier than mine, I would add that, in his very next paragraph, he writes that the African American, knowing "that Negro blood has a message for the world," wants "to be both a Negro and an American, without being cursed and spit upon by his fellows, without having the doors of Opportunity closed roughly in his face" (4). I argue that social justice can be expressed in the language of double consciousness. Smitherman quotes Du Bois and adds that mainstream pressures to assimilate create a "linguistic push-pull," on the positive side of which are "Black folk loving, embracing, using Black Talk" (6).

[5] Certainly in the last four decades scholars of color have inscribed race on the critical menu. Critics such as Lisa Lowe, Paul Gilroy, E. San Juan, Jr, and Elizabeth Cook-Lynn are known to scholars of contemporary and "ethnic" criticism, but they

applied in very broad strokes, and aimed at suggesting new paths for a socially engaged literary and cultural criticism.

It is also geographically restricted, as CPT has been studied so far mostly within the United States, and, as I will focus on U.S. and English-language examples, I would suggest, however, that alternative time consciousnesses are necessarily and inevitably common to the politics and the narratives of racially oppressed peoples everywhere. This is an important point, for oppressed peoples can achieve justice together only after recognizing the common features of their narratives of oppression. Moreover, dominant cultures have imposed upon them impossible and coercive standards based upon their assimilating and their "forgetting" of their old home cultures. Abdul R. JanMohamed and David Lloyd argue that oppressed peoples must reject assimilation and refuse their oppressors' "assumption of the timeless universality of cultural products and of the concomitant tendency to read cultural texts exclusively for their representation of 'aesthetic' effects and 'essential' human values."[6] After all, an assumption of universality is what gives dominant cultures the luxury of assuming "timelessness" in their values. CPT rejects Eurocentric universalities by insisting on very real alternative time consciousnesses born of oppression and resistance. "Coerced into a negative, generic subject-position," write JanMohamed and Lloyd, "the oppressed individual responds by transforming that position into a positive, collective one."[7] Racism within the United States differs in many ways from racisms elsewhere, many of which take the form of repressive colonizations, and yet this trans-

are not among the lions of the grand critical tradition. The very intensity of the "culture wars" represented a massive push by conservatives to keep such critics outside the mainstream, and the extent of their success may be measured in the fact that, even today, specialists in, say, Chaucer may earn their advanced degrees without ever reading works by novelists, poets, or scholars of color. Furthermore, as probably most literary critics of color would argue, most college courses in American literature assume an Americanness that assimilates writers such as James Baldwin and Maxine Hong Kingston while keeping them in an "ethnic" niche. Finally, narrative criticism in the U.S. discusses race but not racism. Race fits easily into discursive and structural categories, while racism eludes apprehension. Not surprisingly, scholars of color are seldom found in the social and professional circles of narrative critics. Perhaps most relevant to this study, I have seen absolutely no discussion of CPT in mainstream literary criticism.

[6] Abdul R. JanMohamed and David Lloyd, "Toward a Theory of Minority Discourse: What Is To Be Done?," in JanMohamed and Lloyd, eds., *The Nature and Context of Minority Discourse* (New York: Oxford University Press, 1990), 10.

[7] JanMohamed and Lloyd, 10.

formation into a collective subject-position suggests that some form of CPT is common to the resistant narratives of racially oppressed peoples. To speak of a "form" or "structure" of CPT is, however, to risk undermining its value. It is both ambient and contingent, both a tool and a blanket, both responsive and assertive, and sometimes all of these, as different situations, communities, and narratives see fit.

Because narratives are delivered in language, I wish to cite another of Smitherman's examples of racialized linguistic temporality from AAL, or African American Language. After citing Toni Morrison's remark about the "five different present tenses" that black children bring to school, Smitherman quotes H. Samy Alim's observations that in AAL the verb *be* indicates that "a speaker attempts to speak an ultimate truth (and often emphatically in relation to one's identity)," and that *be* "equates the subject and predicate nominal in such a way as to reveal what the speaker believes to be a 'realer than real' state of affairs."[8] But this is an evolutionary development, for, in traditional AAL, "*be* as a full verb form conveys the message that a reality is continuing, intermittent, or recurring."[9] Smitherman is not speaking here of conjugated forms of the verb but of the word *be* itself, a word that in mainstream English links social, material, and temporal states but that in AAL contains at least three of the "five different present tenses" that Morrison hears. The very word *be* models an idea of being "in time."

Two preliminary points beg to be stressed. First, CPT is, as Smitherman notes, political and philosophical. It is exclusively neither a fad nor a reaction. It emerges from an ongoing, indefinite condition of oppression. It emerges, in one form or another, across different communities suffering different kinds of racism in different places. While it is more than a coping strategy, it is *also* a coping strategy, so that I would not argue, as many postmodernist critics would, that it is always and necessarily actively resistant. Sometimes the only available resistance is passive. After all, CPT is, at least, a source of jokes inside communities of color.

Mark M. Smith traces CPT to an emergence among slaves on Southern plantations:

> African Americans can adjust to white time sensibilities, which stress punctuality and are future oriented, but they can also reject these same

---

[8] Smitherman, 8, 103.
[9] Smitherman, 102.

> sensibilities as a form of protest against democratic capitalism generally, white bourgeois sensibilities specifically, by eschewing the authority of the clock and adopting presentist and naturally defined notions of time, a tendency that sociologists and the public alike have come to call Colored Peoples' Time, or CPT.... CPT is a useful shorthand to describe how African Americans as a class of laborers resisted planter-defined time during and after slavery.[10]

This history rejects many sociologists' notion that CPT is merely cultural. In the 1960s John Horton, in an essay titled "Time and Cool People," claimed that on the streets of black neighborhoods "watches have a special and specific meaning. Watches are for pawning and not for telling time. When they are worn, they are decorations and ornaments of status."[11] Horton's claim that CPT is known only to middle-class blacks, not to the watch-wearing status seekers, ignores the fact that CPT was invented *because* its inventors were oppressed. Not surprisingly, then, Smitherman hears CPT in hip hop culture.[12] As if to underscore its source, anthropologist Johannes Fabian discusses "temporal relations" between dominant classes and their others: "What makes the savage significant to the evolutionist's Time is that he lives in another Time."[13] And sociologist Pierre Bourdieu analyzes temporal hierarchies in his work on cultural production, so that what he says about artistic movements applies to social inequalities and hierarchies: "To introduce difference is to produce time."[14]

My second preliminary point is that CPT, despite its history and its current usage, is not essentialist. Smitherman's claim for AAL applies usefully to CPT: she argues that "race does not determine what language a child will speak, there is no such thing as a 'racial language,' and no race or ethnic group is born with a particular language," but she adds,

[10] Mark M. Smith, *Mastered by the Clock: Time, Slavery, and Freedom in the American South* (Chapel Hill, NC: University of North Carolina Press), 130.

[11] John Horton, "Time and Cool People," in *Soul*, ed. Lee Rainwater (New Brunswick, NJ: Trans-action Books, 1970), 43.

[12] See Smitherman's discussion of "Black Hip Hop linguistics" on 102, 103.

[13] Johannes Fabian, *Time and the Other: How Anthropology Makes Its Object* (1983; New York: Columbia University Press, 2002), 27.

[14] Pierre Bourdieu, *The Field of Cultural Production* (New York: Columbia University Press, 1993), 106.

> The roots of African American speech lie in the counter language, the resistance discourse, that was created as a communication system unintelligible to speakers of the dominant master class.[15]

I would add to this important claim only that CPT, though it might have emerged on plantations, exists among all marginalized racial communities, as I argued earlier, so that, while in their details "Indian time" may differ from "Samoan time" and from "African People's Time," still David Parker's conclusion about workers in Chinese takeaway restaurants in Britain echoes Smith's conclusion about black slaves:

> The temporal discipline of ceding time through labour to meet the consumption demands of others has been a defining experience for the families owning Chinese takeaways.[16]

Despite differences in its various particular manifestations, racism still generally functions as the exploitation of communities of color. And all varieties of CPT, regardless of their manner of expression or the extent of their active resistance, stake a political and philosophical claim against racism merely by creating a resistant "insider" time-sense.[17]

If CPT is a function of a strategic double consciousness, then, the narratives that come from peoples of color should reflect an additional, different time-sense. And this difference might manifest itself most conspicuously in endings. While my purpose here is to make that

---

[15] Smitherman, 5, 3.

[16] David Parker, "The Chinese Takeaway and the Diasporic Habitus: Space, Time, and Power Geometries," in *Un/settled Multiculturalisms: Diasporas, Entanglements, Transruptions*, ed. Barnor Hesse (London: Zed Books, 2000), 90.

[17] By defining racism largely in terms of exploitation, my argument is vulnerable to an assumption, frequently voiced by students in low-level Ethnic Studies classes, that race may be subsumed by class, or at least that more attention must be paid to the everyday experiences that people of color share with working-class and impoverished whites. Most activists would agree that racism and class oppression are functions of each other—except for the residual issue of race. The wealthiest black man is less likely to get a taxi than a middle-class white man, and, when convicted of the same crimes, blacks and Latinos are almost certain to get much harsher sentences. But, complaining of "whining" people of color, many whites insist that, because they enslave no one and participate in no genocides (and now too because the U.S. has elected a biracial black president), then studies of racism must be subordinated to studies of oppressions that, being more widely dispersed, equally affect whites. But even class oppression affects people of color more deeply than it affects whites, as reflected in a racialized maldistribution of wealth and institutional power. The subject of this argument is "colored people's time," after all, not "oppressed people's time."

claim rather than apply it to particular narratives, I will suggest three African American writers whose fictions may recommend themselves to an examination of time differentials. Toni Morrison's novels riff on history, but I would argue that Morrison constructs a particularly black history rather than a black unit of standard U.S. history, refusing to take readers of, say, *Beloved* back to the age of slavery but rather moving that age forward, bringing it to the present and confronting readers with it.[18] The lyrical ending of James Baldwin's short story "Sonny's Blues" superimposes on a Joycean epiphany the time-sense of musical—and racial—improvisation.[19] And the recent short fiction of Amiri Baraka collapses and explodes time, teasing readers into a notion that Thelonious Monk might actually still be walking the city streets.[20] Smitherman's claim that AAL creates a system that may be "unintelligible" to the "master class" may transfer to a suggestion that CPT, to the extent that it informs the works of these writers of color, renders transparent only the "clock time" of their narratives and makes itself available only to readers sensitive to its differential time.

According to traditional social sciences, primitive peoples—that is, dark-skinned peoples—mark time by the seasons. Having no need of technologies such as clocks that measure linear progression, they live by the migrations of animals and the changing of seasons. Such a circular time-sense frustrates endings and closure. I would argue that this time-sense, to the extent that it even still exists among peoples of color, is not CPT. I would also argue that John Horton is wrong to ascribe CPT to the decision by black men on the street to wear watches only as status symbols. For CPT simultaneously resists racism and refuses to answer to it. It can be flexible, improvisational. A woman of color may be on CPT whether she is five minutes late or twenty minutes late to a meeting. She may be on CPT even if she arrives on time, though she will probably not say so, for the joke would be lost.

[18] Morrison insists on the relevance of slavery to the present. For most people of color, this is agreeable, but it remains vexing for whites who say, "We got rid of slavery, and so racism is not so serious anymore"—a frequent claim made by students in introductory Ethnic Studies classes. My point is that, whereas for whites slavery exists only in the past, for people of color it is still present, and will remain an urgent issue at least until racism is eradicated.

[19] For race's intersections with music and rhythm, see the work of Fred Moten and Michael Golston.

[20] Morrison's and Baldwin's fictions are widely available. For Baraka's recent stories, see his *Tales of the Out & the Gone* (New York: Akashic Books, 2006).

More important, CPT envisions endings. The very point of resistance, whether active or passive, is to achieve justice. Indigenous peoples envision their restoration to the land. Refugees envision a homeland. Slaves envision their liberation. Implicit in all varieties of CPT, then, no matter how silly each instance of it might be, is the satisfactory resolution of a story of injustice. When Martin Luther King, Jr, repeated the exultation "Free at last," he saw freedom and justice as the end-point of the struggle, coming literally "at last."

In important ways, then, CPT's sense of an ending differs from European senses of apocalypse. Traditional apocalypse narratives construct the ultimate triumph of the good as immediately following, even as commensurate with, a near-triumph of evil.[21] These stories say that, no matter how terrible the present situation may be, the future will bring a clash of good and evil and that, though good will surely prevail, evil will wage a fierce fight. Frank Kermode wrote his classic study *The Sense of an Ending* in the 1960s, during the most intense time in the Civil Rights movement, though he wrote from within a British critical mainstream that regarded the Christian Bible as an objective touchstone for literary commentary. For this reason his work recommends itself to a comparison of the Euro-American sense of apocalyptic endings to CPT's very different sense. Kermode concedes that "the End itself, in modern literary plotting loses its downbeat, tonic-and-dominant finality," but still "we think in terms of crisis"[22]—the very kind of crisis that characterizes the near-triumph of evil in traditional narratives. And

> we still feel a need, harder than ever to satisfy because of an accumulated scepticism, to experience that concordance of beginning, middle, and end which is the essence of our explanatory fictions, and especially when they belong to cultural traditions which treat historical time as primarily rectilinear rather than cyclic.[23]

The fact that high school teachers of literature and reviewers of popular films still discuss novels and movies largely in terms of their formal features attests to the abiding temporal structures that contemporary culture attaches to all narratives, even those of history and experience.

---

[21] See chapters 2 and 3 in Frank Kermode, *The Sense of an Ending* (1967; New York: Oxford University Press, 2000).

[22] Kermode, 30.

[23] Kermode, 35–36.

Kermode explains the basic temporal structure of narratives in terms of a *tick* that denotes beginnings and a *tock* that denotes endings. Fictions emerge in the interval between the sounds:

> The clock's *tick-tock* I take to be a model of what we call a plot, an organization that humanizes time by giving it form; and the interval between *tock* and *tick* represents purely successive, disorganized time of the sort that we need to humanize.[24]

What Kermode fails to say, though he hints at it, is that a desire to explain the interval between one *tick-tock* narrative and another is a desire not to create a metanarrative but to contextualize, perhaps even to understand one narrative in terms of another in such a way that demarcations between beginnings and endings may blur, as they do in the modernist fictions that, he says, defy traditional temporal structure. Kermode argues that writers reject simple chronicity as "humanly uninteresting successiveness" and fill the interval with a "significant season:"

> It has to be, on a scale much greater than that which concerns the psychologists, an instance of what they call "temporal integration"—our way of bundling together perception of the present, memory of the past, and expectation of the future, in a common organization. Within this organization that which was conceived of as simply successive becomes charged with past and future.[25]

I would like to argue that what Kermode refers to as "our way of bundling together perception of the present, memory of the past, and expectation of the future" does not result in the "common organization" he prescribes, that historical racism has led not to "temporal integration" but to temporal segregation of narrative communities, and that, to the extent that narrative consciousness helps frame historical experience, CPT is one outcome of this segregation. The endings of apocalypse narratives, then, differ from the freedom-and-justice narratives of peoples of color because Western end-time is not on CPT. That is, for peoples of color the near-triumph of evil has already happened and continues to happen. To be sure, it evolves. Slavery and Indian massacres nominally ended in the U.S. But these endings did not result in the paradise that apocalypse narratives predict. Racism

[24] Kermode, 45.
[25] Kermode, 46.

finds other sites and strategies, components of the evil that continues to happen.

Kermode does clear a possible space for CPT. He identifies phases of cultural change, the first two of which are a desire to learn from the past and a desire to "find patterns in historical time."[26] As a result of the second phase, history becomes "a fictive substitute for authority and tradition, a maker of concords between past, present, and future, a provider of significance to mere chronicity."[27] I would argue that history is, comparatively, a luxury for those who create racism and a necessity—not "a fictive substitute for authority and tradition" but an expression of authority, of subjecthood—for those who suffer it. Thus the sense-making narratives that Kermode refers to as "concord fictions" differ between communities. And the desire in modern fiction to deny the appearance of concord in favor of skepticism represents a move—surely an unwitting one—toward the historical messiness of freedom-and-justice narratives. By contrast, those who create racism enjoy the luxury of filling the interval between *tick* and *tock* according to their means, which is to say that they exercise some control over their own endings.

But, if the near-triumph of evil has already happened and continues to happen to peoples of color, then is the longed-for ending of freedom and justice a mere afterthought and anticlimax? Is it as uninteresting to peoples of color as the idea of paradise is boring to the dominant culture? After all, fictions created by writers of color construct the historical experience of racism but not the dream that Martin Luther King, Jr, dreamed. A traditional explanation would merely claim that conflict is inherent in plot, and that endings, whether happy or tragic, build upon tensions in middles of stories. Some psychoanalytic critics even claim—in a variant on the familiar "the journey is more important than the destination" thesis—that the narrative experience resembles the sexual experience, and that audiences rush through climaxes so that they might relieve, and relive, the building up of tensions.[28] But narratives from the racial margins, because they emerge from the time-sense of CPT, depend not only on internal story-tensions but also on the external tension that historical racism imposes on their

[26] Kermode, 56.

[27] Kermode, 56.

[28] See, for example, Peter Brook, *Reading for the Plot: Design and Intention in Narrative* (New York: Knopf, 1984).

writers and tellers. That is, CPT, as the double consciousness of time, is simultaneously entrapped by and liberated from Western time. It is the time-sense of the struggle. And it may be discarded as unnecessary when the dream comes true, as most peoples of color believe it will.

Until then, the location of the struggle needs to be clarified. My claim that CPT exists both within and outside stories, that it measures story-time and storytellers' time, is not an argument for trendy postmodern border-crossing. Nor is it a plea for understanding CPT through metanarratives. For CPT refuses to recognize narrative borders. And, in most criticism, even metanarratives exist in a space bordered by the writer's immediate context. CPT functions as what I will call "deep context," keeping time alike for the writer, the story, and the consciousness of an oppressed people. It is perhaps in recognition of this function of CPT that recent works by writers such as Jamaica Kincaid, Amy Tan, and Amiri Baraka willfully blur old generic distinctions between essay and fiction. When Baraka writes about meeting Thelonious Monk on a street only months after attending Monk's funeral, the historical and the fictional narrator are as real as the historical and the fictional Monk.[29] Writers of color, including science fiction writers such as Octavia Butler and Samuel Delaney, bring the historical struggle into their narratives. There are no separate worlds, one fictive and the other historical, that CPT straddles. Rather, CPT keeps time for all narratives of the struggle. A young James Baldwin, in a 1947 review of Chester Himes's *Lonely Crusade*, warned that race relations, telling "a far uglier story and with more sinister implications than have yet found their way into print," had evolved into "something which can be measured in decades and generations and which may spell our doom as a republic and almost certainly implies a cataclysm."[30] And yet Baldwin holds out hope for a waking from this historical nightmare, a release from the struggle and a need for CPT. Perhaps he would agree with Ernesto Cardenal's claims, in the *Cosmic Canticle*, that "time is merely / our inability to see everything at the same time," and that, because "the proletariat is king of history," a Final Judgment will inevitably lead to "the destruction of injustice on

[29] See Baraka, "A Monk Story," in his *Tales of the Out & the Gone*, 206–10.

[30] James Baldwin, "History as Nightmare," in Baldwin, *Collected Essays* (New York: Library of America, 1998), 581. This review originally appeared in *The New Leader* of 25 October 1947.

earth."[31] And because injustice will be destroyed on earth, then justice will necessarily replace it on earth, the site of the struggle, and not in a heavenly realm.

Certainly, however, a dream of heaven still holds mainstream U.S. culture in thrall. An "age of information" has not displaced a faith in an afterlife, and, as Kermode observes, even believers who are skeptical of apocalypse narratives may find in those narratives a hope of a happy ending. The problem is that, in the European tradition, that happy ending is a reward for being good through the near-triumph of evil, for being, that is, a virtuous individual. Happy endings are thus personal redemptions, just as the struggle with evil has been a personal struggle. But this is not the struggle that engages peoples of color. Racism results from public policy, practice, and behavior. The struggle for peoples of color is communal. Traditional apocalypse narratives are possible because they reflect a naturalization of narratives of personal struggle and redemption. Freedom-and-justice narratives, on the other hand, are possible only because of a common belief—though by no means a naturalized belief, as the condition of oppression thwarts affirmative naturalization—in communal action. And CPT is the time-sense of this communal action. Peoples of color cannot afford the luxury of diffuseness, multiplicity, and fragmentation that postmodernism often celebrates, a sense of differentiation that reflects the division of the planet, in the late nineteenth century, into time zones. What could be more fashionably postmodern than a watch that keeps accurate time across time zones, simultaneously recognizing unitary order and difference? But this is not CPT. For peoples of color, having already suffered the evil of racism, the end-time might appropriately begin in justice meted out to those who profited from that racism and end in true communal freedom. On CPT, the revolution will come too late to be televised.

Gary Saul Morson offers a reading of time that may be useful to an understanding of CPT and its relationship to closure. He advocates a literary construction, "a particular conception of open time," that he calls *sideshadowing*.[32] This is not exactly an opposite, nor is it mutually exclusive, of the more familiar foreshadowing, which constructs a

---

[31] Ernesto Cardenal, *Cosmic Canticle* (1989; Willimantic, CT: Curbstone Press, 2002), 472, 477.

[32] Gary Saul Morson, *Narrative and Freedom: The Shadows of Time* (New Haven, CT: Yale University Press, 1994), 5.

world of closed time and involves "backward causation, which means that, in one way or another, the future must already be there, must somehow already exist substantially enough to send signs backward."[33] Foreshadowing creates prolepsis or the nebulous causation that is named in fashionable critics' adverb phrase as "always already." Side-shadowing, however, exists in a world of possibility:

> [S]ideshadowing admits, in addition to actualities and impossibilities, a *middle realm* of real possibilities that could have happened even if they did not. Things could have been different from the way they were, there were real alternatives to the present we know, and the future admits of various paths. By focusing on the middle realm of possibilities, by exploring its relation to actual events, and by attending to the fact that things could have been different, sideshadowing deepens our sense of the openness of time. It has profound implications for our understanding of history and of our own lives while affecting the ways in which we judge our present situation. It also encourages skepticism about our ability to know the future and the wisdom of projecting straight lines from current trends or values.[34]

While foreshadowing is so artificial that it is easy to recognize and thus easy to teach in high school literature classes, recommending itself, says Morson, to beliefs in omens and "laws" of history,[35] sideshadowing risks the limitations of audiences' subjectivities. Peoples of color, lacking access to institutional power and archives, are often accused of making claims lacking documentation and objectivity. For example, in Ethnic Studies classes, I present undergraduates with the "news" that, mere days after the U.S. dropped its atomic bombs on Japan, Langston Hughes and other writers in the black press suggested that the bombs would never have been used on Germans, a white enemy, that the bombs were a race weapon reserved for dark-skinned enemies.[36] While historians may never know the extent to which racism factored into the decision to use the bomb, or even whether it factored at all, the important point is that Hughes and many other thoughtful peoples of color read racism into the decision. Students need to understand why

[33] Morson, 7.
[34] Morson, 6.
[35] Morson, 7.
[36] Carole Doreski, *"Kin in Some Way": Reading Citizenship, Reading Relocation at the* Chicago Defender (manuscript, 1999), 25. For other black journalists' reactions, see Reginald Kearney, *African American Views of the Japanese: Solidarity or Sedition?* (Albany, NY: State University of New York Press, 1998), 122.

this "news" is important. The subjectivity of the racially oppressed, relying on painful historical memory, rejects standard mainstream narratives. Hughes's suggestion is thus not paranoia or "playing the race card" but is either historically true or at least an example of sideshadowing. Morson writes:

> Sideshadowing...calls attention to the ways in which narratives, which often turn earlier presents into mere pasts, tend to create a single line of development.... Alternatives once visible disappear from view and an anachronistic sense of the past surreptitiously infects our understanding. By restoring the presentness of the past and cultivating a sense that something else might have happened, sideshadowing restores some of the presentness that has been lost.[37]

I should note that while, for Morson, sideshadowing is a predominantly literary construct, it can provide for peoples of color glimpses into alternative historical narratives whose temporal settings are on CPT. This ranges far beyond the conceits of novelists and speculative historians in books imagining, say, a postbellum U.S. after a Civil War victory by the Confederacy. The narrative of King's dream also exists in a sideshadowing, and the thought abides in peoples of color that the parallel track on which sideshadowed narratives exist will someday merge with living history in a longed-for end-time beyond a need for CPT.

In January 2007 the Board of Directors of the *Bulletin of the Atomic Scientists* announced that they were moving up their Doomsday Clock to five minutes before midnight.[38] The Clock was created in 1947 as a symbolic indication of the urgency of nuclear perils. At that time it was seven minutes to midnight, and, in the decades since, the Clock has fluctuated between two minutes to midnight in 1953, when the U.S. and the U.S.S.R. were developing and testing hydrogen bombs, to seventeen minutes to midnight in 1991, when superpowers reduced their nuclear arsenals and the readiness of their intercontinental ballistic missiles. What prompts the new urgency in resetting the Doomsday Clock is unprecedented in the Scientists' timekeeping:

---

[37] Morson, 6–7.

[38] Except where noted, information and quotes in this paragraph derive from *The Bulletin Online*, the Web site of the Bulletin of the Atomic Scientists, accessed 3 July 2007 at www.thebulletin.org/minutes-to-midnight/board-statements.htm and www.thebulletin.org/minutes-to-midnight.htm/timeline.htm.

> We have concluded that the dangers posed by climate change are nearly as dire as those posed by nuclear weapons. The effects may be less dramatic in the short term than the destruction that could be wrought by nuclear explosions, but over the next three to four decades climate change could cause drastic harm to the habitats upon which human societies depend for survival.

This decision is, however, more than a resetting of the Clock. It represents as well the superimposition of the long time-frame of environmental ruination on the short time-frame of deployed nuclear weapons. In other words, the current setting of five minutes to midnight is really a synecdoche for at least three or four decades. The two time-scales need to be reconciled, even if only by parsing them into different apocalyptic time zones.

As the expression of a military-industrial-national complex, the Cold War created its own time-sense, and the Doomsday Clock is the critical timekeeper of its apocalypse narrative. Given the history of racialized access to institutional power, I could even argue that it keeps White People's Time, except that standard "universal" methods of timekeeping are already a white people's time. Yet the Doomsday Clock does not keep the time that slaveowners forced upon slaves and that railroad barons forced upon a modernizing society. It wants permanently to avoid an ending. It defies the time-sense of linear progression, urging a turning back of the clock. It also defies the capitalist equation "time is money," an equation built on an assumption of linear progression. In his analysis of what he calls "time-space compression," David Harvey speaks less of "time and space" than of "money, time, and space."[39] After all, money can accumulate upward while time pushes forward. Harvey argues that "shifts in tempo or in spatial ordering redistribute social power by changing the conditions of monetary gain (in the form of wages, profits, capital gains, and the like)."[40] The Doomsday Clock could effect such a redistribution if it advanced, subjectively, a shift in tempo, if it imagined a new narrative with closure. But the purpose to which the Atomic Scientists put it is merely to witness, not to analyze and dream. And what it witnesses, in any given moment, is the present, the terrifying now.

---

[39] David Harvey, *The Condition of Postmodernity* (Cambridge, MA: Blackwell, 1990). See especially Part III, "The Experience of Space and Time."

[40] Harvey, 232.

CPT is both witness and vehicle of hope. In the United States today, few educated nonscientists fuss with contradictions and accept time as both linear (a function of aging toward an inexorable personal demise) and circular (seeds of the tomatoes that refused to grow this summer can be planted in a different soil next summer). But the many nonscientists who might even have heard of CPT may not recognize its function as a vehicle of communal hope. Certainly the multicultural turn of recent decades has enabled a recognition of CPT. But, concerned more with identity than with justice, multiculturalism and institutional diversity have failed to distinguish among various narratives of racial oppression and various time-senses. Critics Robin Kelley and Vijay Prashad advocate for what they call polyculturalism, a politics by which differently oppressed peoples work toward a common justice and which seems to permit a sense of time as a political frame for lived experience.[41] There is no irony in the fact that the privileged classes who have created the conditions of oppression are also creators of a scary and necessary Doomsday Clock while the oppressed classes are creators of the only time-sense that constructs justice as a communal reward. Indifferent to the formalisms that concern narrative theory, CPT nevertheless envisions a closure that rejects apocalypse and embraces justice.

And, until that justice is realized, it gives hope. Eduardo Galeano tells the story, appropriately titled "Time Takes Its Time," of an old man, a miner "knee-deep in the mud, grinding stones and scraping sand in the abandoned mine that doesn't have a cemetery because even the dead don't want to stick around."[42] The man has worked the mine for a half-century and is now worn and tired, stirring "the sand day after day, seated beside his pan, under a tree even skinnier than he that barely offers any defense against the biting sun." The photojournalist Sebastião Salgado visits the mine, and the old man shows him a blurry, dog-eared, and very old photograph of his wife—as old as the man's labor and yet as immediate as its effects, in the way that slavery remains immediate to people of color today. "'She's waiting for me,'"

[41] See Vijay Prashad, *Everybody Was Kung Fu Fighting: Afro-Asian Connections and the Myth of Cultural Purity* (Boston: Beacon, 2001), especially chapter 2, "The American Ideology."

[42] Eduardo Galeano, "Time Takes Its Time," in *Voices of Time: A Life in Stories* (New York: Metropolitan Books, 2006), 7. All quotes in this paragraph appear in this story, on page 7.

the man tells Salgado. In the photograph, writes Galeano, the wife is twenty years old: "For half a century, she's been twenty, somewhere in the world."

References

Baldwin, James. "History as Nightmare." In *Collected Essays*, by James Baldwin, 579–81. New York: Library of America, 1998.

Baraka, Amiri. *Tales of the Out & the Gone*. New York: Akashic, 2006.

Bourdieu, Pierre. *The Field of Cultural Production*. New York: Columbia University Press, 1993.

Brook, Peter. *Reading for the Plot: Design and Intention in Narrative*. New York: Knopf, 1984.

Bulletin of the Atomic Scientists. *The Bulletin Online*. http://www.thebulletin.org/.

Cardenal, Ernesto. *Cosmic Canticle*. 1989. Willimantic, CT: Curbstone Press, 2002.

Doreski, Carole. *"Kin in Some Way": Reading Relocation, Reading Citizenship at the* Chicago Defender. Manuscript, 1999.

Du Bois, W. E. B. *The Souls of Black Folk*. 1903. Millwood, NY: Kraus-Thomson, 1973.

Fabian, Johannes. *Time and the Other: How Anthropology Makes Its Object*. 1983. New York: Columbia University Press, 2002.

Galeano, Eduardo. "Time Takes Its Time." In *Voices of Time: A Life in Stories*, by Eduardo Galeano, 7. New York: Metropolitan Books, 2006.

Golston, Michael. *Rhythm and Race in Modernist Poetry and Science*. New York: Columbia University Press, 2007.

Harvey, David. *The Condition of Postmodernity*. Cambridge, MA: Blackwell, 1990.

Horton, John. "Time and Cool People." In *Soul*, edited by Lee Rainwater, 31–50. New Brunswick, NJ: Trans-action Books, 1970.

JanMohamed, Abdul R., and David Lloyd, "Toward a Theory of Minority Discourse: What Is To Be Done?" In *The Nature and Context of Minority Discourse*, edited by Abdul R. JanMohamed and David Lloyd, 1–16. New York: Oxford University Press, 1990.

Kearney, Reginald. *African American Views of the Japanese: Solidarity or Sedition?* Albany, NY: State University of New York Press, 1998.

Kermode, Frank. *The Sense of an Ending: Studies in the Theory of Fiction*. 1967. New York: Oxford University Press, 2000.

Morson, Gary Saul. *Narrative and Freedom: The Shadows of Time*. New Haven, CT: Yale University Press, 1994.

Moten, Fred. *In the Break: The Aesthetics of the Black Radical Tradition*. Minneapolis: University of Minnesota Press, 2003.

Parker, David. "The Chinese Takeaway and the Diasporic Habitus: Space, Time, and Power Geometries." In *Un/settled Multiculturalisms: Diasporas, Entanglements, Transruptions*, edited by Barnor Hesse, 73–95. London: Zed Books, 2000.

Prashad, Vijay. *Everybody Was Kung Fu Fighting: Afro-Asian Connections and the Myth of Cultural Purity*. Boston: Beacon Press, 2001.

Smith, Mark M. *Mastered by the Clock: Time, Slavery, and Freedom in the American South*. Chapel Hill, NC: University of North Carolina Press, 1997.

Smitherman, Geneva. *Words from the Mother: Language and African Americans*. New York: Routledge, 2006.

Warm Water, Luke. *On Indian Time*. Rapid City, SD: Trickster Arachnid, 2005.

RESPONSE 1

Robin Lucy

Colored People's Time is linked to a body of political, psychological, and philosophical strategies of resistance to oppression handed down through time and utilized as long as the "dream" of social justice is not realized. These practices look to the past and toward the future. What I explore further here is that African American texts and orature have always themselves theorized the experience of time for a community subject to coercive material and legal practices that attempt to render the group powerless and invisible and to produce an intellectual and spiritual dispossession of which Frederick Douglass writes, "The thought of only being a creature of the *present* and the *past*, troubled me, and I longed to have a *future*—a future with hope in it. To be shut up entirely to the past and present, ... is to the soul what the prison is to the body...."[1] Douglass escapes to a future in the North. Harriet Jacobs liberates her children and describes this event in metaphors that evoke Douglass's "glorious resurrection, from the tomb of slavery..."[2]: "The darkest cloud that hung over my life had rolled away..... my little ones were saved."[3] The Black Jeremiad, too, envisioned a future in which a just God would destroy the oppressor and liberate the oppressed so that sacred and secular time would converge.

Douglass describes the African American spirituals as "a testimony against slavery, and a prayer... for deliverance from chains."[4] For Du Bois, these songs were also evidence of a culture and politics of liberation struggle and of African-derived practices and beliefs that subverted the "forgetting" imposed by the masters. Out of these emerged double consciousness and CPT. CPT being rooted in Africa is evident,

---

[1] Frederick Douglass, *My Bondage and My Freedom*, ed. John David Smith (1855; New York: Penguin, 2003), 200; emphasis in original.

[2] Frederick Douglass, *Narrative of the Life of Frederick Douglass, an American Slave, Written by Himself*, ed. David W. Blight (1845; Boston: Bedford / St. Martin's, 2003), 89.

[3] Harriet A. Jacobs, *Incidents in the Life of a Slave Girl, Written by Herself*, ed. Jean Fagan Yellin. (1861; Cambridge: Harvard University Press, 2000), 109.

[4] *Narrative*, 51.

Jo Alyson Parker, Paul A. Harris and Christian Steineck (Eds), Time: Limits and Constraints, pp. 236–237

for example, in the presence in the diaspora of the Ki-Kongo cosmogram, an ideogram that represents the continuity of human experience in Great Time, the intersection of the living and dead, the human and divine, the spatialization of a past through which all things pass, and the abiding presence of the ancestor. The cosmogram represents a world distinct from that defined by Western beliefs in linear time and the strict separation of the material and the spiritual. Toni Morrison, for example, employs Great Time to locate the individual's place in the continuing collective struggle. As Dr. Streamas argues, in writing shaped by CPT, history (Great Time) and fiction are not "separate worlds." Great Time also gives rise to "sideshadowing," to the recovery of histories and memories that disrupt the masters' narratives.

CPT is part of the political and cultural repertoire of oppressed peoples of color worldwide. Du Bois, too, recognized multiple forms of oppression and envisioned a pluralistic, even "polycultural," America in which the spirituals' "faith in the ultimate justice of things" would be realized.[5] This vision—this dream—is grounded in the process of moving toward this future, toward collective triumph, while at the same time in recognizing past and present practices that have produced and profited from oppression and exploitation. To assert this vision in our teaching, scholarship, and activism is to refuse to reify and reproduce the very forces that have subverted the movement toward the "end-time" of CPT: liberation. Dr. Streamas's inclusion of indigenous peoples, refugees, and the working class as participants in the *work* of CPT points to the direction this larger collective struggle might take.

References

Douglass, Frederick. *Narrative of the Life of Frederick Douglass, an American Slave, Written by Himself.* Ed. David W. Blight. 1845; Boston: Bedford / St. Martin's, 2003.

———. *My Bondage and My Freedom.* Ed. John David Smith. 1855; New York: Penguin, 2003.

Du Bois, W. E. B. *The Souls of Black Folk.* Ed. David W. Blight and Robert Gooding-Williams. 1903; Boston: Bedford / St. Martin's, 1997.

Jacobs, Harriet A. *Incidents in the Life of a Slave Girl, Written by Herself.* Ed. Jean Fagan Yellin. 1861; Cambridge: Harvard University Press, 2000.

[5] W. E. B. Du Bois, *The Souls of Black Folk*, ed. David W. Blight and Robert Gooding-Williams (1903: Boston: Bedford / St. Martin's, 1997), 192.

RESPONSE 2

DEIRDRE H. MCMAHON

In "Closure and 'Colored People's Time,'" John Streamas offers intriguing insights into the whitewashing of popular and academic understandings of both time and narrative form. Streamas positions "Colored People's Time" (CPT) as the temporal dimension of W.E.B. DuBois's germinal theory of African-American double consciousness—a divided or dual state of mind in which one is both subject and object, both looking and being looked at. For DuBois writing in 1903, double consciousness expresses the tensions that accompany one's subject position as a racially marginalized figure within dominant (white) culture. Streamas suggests that exactly in its refusal to internalize the supposedly objective demands of the clock, "Colored People's Time" operates as a political strategy and philosophy that enables a better future, free of racism, to be imagined even as racism as a phenomenon continues, takes new forms, and cannot yet be placed into the past tense.

"Closure and 'Colored People's Time'" fruitfully connects historian Mark M. Smith's discussion of CPT as a means of protesting the power of the clock in capitalist production with literary theorist Gary Saul Morson's concept of "side-shadowing," which allows for a range of alternative presents that could have occurred even if they did not. In doing so, Streamas argues cogently and productively for more open notions of time and political possibility. Future work on the ways CPT might influence the construction or shape of time as a function of narrative closure, however, would benefit from analysis of the substantial scholarship that has followed Frank Kermode and Peter Brooks, Streamas's touchstone critics, in the decades since their publication. Given well-established and more recent work on the imbrications of race and narrative via scholars such as Kwame Anthony Appiah, Bernard Bell, Hazel Carby, Rey Chow, Wai-Chee Dimock, Laura Doyle, Henry Louis Gates, Jr., Sander Gilman, Hortense Spillers, Werner Sollors, Claudia Tate, and Cornel West, extended work on CPT's imaginative potential through text would profit from as well as contribute to the diversity and complexity of literary critical race studies.

Jo Alyson Parker, Paul A. Harris and Christian Steineck (Eds), Time: Limits and Constraints, pp. 238–239

The key contentions of Streamas's piece—that CPT exists for all racially marginalized peoples and fuels the communal action of freedom-and-justice narratives—encourage further extrapolation. If CPT exists across communities or individuals, for example, it may provide common ground for strategic solidarity among disparate groups advocating social change. Similarly, an alternative or resistant sense of time as an expression of communal identity and political agency need not be confined to an essentialist definition of race. Additional analyses of CPT might take into account how the time sense Streamas describes might be distinguished from the experience of time for economically dispossessed workers or politically oppressed peoples across supposed racial or national divisions. How a rejection of linear, progressive time leads toward communal justice and the elimination of racism requires additional elucidation. While further engagement with critical race studies in the academy, much of which addresses double consciousness directly and issues of time tangentially, would help to situate CPT and side-shadowing amongst the many strategies for resistance deployed by those racially marginalized against dominant power structures, John Streamas's essay remains valuable for discussing how more open notions of time might facilitate communal justice on any number of levels. The provocative re-imagining of time provided by this piece will enrich the scholarly conversation about race, power, labor, and the potential for social transformation afforded through narrative.

CHAPTER TEN

DRAMATIC TIME: PHENOMENA AND DILEMMAS

Carol A. Fischer

Summary

The stage is a unique space on which time plays. The linear time of a theatre event, from the audience's entrance to the exit of the last spectator, frames a performance: the start designated by dimming houselights and the show proceeding minute by minute to the final stage cue. But within this frame is the spacetime of the play, which could present or represent historical time, future time, anytime, or time outside of time. On a stage, time is apt to go backwards or jump around undisciplined, suddenly stopping or racing ahead, becoming a character with its own energy. Actors and directors are concerned with rhythm of speech and action, beats of plot development imposing limitations on their embodied creativity, but all such staging is manipulated by the playwrights who bend the event with the pressure of time in one of its many forms. In this essay I explore the interplay of meaning and function regarding time within the dramatic art form.

The twentieth century experienced shifts in the consciousness of the culture as new scientific ideas about time became assimilated into aesthetics of art. The historical theorist Michel Foucault focused on disjuncture and ruptures, which influenced theatre historians and critical analysts as they searched out alternate methods to Aristotle's *Poetics*, the incipient guide to dramatic criticism. I use such theories to discuss various ways contemporary playwrights use or abuse time in their works. The playwrights included are Tom Stoppard, Marsha Norman, Steven Dietz, Liz Duffy Adams, Penny Penniston, José Rivera and Neil LaBute.

As Tom Stoppard's play *Rosencrantz and Guildenstern Are Dead* begins, Rosencrantz and Guildenstern are flipping a coin, which has turned up "heads" ninety times in a row. Guildenstern attempts to rationalize the phenomenon and wonders if time itself has stopped dead. The two characters struggle to experience a particular timeframe and thereby a particular existence. Memories of past events are fuzzy, and the future is blank.

> GUILDENSTERN. One must think of the future.
> ROSENCRANTZ. It's the normal thing.
> GUILD. To have one. One is, after all, having it all the time… now…and now… and now…
> ROS: It could go on for ever.

Jo Alyson Parker, Paul A. Harris and Christian Steineck (Eds), Time: Limits and Constraints, pp. 241–256

.....

ROS: Eternity is a terrible thought. I mean where's it going to end?[1]

These men exist in chronological time only when their space is invaded by scenes from *Hamlet*, in which they are Shakespearean participants. As purely (and doubly) created characters, they exist outside a normal relationship to time, a place where eternity is a terrible weight. But this is not the case for us who live offstage.

In quotidian reality, time and timing certainly can emphasize the theatricality of our lives in subtle and not-so-subtle ways. There can be much drama in getting children ready for school in a short amount of time, an entire play being created by a small person with a determined teleology. A wannabe wedding groom will go to expensive lengths to ensure perfect timing in order to get an unqualified "yes" from his intended bride. The physical expression of these events is a heightened dramatic episode because of subjectively experienced time. The specificity of location and timing is crucial to the emotion of the participants fully involved in the *now*.

As Rosencrantz and Guildenstern exemplify by their confusion, time displayed in a formalized theatrical setting is unique from such "life" drama because of the text in performance. The performance of time as presented in the material of a play can suggest universal myth, with its similarity to repeatable material as expressed in ritual, as well as the singular meaning of a particular production. Time is integral to the truth of the text and audience reception in the large, theoretical sense and to the cultural context and experience of an individual involved in a particular performance in a narrow, subjective sense. Exploration of stage time-ness is complex because it is vital; the quality of liveness on the stage energizes or depresses response from spectators and then circles back to the actors. The time presented by the play material itself, the time of the presenting, and the timing of those presenting enfold one another in a hologram of experience so that any person's consciousness regarding his or her perception of time is in constant fluctuation. In this essay, I delve into the interaction of theatre scholarship, playwriting, theatre practice, and approaches to time. Theatre warps normative expression of linear time. To begin to understand how that happens, I look at cultural and performance analysis that is

[1] Tom Stoppard, *Rosencrantz and Guildenstern Are Dead* (New York: Grove Press, 1967), 70–71.

shifting some traditional methods of analysis about dramatic time and explore works of playwrights who have agitated the sense and representation of time on stage.

Theatre criticism that grapples with issues of modernism, postmodernism, realism, anti-realism, structuralism, poststructuralism, and other twentieth century *-isms* has been influenced by the thought of Michel Foucault that inverts, or at least distracts from, conventions of analysis based on linearity and causality. The types of questions Foucault asked in his pursuit of historical cultural knowledge are pertinent to analysis of stage phenomena involving time: "How is one to specify the different concepts that enable us to conceive of discontinuity (threshold, rupture, break, mutation, transformation)? By what criteria is one to isolate the unities with which one is dealing?"[2] How does one specify the ruptures on stage involving time? What conversation takes place among the representations of and vital movements of time on a stage? Additionally, theatre historians have taken Foucault's questions to heart. Rather than manipulating a cause/effect relationship, some historians seek for the situations that undermine normal searches for trajectories, understandings, or discourses. Those types of questions reposition the angle of perception.[3]

In a visible shift from conventions for dramatic epistemology, newer anthologies for introductory courses to dramatic arts attempt to reframe the student's perspective. The ten editions of Oscar Brockett's *History of the Theatre*, which follow theatrical traditions from possible origins in ancient ritual, meander its linear-time-oriented way through Western/European dramatic practices, briefly including some non-Western theatre traditions. Well deserving the fine reputation assigned to it, Brockett's book has been a solid bible reference for many teachers and practitioners in its forty years of service. But a differently conceptualized volume, *Theatre Histories*, published in 2006 by a collaborative editorial team, looks at theatres globally from multiple perspectives: "We consider many cultures and multiple methods of interpreting performances, theatre, and drama from around the world.

[2] Michel Foucault, *The Archeology of Knowledge* (New York: Pantheon Books, 1972), 4.

[3] Rosemarie Bank and Michal Kobialka particularly have worked with Foucault's thought in their areas of expertise as theatre historians. See "SUPPLEMENT: Physics and the New Historiography," *Journal of Dramatic Theory and Criticism* 5:2 (Spring 1991): 61–100.

We do not privilege chronologies or lists, [and aim] to stimulate the appreciation and interpretation of many kinds of performance and theatre, and many kinds of critical thinking about them."[4] In broadening the location of theatre history from Europe to the world, these authors note that a linear, deterministic viewpoint is only one of several possible historiographies and re/consider varying meanings according to specific locations and circumstances.

The authors further explain that their shift in historical method relates to shifts in scientific method as expounded in *The Structure of Scientific Revolutions* by Thomas Kuhn, whose study of historical paradigms renders untenable a claim for pure objectivity, including any discovery of an unquestionable universal truth.[5] In his own scholarship, Foucault exchanges one type of historical perception for another and utilizes the science and philosophy of quantum physics in developing his emergent historical and cultural theories.[6] Such cultural and scientific conceptual shifts support the analysis of drama from a variant perspective, a different point of view. Space and time overlay and interact with the space-time and energy-mass identified as theatre both in dramatic literature and in stage production. Playwrights and critics can discuss the interactions that create theatre with renewed vigor from a shift in presuppositional thought.

H. Porter Abbott writes that "narrative is the principle way in which our species organizes its understanding of time.... As we are the only species on earth with both language and a conscious awareness of the passage of time, it stands to reason that we would have a mechanism for expressing this awareness."[7] Despite other available modes of telling time from the seasons' cycles, the daily passage of the sun or phases of the moon, we humanoids like to organize our histories through the

[4] Phillip Zarrilli, Bruce McConachie, Gary Jay Williams, Carol Fisher Sorgenfrei, *Theatre Histories: An Introduction*, (New York: Routledge, 2006), xvii.

[5] Ibid., xxiv.

[6] Pamela Major-Poetzl explains at length the connection of Foucault and quantum physics in her book, *Michel Foucault's Archeology of Western Culture* (Durham: University of North Carolina Press, 1983). "Foucault's works might be seen as a series of 'field histories' in which there are no material actors and which deal with the web of relationships in an immediate neighborhood. He, too, avoids causal theories connecting distant events by big steps and explores instead a multitude of small events that he views in spatial terms" (64).

[7] H. Porter Abbott, *The Cambridge Introduction to Narrative* (Cambridge: Cambridge University Press, 2002), 3.

personal narration of a succession of events. Mothers might remember the time of a particular political event or natural disaster because of which baby they were pregnant with then rather than the specific numerical year. Men (at least according to a general stereotype) mark time by noting how a particular World Series played out or by remembering how television coverage switched to a movie just before the dramatic upset of the "Heidi Bowl" football game. This is not a gender point here but an acknowledgement of the functional reality of Abbott's statement. Many today will remember 9/11 by recalling what they personally were doing when the news of the World Trade Center attack reached them. People consistently mark the passage of time in their lives through narrative.

The stories that are told on stage evolved in similar fashion. What was important for humans to know about themselves and their environment became woven into narratives of relationships—man or woman and power, humans and knowledge, humans and the gods, man and woman—and then was presented for observation, association, comment, and response. Aristotle's *Poetics* outlined the most congenial production of a story for an audience of his era, beginning with imitation, *mimesis*, as part of nature, and expanding on the details of tragedy and epic poetry in accord with his understanding of correct audience reception, all of which included a formalized unity of time, place and story. The contemporary theatre scholar Hans-Thies Lehmann concludes a discussion about Aristotelian unity of time by stating that the tradition's aim is "to prevent the appearance of time as time.... [It] points to a much more profound fantasy image, a phantasm of continuity."[8] But on a number of contemporary stages, causal narrative is less essential to the experience. The orientation of the substructure might still function from a position relative to a normalized experience of time, but the consciousness of time in the world of a play de-familiarizes that experience, wakes an audience up to a previously unperceived relationship, inverts expectations, surprises, resurrects deadened responses, and mythologizes because of time's interactions with characters and story. Collisions of energy fields create an observable "event" (to use vocabulary associated with events described in

[8] Hans-Thies Lehmann, *Postdramatic Theatre*, trans by Karen Jürs Mundy (New York: Routledge, 2006), 161–162.

quantum physics). A play in its specificity provides the location for this de-familiarization, offering a "playing" field, so to speak, for analysis. As I suggested in noting current shifts in theatre historiography, some scholars of dramatic study and analysis as well as playwriting artists are also reconsidering, re-envisioning time on stage.

Stage time *resembles* human time in the sense used by Abbott. Audience attending the theatre could tell a sequential narrative about their evening: getting dressed, driving to a restaurant for dinner, walking to the theatre, seeing the show, stopping at a bar before heading home. And generally a play script is similarly organized with linear designations: Act 1 scene 1, Act 1 scene 2, etc. The playwright produces a pile of papers representing the hours he has worked—pages presented in normal numbered sequence. Some playwrights designate that their plays be performed without an intermission so that there is no disruption to the story that sequentially builds tension towards a climax and resolution (or with inverted intention, maintains a static sense of entrapment or boredom), not wanting to remove the consciousness of spectators from the space-time of the play before its completion.

Many books ground the theatrical experience in space and in matter, the interaction of location and of bodies.[9] In such analyses, the illusive nature of time gets transformed into a solid body's action or a space's transformation as representative of time passing—time as movement or change—effectively keeping time invisible.[10] However, even within a traditional narrative there is something more to consider. The same play (virtually) is performed twice or 300 times, perhaps counting down nearly the exact same clock time. "Yup, 1 hour and 56.5 minutes again!" the stage manager claims when pressing the button on his stop-watch. Even with different actors, the play is considered the same play because the text is the same and the action is the same to a great degree. But the day on which the "same" play is performed is a different day—not yesterday nor tomorrow. The *now* of being, the

[9] The following are a few of the books that focus on the physical interaction of the space and the actor: Peter Brooks, *The Empty Space* (New York: Touchstone, 1968); Deborah Saivetz, *An Event in Space, JoAnne Akalaitis in Rehearsal* (Manchester, NH: Smith and Kraus, 2000); Una Chaudhuri, *Staging Place, The Geography of Modern Drama* (Ann Arbor: University of Michigan Press, 1997); Una Chaudhuri and Elinor Fuchs, *Land/Scape/Theater* (Ann Arbor: University of Michigan Press, 2002); Bonnie Marranca, *The Theatre of Images* (Baltimore: The John Hopkins University Press, 1996).

[10] "Space" is understood here as an area delineated by the materials that create location for a play: the set or the curtains or borders enclosing an open acting area.

was of having been, and the *will be* of becoming are enfolded in such reiterated performance. One possible way to discuss this labyrinth of time in performance is through the work of a theorist contemporary to Foucault, Michel Serres.

Michel Serres dissects the word "circumstance" in his writings "Les cinq sens" and "Statues." In her analysis of Serres's work, Maria Assad explains that "circum-" refers to the dynamical and "-stance" to the static. The linguistic bisection paradoxically separates the static from the dynamic even while the word unites them into one concept. "The statue as circumstance is no longer either purely stable or utterly chaotic; it is both circum- and -stance, an oxymoronic 'dynamical statue'."[11] This is a useful concept to apply to the discussion of the time-sense of repeatable theatre productions. The circum-stance of a particular performance event includes the static nature of the text reiterated within the dynamical event of live actors in front of a live audience. An audience member knows that he saw the "same" play as his friend did last week, but conversation with that friend discerns where the performances and audience responses, as well as their own personal reactions, differed. The -stance/stasis surrounds and offers freedom to create and to experience the circum-/dynamic. This idea about the interaction of the static and the dynamic leads us to some specific anomalies about how time can function on stage, how it is dynamic and not invisible.

Walter Benjamin comments in his essay "The Work of Art in the Age of Mechanical Reproduction" that "a clock that is working will always be a disturbance on stage. There it cannot be permitted its function of measuring time.... Astronomical time would clash with theatrical time."[12] In writing about the phenomenology of theatre, Bert O. States responds that stage designers usually remove the minute hand or mask the clock face in some way; but, in contrast to Benjamin, he wonders if the disturbance to an audience would be minimal.[13] And yet there is that slightly uncomfortable awkwardness of a real clock measuring theatrical time with the parameters of real time—the clock in that case

---

[11] Maria Assad, *Reading With Michel Serres* (Albany, NY: State University of New York Press, 1999), 104.

[12] Walter Benjamin, *Illuminations*, ed. Hannah Arendt (New York: Schocken Books, 1988), 247.

[13] Bert O. States, *Great Reckonings in Little Rooms* (Los Angeles: University of California Press, 1987), 30.

isn't pretending to be a clock in the same way that a stage set pretends to be a room or an actor pretends to be a character.

Indeed, directors take the presence of a clock on stage very seriously. Because a stage contains a man-made universe, everything put there assumes extraordinary meaning. With such excess of meaning, a material working clock *does* become as a character stepping forward and speaking an indicative line, although the pointing hands physicalize time in a very different mode than a human Shakespearean character stepping forward to announce the day or season or hour. Time's costume of a clock body turns the invisible man visible.

In the play *'night, Mother*, author Marsha Norman deliberately designates at least two functioning clocks on stage. The play is a countdown to a suicide, and Norman instructs the director to have the performance begin about 8:15 in the evening so that the suicide happens about 10:00. The minutes, not just any minutes, but those between 8:15 and 10:00 p.m., are integral to the dialogue and the story. Time is tapping its foot in the corner, visible and inexorable. At least one critic agreed with Benjamin, however, and was disturbed by the prominent display of real time in Norman's play:

> In yet another deadening display of Idiot Realism, dramatic time matched real time as the timepiece [a big clock hanging conspicuously on the back wall of the single set] ticked away till the end of the play.... The grinding inevitability of that clock is not so much tragic as pathetic, realism at its most unrealistic.[14]

Despite the pejorative intent, the comment captures the dilemma of a working clock on stage, the clash of the real*istic* and the real. This critic represented an aesthetic of postmodernism with more appreciation for the disjuncture and ruptures presented in art than for realism. A surrealist painter of the mid-twentieth century can help to clarify the artistic culture's growing awareness that the passage of time is other than the ticking of a clock.

Remedios Varo illustrates the shift in man's perception of time because of the entrance of Einstein's relativity in her painting, *Revelation* or *The Clockmaker* (1955). The painting shows many grandfather-type clocks in a Medieval-looking room, all similar, all hands pointing to the same time. But a new essence spins in through the window, visible but not solid, and shocks the clockmaker's long-standing per-

[14] Robert Asahina, "The Real Stuff," *The Hudson Review* 37:1 (Spring 1984), 100–101.

ception regarding the substance of his work. The ethereal circularity and energy balls of his vision resemble the round clock faces and the spherical gears that are falling onto the floor off the clockmaker's table, but they also have a sense of dimensionality that the gears and clock faces do not. The clockmaker has been surrounded by ticking; all the clocks' pendulums are at the farthest-out point of their swing. Time has always moved on a horizontal plane of pendulum action for him, but this new essence is a wheel of whirling energy. Varo described the painting as a fundamentally new vision of how time works:

> The clockmaker represents our ordinary time…operating in a clockwork universe, a system of absolutes in which the flow of time is uniform and unchanging.… The vision represents the Einsteinian revelation that time is relative…not a fixed moment to be trapped within a clock.[15]

The grandfather clocks (presenting "old") have curtains (rather than doors) that look like tied-back stage drapes revealing people in period clothing, and at least one character clearly has a Shakespearean cast to him. Each clock has an inside window with a peek of blue daylight outside, even though the clock bodies stand inside a building with dark night outside. I perceive these blue windows as something metaphysical. Could it be that the heavens—symbolically God's plan believed through many centuries—has entrapped time in a teleological format enclosing man's history into rooms with bars on the windows? Einstein's introduction of time as a relative phenomenon moves in or out of the room freely. Close observation reveals that the pointed peaks of the clocks in the painting are holding up the ceiling above the clockmaker. This world has maintained itself because of the artificial mathematical precision of tick-tock, but the ceiling is collapsing around the clockmaker like a circus tent being pulled down to make room for the next show.

Aristotelian poetics, the mainstay historical support for the function of time in drama, holds up a softening roof similar to the clocks in Varo's painting. Today the clearly defined borders between comedy and tragedy fluctuate and blur so that "tragi-comedy" has become a common subtitle; a story does not always have a beginning-middle-end, may turn back on itself, or not be a story in the sense of narrative at all. Characters might revisit situations in their minds but do so

[15] Janet A. Kaplan, *Unexpected Journeys: The Art and Life of Remedios Varo* (New York: Cross River Press, Ltd., 1988), 174–175.

physically in an audience's presence—an audience who then wonders what part is real, what part memory, faulty memory, or only imagined. But is anything on stage truly "real"? This *is* performance after all. In the play *This Is How It Goes* by Neil LaBute, the character named "Man" also acts as narrator directly addressing the audience. He stumbles over his words in attempting to explain the time-linear goings-on, realizing that the time frame of telling a story on stage, the relativity of tenses, is not as simple as it might at first seem:

> Okay. This how it goes, I mean, went. This is the way it all played out, or is going to. Or is ... right now. Doesn't matter; you'll figure it out. I think. No, you will ... sure you will! No problem. (beat) What you need to know for now, I mean, *right* at this moment, is that there was a girl.[16]

And later:

> And, so, that's how I got the apartment over the garage. I mean, I may've walked her out to her car, or am going to—I still don't get the time thing here!—but it basically happened that way.[17]

Narration on stage is different than narration in a written story. The Man here is telling in the present while showing the past, but the past is the present because we, as audience, are in the now moment seeing it unfold on the stage. The odd thing in this particularly self-conscious play is that at the end one may not have seen what "really" happened anyway. Is the narrator, who claims to be a neophyte writer, making up the whole story as a writing project? And, if he is, then the time frame containing the story that he offers us as memory never existed. Theatre spectators generously suspend their own reality for the sake of the fictional stage world, but this play's ending uncomfortably suggests that the audience has been flimflammed, victimized by lies. The story, claimed as a real history belonging to the narrator's character, ends up floating in a time-less, space-less place for all its resemblance to some reality. At the particular performance by the Ensemble Theatre Company of Santa Barbara,[18] the audience was awkwardly silent as lights faded to black, unsure if the performance was finished, and unsure what their response should be. Applause did not begin until

---

[16] Neil LaBute, *This Is How It Goes* (New York: Faber and Faber, Inc., 2005), 5.

[17] Ibid., 36.

[18] Neil LaBute, *This Is How It Goes*, dir. Jonathan Fox, Ensemble Theater Company of Santa Barbara, CA, 12 June 2007.

the cast was back onstage and lights were up for curtain call. For there to be a sense of a story's existence in time, a sense of shared existence in space is needed, but the blurriness offered by the narrator leads us to question if the events happened even in the presented imaginary or existed only in his computer's memory, changing even as they were being played as an actuality. Any possible grounding was impolitely yanked out from under audience members. Nonetheless, and perhaps precisely because of the "sting" operation, the play provided an energizing night of theatre.

Some plays use time as a focused subject in the plot as opposed to the ether that a story lives in. That is the case in *The Ruthless Reckless Brutal Charge of It or The Train Play* by Liz Duffy Adams (2005). A female scientist is on a train because, in the midst of other research, she has uncovered Nature's apocalyptic plot to end time soon—very soon—and she is physically distancing herself from the angst of the discovery as well as travelling to see her girlhood town one last time, if the end holds off long enough. Also on the train is a twelve year-old girl who is running away from home, has named herself "Leopard-girl," and is convinced she has an inchoate superpower. After some trial and error, Leopard-girl discovers that she can make time stop (which inadvertently gives Nature a way to figure out how not to implode its existence—a nice counteraction to the cause/effect entropy that the scientist had uncovered). Additionally, the earth goddess Gaia is onboard; she informs us that when she exits time for awhile (takes a nap or something), disasters and wars happen. A reader/audience's perception is tickled by such performances of time as something that can end or be stopped at will, or that plays tricks. Time has become a visible character itself (but far different than the clock face of *'night, Mother*), clown-like with somersaults and multiple personality traits. Part of the audience's enjoyment is in thinking of time doing something other than make a clock tick. We like the fact that we don't have to succumb to an assumed, inevitable absoluteness and that the uncontrollable might yet be controlled by some superpower. Note, too, this play takes place on a train, one of those fascinating vehicles that can fool you into thinking it is moving when it's not and whose speed played integrally into the thought experiments of Einstein as he developed his theory of relativity.

The current theatre culture's propensity to update Shakespeare, Greek tragedies, and other ancient mythologies is another point of discussion regarding time on stage. How do we examine the motivations to stage *Romeo and Juliet* in Elizabethan style or in Harley leathers with motorcycles? Is Antigone mourning her brother during a war in Thebes or on the streets of Baghdad? The musical *Rent* (1996) re-plays and updates the decadence of *La Bohème*,[19] but some critics already consider it valuable only as a historical piece that refers to the drug and homeless scenarios in the broadening recognition of the AIDS epidemic, the mixture of sexual orientations and ethnicities being regarded as overtly didactic. However, we are currently seeing new productions of *Hair, The American Tribal Love-Rock Musical* (1967), that re-capture the '60s with the purpose of saying something current about war and peace and community, of making that time relevant to now. Richard Parison, director of a 2007 production in Philadelphia commented that

> every one of the messages in *Hair* is equally resonant today... [it] is about the environmental movement, it's about the antiwar movement, it's about the gay-liberation movement, it's about the feminist movement... the black-power movement.[20]

For these plays the relationship of timeframes and relevancy gets worked out visually in the technology, the designs of costumes and scenic elements, but the success of the production is in the effective translation of the universal that is *other* than narrative history, that is relatively timeless and can be costumed in many fashions. Such a message might have been once dressed in doublet and hose and now is in distressed jeans or in nothing at all, but it contains an essence of reiterability, of circum-stance/dynamic-stasis, while on stage. It seems possible that *Rent*, being based as it is on Puccini's century-old opera *La Bohème*, may also find, after more years have passed, updated theatrical relevance similar to what *Hair* is accomplishing.

Playing with linearity and updating older works are not the only ways that playwrights imagine time outside the quotidian. In considering another style of contemporary playwriting, I see an interesting, possible development of a model somewhere between the philosophy of quantum physics and theatre theory that could connect those two disparate disciplines (similar to the influence of Foucault's model on

---

[19] Puccini's *La Bohème* was first performed 1896.

[20] Howard Shapiro, "*Hair* still grooves for those who are losing it," *The Philadelphia Inquirer*, 10 August 2008, HO1. A Broadway revival of *Hair* opened at the Al Hirschfield Theatre, NY, in March 2009.

theatre historiography). Theatre criticism can get bogged down in traditions of analysis that work well for realistic and naturalistic plays or in the distinctive modes of criticism used when regarding the experimental nature of avant-garde works that often emphasize what the play is not. But some works resist both approaches as they highlight something hovering at the fringe, an emotion or relational aspect or psychosis or relative thought that has been ignored by theatre and/or by current culture. Foucault encourages us to pay attention to what has been previously censored or submerged. Phenomenology helps more than a method such as semiotics in finding a way to speak of these new plays, but our culture has been, is being, infiltrated by the weirdness of quantum physics (and its corollary, the pseudo-quantum physics of pop culture) through fiction, pop physics, and movies, and playwrights dabble in it. The perception of time in these plays is one way that the assimilated influence of twentieth-century sciences reveals itself. The scripts don't necessarily try to explain time or quantum theories (although some do), but it is useful and perhaps even mandatory for a researcher to include the science as part of a critical approach.

Now Then Again (2001) is such a play. Penny Penniston writes about a love triangle, but she develops it through the concept of quantum field theory and time reversal. She acknowledges John Cramer's book, *Transactional Interpretation of Quantum Mechanics*, as her main source for the physics. The first act moves forward in time: now, now plus one day, now plus two weeks, etc. She sets the story in the Fermilab National Accelerator Laboratory in Illinois, with two twenty-something physicists and a clairvoyant janitor who claims to be more interested in possibilities than in reality. By the end of Act One (now plus 13 weeks), the female has married her childhood sweetheart and is going to give up her science pursuits even though she has won a prestigious physics prize; the janitor has died of a brain tumor. But Act Two reverses the process: now plus 13 weeks, now plus 10 weeks, going backwards in time, the janitor living again, each step revealing a different choice, decision, or observation until we reach the beginning "now," which is and is not the same as Act One's "now." In another reversal in the last scene a new forward is initiated, a future for the female to be with the bumbling but cute and very smart male physicist who will encourage her research and personal career satisfaction. Analyzing the aspect of time in the play is enticingly tricky. The audience experiences the hour that moves the story backwards in time as forward in the play and in their evening out. The interesting part is that the play does not merely look backwards into history to uncover some linear cause and effect as one would find, for example, in Harold Pinter's play *Betrayal*. Time has not simply reversed for the sake of dramatic understanding, a concept that can be smoothly appreciated and fits a variation of traditional parameters of dramatic writing. But in light of the play being a reiterable text, there is also a determinism inherent, that of the author determining to capture through imagination what the quantum world of possibilities might be like if played out in human dimensions by using time reversal for a re-envisioning. (Also applicable for analysis perhaps, might be the multi-paths of the many worlds theory.) Looking at *Now Then Again* with an attitude of being open to unusual possibilities, with observation of events jumping into and out of focus, and with the interaction of fields of energy being of primary concern, a reader/spectator gains a fuller appreciation of the play as well as some understanding of the discontinuities (Foucault's "threshold, rupture, break, mutation, transformation") in the subatomic world.

The idea of certain historical worlds in direct communication with each other even though years and centuries apart, is another location of fruitful work and thought about time by playwrights. Tom Stoppard's *Arcadia* (1993) and Steven Dietz's *Inventing Van Gogh* (2001) both conjoin time-distant worlds and people. *Arcadia* connects the timeframes within one setting: a sitting room with a garden view on a large English estate. The stage dressing—books, table, chairs, tortoise—belong to both 1809 and the present day. Some of the "now" people are descendants of the earlier ones, but all who live in the present are interested in researching what happened in the house in 1809. The characters of the different centuries never directly interact, although a costume ball in the final scene has those of both eras dancing alongside each other. Stoppard's clever time-weaving includes a cautionary message: today's researchers cannot fully know what happened earlier since hints and evidence are only that and not a contemporary videotaping—alternative stories, easily conjectured, may or may not be true to the original scenario.

Inventing Van Gogh requires a different means of perception. Rather than using the time travel of Stoppard, Dietz has the painter Vincent van Gogh and some of his compatriots show up in a present day artist's studio. Van Gogh and Patrick Stone, a frustrated young modern artist, are played by different actors, but all the other actors play two characters: one character related to Patrick's world today and a parallel character related to Van Gogh's world in the 1880s. The location in time represented through the change in character is accomplished by a quick on-stage shift in costume. Van Gogh himself acts as a teacher/mentor for Patrick, and Patrick's life changes because of the personal exchange. Adding to the complexity of times displayed is the re-enacting of moments in the histories of Patrick's life and of van Gogh's life. The psychosis of the Dutch Master, which was apparent in the 1880s, has parallels in Patrick's situation in the present. Dietz's notes for the setting have a sense of multiple timeframes as well as stage space: "Evocative, not representative. Aggressively theatrical.... 'Exaggerate the essential; leave the obvious vague,'"[21] a reference to van Gogh's personal painting theory. Differently from *Arcadia*, this play asks questions regarding the strength of influence of one time on another. Stoppard's characters are like archeologists digging up the bones of past lives; Dietz wrinkles time so that what was far apart is in immediate juxtaposition. Neither is simply an exposition of historical cause and effect.

A last play I'm considering in looking at anomalies of time as representated on stage is *Cloud Tectonics* (1997) by José Rivera. In his play, the female character, Celestina del Sol, very pregnant, is hitch-hiking in the rain and gets picked up by a sympathetic Anibal de la Luna. With such names, we immediately assign a semiotic understanding of the situation—she's the sun and he's the moon. But when the two enter Anibal's house, all the clocks stop. What is discovered as the play progresses is that Celestina remains pregnant for over two years—that is, two conventionally understood years—and when she sees Anibal forty conventional years later, the baby is only six months old. The interplay of two extremely divergent movements of time is eerie and disconcerting. Celestina has no concept of what time is or does: she asks Anibal to imagine: "What if there are people born who don't have that sense? Don't have that inner clock telling them when a moment has passed, when another has started...."[22] Her lack of normal aging reveals more than a mere difference in genetic-based perception, however; it is as if she is living on a light beam—a sun-beam—and looking out at the world. Clocks don't tick when one is traveling at lightspeed; in the moments when she slows down, the world has radically changed about her.

---

[21] Steven Dietz. *Inventing Van Gogh* (New York: Dramatists Play Service, Inc., 2004), 7.

[22] José Rivera, *Cloud Tectonics* in *Marisol and Other Plays* (New York: Theatre Communications Group, 2004), 153.

Conversation regarding stage and time is more complex than a literary analysis of realistic or quirky moments or of stories that engage some physics theory. In their imaginaries, those creative places from which plays emerge, contemporary playwrights are questioning human relationship to time with its multiplicities and disguises. Audience and theatre clock time—chronos time—enfolds the imaginative, experienced kairos time of the play's situation, which for these playwrights is a re-consideration of their perceptions in light of science's continuing explorations and the aesthetics of the culture. Time on stage coagulates into a character with its own distinct personality interacting with other characters, demanding a reaction or response. The current proliferation of new plays working with anomalies of time suggests that time, a character itself, is asking us to re-consider our perceptions of its presence and the meaning that it carries to our stages.

References

Abbott, H. Porter. *The Cambridge Introduction to Narrative*. Cambridge: Cambridge University Press, 2002.

Adams, Liz Duffy. *The Reckless Ruthless Brutal Charge of It or The Train Play*. New York: Playscripts, Inc., 2005.

Aristotle. *Poetics*. trans by James Hutton. New York: W. W. Norton & Co., 1982.

Assad, Maria L. *Reading With Michel Serres*. Albany, NY: State University of New York Press, 1999.

Dietz, Steven. *Inventing Van Gogh*. New York: Dramatists Play Service, Inc., 2004.

Foucault Michel. *The Archeology of Knowledge*. New York: Pantheon Books, 1972.

Heise, Ursula K. *Chronoschisms*. Cambridge: Cambridge University Press, 1997.

Kaplan, Janet A. *Unexpected Journeys, The Art and Life of Remedios Varo*. New York: Abbeville Press, 1988.

LaBute. Neil. *This Is How It Goes*. New York: Faber and Faber, Inc., 2005.

——. *This Is How It Goes*, dir. Jonathan Fox, Ensemble Theater Company of Santa Barbara, CA, 12 June 2007.

Lehmann, Hans-Thies. *Postdramatic Theatre*. Trans. Karen Jürs Mundy. New York: Routledge, 2006.

Major-Poetzl, Pamela. *Michel Foucault's Archeology of Western Culture: Toward a New Science of History*. Durham: University of North Carolina Press, 1983.

Norman, Marsha. *'night, Mother*. New York: Hill and Wang, 1983.

Penniston, Penny. *Now Then Again*. New York: Broadway Play Publishing Inc., 2001.

Rivera, José. *Cloud Tectonics* in *Marisol and Other Plays*. New York: Theatre Communications Group, 2004.

Shapiro, Howard. "*Hair* still grooves for those who are losing it," *The Philadelphia Inquirer*, 10 August 2008, HO1.

States, Bert O. *Great Reckonings in Little Rooms*. Los Angeles: University of California Press, 1987.

Stoppard, Tom. *Arcadia*. New York: Faber and Faber, Inc., 1993.

——. *Rosencrantz and Guildenstern Are Dead*. New York: Grove Press, 1967.

RESPONSE 1

Laura Pattillo

Fischer does a fine job of illustrating and examining various theoretical and practical ways in which time functions and can be explored in drama. She provides a worthwhile discussion of how time and space function in the theatre, both onstage in the world of the play (through act and scene designations, clock time, etc.) and in a meta-theatrical sense in terms of repeated performances of the same script in which the static words on the page intersec in different ways each time with the dynamic forces of actor, audience, etc. I might note that the narrative of the larger evening out (beyond the play itself) that Fischer proposed for the audience members applies to the performers as well, since their days are arranged around call time, curtain time, and post-show activities or meals, not just the playing time itself.

Fischer also addresses the updating of classics and the revivals of supposedly "period" pieces such as *Hair* (and, perhaps someday, *Rent*) as speaking across time to present-day issues. The question of how play revivals speak differently to audiences than they did when first produced is of immediate interest since the recent economic downturn has led many theatres to rely on revivals of older crowd pleasers as "safe bets" in terms of ticket sales; however, some revivals, like *Hair*, communicate across time in interesting new ways at the present moment, including the 2007 Broadway production of the World War I play *Journey's End* (which resonated powerfully with audiences who were viewing it through the lens of current wars), and the 2009 Broadway revivals of *Finian's Rainbow* (with its discussions of race, exploitation, and the dangers of living on credit) and *Ragtime* (with its sweeping look at race, labor, and other issues, echoed by the 2008 presidential race in many ways).

Fischer presents evidence of new developments in the organization of theatre history texts, which serve as a telling measure of how our concepts of linearity are shifting with regard to drama. The discussion of clocks on stage (further expanded by discussion of a surrealist painting) is also of special interest because it speaks so well to practical issues of dramatic time in performance.

Jo Alyson Parker, Paul A. Harris and Christian Steineck (Eds), Time: Limits and Constraints, pp. 257–258

Fischer examines how Aristotle's unities are adhered to strictly by plays such as *'night Mother* yet disrupted completely, often in ways that reflect theories of quantum physics, by other texts. The choice of plays provides a wide range of rich, useful examples for the discussion of dramatic times. Examining works by LaBute, Adams, Penniston and Dietz, alongside such iconic texts (in terms of dramatic time) as Norman's *'night Mother*, Stoppard's *Arcadia* and Rivera's *Cloud Tectonics* makes for a series of excellent illustrations of dramatic time. The mention of Pinter's *Betrayal* is also a good choice, as its "reverse chronology" is familiar to many people either through the play itself or the film adaptation. More mention might be made of other works in which the same playwrights discussed also address the issue of time. For example, one might simply note that Stoppard has an ongoing interest in exploring dramatic time that is not limited to *Arcadia* or *Rosencrantz and Guildenstern*. And, of course, while Fischer cites several excellent examples of recent plays that handle time in interesting ways, these are but the tip of the iceberg, which should inspire further study.

RESPONSE 2

Katherine Weiss

Time has always been a central part of theatrical productions. A playwright, after all, marks time on the page by including both a descriptive of when the action takes place and by organizing the work according to acts and scenes. However, many contemporary playwrights, as Carol Fischer's contribution reveals, challenge Western concepts of linear time and causality. Exploring the works of Neil LaBute, José Rivera, and Penny Penniston, among others, Fischer demonstrates that these playwrights veer radically from thinking of time in a traditional cause and effect manner. Using Michel Serres' dissection of the word "circumstance" to examine these new trends, Fischer shows that time in theatre is, indeed, a "dynamic statue" as performances include "the static nature of the text that is reiterable within a dynamical event of live actors in front of a live audience."

For contemporary playwrights, the science and philosophy of quantum physics has opened new avenues to explore time and space on stage. Indebted to quantum physics among other scientific theories that challenge causality, Penny Penniston, Tom Stoppard and Steven Dietz have given audiences dynamic works that push the limits of space and time. Penniston's *Now Then Again* (2001) invites its audience to imagine "what a quantum world of possibilities might be like if played out in human dimensions," whereas Stoppard in *Arcadia* (1993) and Dietz in *Inventing Van Gogh* (2004) use characters from the past and present, allowing the two time periods to interact on stage together in their rupture of "historical cause and effect."

Fischer's contribution does more than highlight trends in contemporary theatre. Her insights allow us to reconsider the ways in which plays from as early as the nineteenth century have challenged causality. Georg Büchner's *Woyzeck*, about the murder of a young woman, is a case in point, although perhaps only accidently. Büchner's manuscript, discovered after his death in 1837, was in disarray; the scenes were neither numbered nor neatly placed in a particular order (perhaps because his manuscript was incomplete), leaving theatre professionals with the task of determining the events. Consequently, every production is

Jo Alyson Parker, Paul A. Harris and Christian Steineck (Eds), Time: Limits and Constraints, pp. 259–260

different as each director unravels the play's events differently. Samuel Beckett's *Waiting for Godot*, which influenced Stoppard's *Rosencrantz and Guildenstern are Dead*, a play Fischer discusses at the onset of her essay, also questions chronological time and causality. Beckett's two tramps wait each day, uncertain as to what day it is and whether Godot will arrive. What Beckett accomplishes in this play is to show that, despite the potential that Godot may one day arrive, he never does and nothing out of the ordinary (or what is determined as such by the tramps) ever happens. Rather they, like actors, live by perpetual repetition.

Fischer's contribution allows us to re-see the ways in which playwrights of the past and present break the spatial and temporal confines of the stage. While time is marked on the page, what the audience often sees on the stage is a breaking down of linear time and a breaking away from deterministic causality.

CHAPTER ELEVEN

EVENTAL DISTENSION: RESTLESS SIMULTANEITY IN STEVE REICH'S *PIANO PHASE*—TOWARDS A REHABILITATION OF THE REAL

Marc Botha

Summary

This work argues that a contemporary resurgence of realism rests on what might seem little more than an intuitive observation: that what is Real is contingently so and that this contingent existence unfolds in a time which is irreversible. If this contingency was not, in fact, a necessity, or if the time in which a Real object existed was actually reversible, then the Real itself would dissipate. Yet aesthetics problematizes this realism in numerous ways, none more clearly than in the case of minimalism. By turning to Steve Reich's epochal minimalist composition *Piano Phase*, this piece traces aesthetics fraught relationship to the real to the point at which a new entity emerges—an event, following the vocabulary of Alain Badiou. Tracing the temporal ambiguity which an aesthetic experience of *Piano Phase* effects—the inability precisely to determine beginning or ending, forward or backward direction—this essay argues that Reich's composition effects an evental distension, an apparent stretching of the event from within its own emergence. While this in no sense diminishes the Real, it takes us a step closer in understanding our fraught contemporary relationship to realism and reality.

1. A Resurgent Realism

1.1 I begin with a manifesto of sorts: five points offered in consideration of how to tie a resurgent philosophical realism to the experience of aesthetic multiplicity, and, if the proof on the basis of which it is forwarded holds true—Quentin Meillassoux's *Principle of Factiality*—then the realism it endorses is neither naïve, nor the assertion of some primary quality that is pervasive in certain entities and absent in others. From the perspective of the post-dogmatic realism I adopt here,[1] there simply can be nothing in being that is genuinely beyond the Real, that could not be Real in some possible world. The work that

[1] I attempt in this formulation a partial synthesis of some points raised by Alain Badiou and Quentin Meillassoux.

Jo Alyson Parker, Paul A. Harris and Christian Steineck (Eds), Time: Limits and Constraints, pp. 261–288

follows this manifesto is an attempt to come to terms with a small part of the sphere that so often seems at odds with such a realism, despite the fact that, technically, there is simply nothing with which to be at odds in the Real. It is a sphere that, in the multiplicity with which it attaches itself to questions of meaning and identity, seems to belie the absolute aspect of the Real. Broadly named, I refer here to aesthetic experience. What is begun below is the recuperation of the Real in a manner that reinstalls it neither as a quasi-transcendental condition of being, nor as a metaphysical dogma.

Because commencing such an enterprise is often itself the prime difficulty, this manifesto sets itself in relation to the notions of beginning and continuation, or, in the terms I develop here, event and evental distension. I offer this latter denomination to account for a peculiar temporal retracement of the event, an internal expansion of evental properties, that is exposed in holding the aesthetic experience encountered in Steve Reich's epochal composition of musical minimalism *Piano Phase* to the revealing, if discomforting, light of a contemporary realism.[2]

1.2 I offer the following five points:

i) Being does not begin; so what we call beginning takes place within the conditions of Being. Being is pure multiplicity and as such has no conditions to which it is tied.[3] Such Being without condition is Absolutely Real, inasmuch as the Real is the mark of that which is beyond any necessary positing, access or interpretation. The only necessity implicit in the Real is the necessity of contingency.[4] Thus

[2] Steve Reich, *Piano Phase (for two pianos or two marimbas)* (London, Universal Edition, 1980). All references to the score are to this edition.

[3] Alain Badiou, *Being and Event* (Continuum: London, 2005), 40–8. Hereafter *BE*. For further explanation, see Peter Hallward's, *Badiou: A Subject to Truth* (Minneapolis: University of Minnesota, 2003), 61–3.

[4] This is the principal maxim to be extracted from Meillassoux's critique of what he terms correlationism—"the idea according to which we only ever have access to the correlation between thinking and being, and never to either term considered apart from the other" (Quentin Meillassoux, *After Finitude: An Essay on the Necessity of Contingency*, trans. Ray Brassier [London: Continuum, 2008], 5). Hereafter *AF*. The many versions of correlationism share a rejection of the Absolute, which Meillassoux argues can be recuperated (constituting the basis for a post-naïve realism) by recognizing that there exist no necessary, universal laws, with the sole exception of the law

any beginning always takes place in the Real. Here the term Real implies the pervasiveness in being of the necessity of contingency and the irreversibility of time (the latter coincidental with such taking-place).

ii) So, we begin with the force of an appearance—a contingent and irreversible occurrence—within the Real. This occurrence may be reiteration, or it may be absolutely new, in which case, following Badiou, it might be termed an event. Badiou identifies the event with the inauguration of something totally new and rare, a new subject in being, "a point of rupture with respect to being [that] does not exonerate us from thinking the being of the event itself."[5] So the event is something proper to being, yet "which is not being *qua* being... which subtracts itself from ontological subtraction."[6] It is this novelty that interests us here. Immediately we must stress that Badiou limits the definition of truth and subject as direct consequences of an event, but that the eventality of the event is irretrievable as such. Much more common is the quasi-evental status of most beginnings—the situation of rearrangements, reattachments of existing elements, those entities one might hesitantly denominate weak subjects and objects—the occurrence of something relatively new, new enough to term a beginning, but without any great self-sustaining intensity in existence, and that must be directed towards, rather than followed by, activity or decision.

iii) We proceed with a distinction that marks existence from being, but is within Being (pure being). Here, of principal significance is the strictly ruptural, non-epiphanic quality of an Event, and the insistence on what is independent of human existence, yet is Real. The distinction with which we proceed in this light, is the bar that

---

that affirms that there is no necessary universality to any law. So all is contingent, except contingency itself, which is absolute (*AF*, 78–81).

[5] Alain Badiou, "The Event as Trans-Being," *Theoretical Writings*, ed. Ray Brassier and Alberto Toscano (London: Continuum, 2006), 100.

[6] Ibid., 101. Paradigmatic of trans-being, the event is both proper to being, while still subtracted from it: "[i]n effect, an event is composed of the elements of a site, but also by the event itself, which belongs to itself" (Ibid., 103). That an event is genuinely rare is based on the mathematical fact that, with the sole exception of the event, every multiple (entity) is founded on an element that it cannot also contain (Axiom of Foundation). According to Badiou, this makes the event a multiple of a multiple, and so, strictly speaking, without foundation—unpredictable, transecting being (Ibid., 102–3; *BE* 173–6, 185–7).

separates existence from inexistence, appearance from disappearance, entities from non-entities. It is thus absolutely coincidental with the emergence of entities. If we can say something exists, then it is an entity.

iv) Accepting these three initial propositions, regarding the Real, the act of beginning, and the necessity for subsequent distinction, it is possible to begin again. We begin again with a discernment within existence—namely the distinction of identity and content, the definition of particularities and singularities in the undifferentiated mass of information that constitutes existence.

v) We continue with decision, a decision that Badiou suggests is the very act upon which rests the possibility of subjectivity and universality. We decide with regard to an event that has taken place in the Real; we decide as far as the association of entities, objects and identities are concerned, tying them to certain truths, subjects, meanings, significances, consequences; and we extrapolate such decisions in specific discourses developed for this purpose.

2. Beginning again: Reich's *Piano Phase*

2.1 My formative musical experiences were highly eclectic, split between the course one invariably must follow in an apprenticeship as a classical saxophonist—endless scales and etudes, the Baroque composers, the French neoclassical literature—and the music that shattered this order, the music I listened to and really aspired to play—what seemed to me free music, from John Coltrane to John Cage. This drift to the delightful anarchic underbelly of the avant-garde was unexpectedly interrupted by a chance encounter with minimalism in a music library. I could not have predicted this, nor could I have anticipated the sublime interplay of force and simplicity that in Glass' *Einstein on the Beach* overpowered me—less still the concentrated fascination of Steve Reich's *Piano Phase*. I am unable to recreate with conviction or description the precision of my experience, so I offer only a few points that might explain its centrality to what follows. For better or worse, I am a highly rhythmic musician. It has always been relatively easy for me to reproduce complex rhythmic patterns and to manipulate beats and rhythmic figures within these. I have a fascination for perplexing irregularity that nonetheless remains within a certain frame of consistency.[7] Anybody familiar with *Piano Phase*, its aural effects and temporal technicalities (which I describe subsequently), will find it no surprise that to someone who instinctively and obsessively overdetermines the beat in music, the process of phasing is one of the most stimulating, if potentially unnerving, musical experiences imaginable. I remember initially thinking something had gone terribly wrong as the melody, forceful and ancient in its modal simplicity, began to pull apart. And the further it pulled apart, the more fervently I attempted to impose my rhythmic will, until, at a certain point of maximum disorder, I stopped trying to force these multiple regularities that I was feeling into a unity and accepted that what was Real in this piece was this temporal multiplicity, this simultaneity of times. Over the years, I have encountered opinions as extremely negative as mine was positive.

It is perhaps out of a misplaced desire to keep inviolable my own experience of minimalism that, to this day, I have not dared to perform minimalist music. Performance would be the only way to be closer even than aural perception to this music. Yet, I fear it might be too close. So, on one level, my instinct is to leave this music explored only through the pure solipsism of individual aesthetic experience or through the technical language of music theory and often drab historical contextualization. But I cannot in good conscience reduce this encounter to these limited paths, whether celebratory or denunciatory, for my increasing conviction is that through *Piano Phase* the listener runs up against the exemplification of genuinely ontological limits that transcend his or her singularity. So I approach the exemplarity of this aesthetic with the only other tools I can imagine appropriate, however blunt they may be or inadequate my application of them: philosophy and cultural theory.

2.2 Most notable amongst the ontological limits exemplified in *Piano Phase* is the question of time. More particularly, there emerges an apparent conflict between a time that proceeds monodirectionally from past to future—which might be called intrinsic, irrefutable, consistent, objective, scientific, real, or, borrowing Meillassoux's term, *ancestral*

[7] The apparent contradiction of accommodating irregularity within consistency is identified here with a post-dogmatic realism, an apparent contradiction that befalls the Real.

time[8]—and the many ways in which it is possible to divide, mark and collect this real time into a host of subjective temporalities. Is it not precisely through the exemplarity of the aesthetic that it becomes possible not only to imagine these two times as coextensive but actually to experience them as such? To be clear, I am not suggesting that from one perspective time is real and singular and from another unreal and multiple. In a simplistic borrowing from McTaggart's philosophy of time, it is never a question of choosing to emphasize an A series, a B-series or a C-series.[9] Predicated on questions of perception and access, these seem ultimately to be arguments that derive from what Meillassoux names the Correlationist assertion in philosophy, "that we only ever have access to the correlation between thinking and being, and never to either terms considered apart from the other,"[10] and I am content with Meillassoux's refutation in this regard.[11] If the defeat of over-generalized perspectivism can be granted, then the central issue changes from the dispute as to whether or not time is real or unreal to the task of identifying the proper time of the Real.

The difference may seem trifling at first, but it is undoubtedly of profound significance. It is, in essence, the difference between installing time as an identity, in which case it will always be exposed to the threat of being reduced to a matter of perspective, and recognizing in a necessary relation to time something strictly infinite. If time is strictly infinite, it cannot be reduced to a finite quantity, some sort of unity, contingent or otherwise, to which access can be gained. If the inapplicability of access to time *qua* time is granted, it is difficult to imagine a set of conditions in which the reversal or a genuine direc-

[8] Meillassoux offers the term *ancestrality* to counter correlationism (see brief description above) and discusses its temporal aspects in some detail. *AF* 10, 14–7, 20–2.

[9] Borrowing here McTaggart's division from his epochal essay "The Unreality of Time" (J. Ellis McTaggart, "The Unreality of Time," *Mind*, New Series, 17:68 (Oct., 1908): 457–74). The A series is marked by relative positions, "from the far past through the near past to the present, and then from the present to the near future and the far future" (Ibid., 458). The B series involves a more fluid progress "from earlier to later" (Ibid.). The C series, McTaggart claims, "is not temporal, for it involves no change, but only an order" (Ibid., 460–1). What is here invoked as Real should not, however, be understood simply as a conjunction of these series. The Real is distinguished precisely by its conjunction of change (implicit in contingency) and time, independent of perception. Hence an instantiated Absolute—Meillassoux's arche-fossil—seems significantly to unify without synthesizing these temporal series (*AF* 9–10, 16).

[10] *AF*, 5.

[11] See notes 4, 8–10.

tional uncertainty of time would be actually possible—except, perhaps, as a rhetorical device to support the assertion that the Real is merely a matter of perspective. Moreover, what is suggested in this association of infinity and time is not that infinity is something *in* time or that time *is* infinite. If time and the Real are mutually implicit, as I suggest above, then time cannot be reduced to a predicate or an identity, nor can it be seen as a strict correlate of any other predicate or identity. In other words, time is not some sort of entity, and it does not inhere in any entity: its identity cannot be named, and it cannot be assumed to be a simple constituent of anything that can be named. The failure of language to translate this realist proposition adequately—the fact that time is a noun, abstract, but a noun nonetheless, and that we are compelled furthermore to articulate these problems by phrases such as *time is*, clearly demonstrates that this condition is pervasive—should not be mistaken for an adequate refutation, which might well be the habitual assertion of certain strands of ordinary language philosophy.

Invoking a proper time of the Real is thus a matter of asserting that time is infinite, irreversible and non-predicative with regard to the earlier description of the Real. In other words, the necessity of contingency—the non-predicative, irreversible, infinity of time—and the Real are mutually implicit. Meillassoux asserts that the Absolute re-emerges within, and also reauthorizes, a speculative realism, and that this happens the moment we recognize that the sole necessity is contingency (raising contingency to a position of genuine immanence). In this light, I claim that such contingency implies the Real on the one hand and infinite non-predicative, irreversible time on the other. Similarly, time implies contingency and the Real, and the Real implies contingency and time. Purely for the sake of convenience, for the sake of putting a name to things, I refer to this relationship of mutual implication under only one of its constituents, the Real.

To the way in which the independent and mutually implicit terms of the Real confirm the infinite, we might add a properly ontological association. Being *qua* being, as pure multiplicity, does not appear nor disappear (rather, this is the characteristic of existence within being) but is expressed through the mutual internal implications of the Real. But if this double guarantee of infinity explains how one might legitimately, or at least consistently, claim that there is only one time—and this time is not real, but rather the time of the Real, that it is infinite and irreversible—it fails to account for the many patent examples where time appears to function in a distinctly surreal manner. Properly to account for these, we would have to consider the limitations of our phenomenological access to the world, our subjective intervention in navigating through memory and sense to project possible futures and pasts for the entity situated in the present.

As soon we assert a mimetic part of our relation to the Real, it becomes vital to recognize how these selections take place—permutations and failures of time and memory that shape and misshape the fictional and experiential world of writers, composers, artists and readers, the aesthetic media in their world-constituting aspect with temporal leaps, sudden halts, and blinding accelerations. In music, the ambiguity presented to the listener by the composition currently under consideration, *Piano Phase*, illustrates an intense example of this upset temporal world, disrupting the consecutive moments we conventionally take as given in auditory experience and throwing them into the shadows of indiscernibility. The principal point to keep in mind, however, is that despite the contradictoriness that clearly inhabits the processes of differentiation, such subtractions from the so-called reality of time do not reduce the Real.

3. Evental Distension

3.1 Although we might assert faithfully the significance of a contemporary rehabilitation of realism, and affirm, along with Meillassoux, Badiou and many others, its irrefutability, it seems an unnecessary violence to categorize such multiple temporal experiences as illusionary and simply to move on. It is also inaccurate because there is no subtraction from being, no number of realities, capable of exhausting the Real. The simultaneous accommodation of the irreversible temporality of the Real with multiple and contradictory temporal experience from the perspective of existence (illustrated here in aesthetic terms) recognizes precisely the way in which the Real extends itself across contingent and contradictory multiples, across the innumerable fluctuating metabolic times of every organism, just as much as it does in a work of art.

The crucial point in the relation of apparent multiple realities to the Real is that there is no real necessity adherent or inherent in any entity in existence to attest to the Real or to being *qua* being. Being and the Real, as we have seen, move through every entity as its innermost potentiality. The Real *simply is* in every existent entity, and art, like anything else that takes place in existence, is under no ontological obligation to offer itself as witness to the Real, nor to reflect on reality *qua* the Real, for the straightforward, if somewhat crude, reason that, as an existent, *art is real*. If these gestures are not obligatory, this in no sense reduces their prevalence, particularly as one of the pivotal moments of self-reflexivity—the moment when a self-reflexive (or self-reflecting) entity knows that it is engaged in self-reflexive activity—is marked by an assertion of self-sufficiency and independence as existential indicator of the Real.

In reflecting on the Real, and in its contrapuntal engagement with the Real, such self-reflexive art holds particular interest precisely because it posits itself as exceptional to the Real. To accomplish this, such art would have to be able to posit and reflect on its own exteriority to the Real, which, in the present formulation is in the strictest sense impossible. The Real, to recall, is both non-predicative, and non-correlative with any predicate. Hence, what these works attest to is in fact that which in existence approximates the Real, or what we habitually imagine must be the presentation of the Real in existence. I propose that this point is nothing other than the moment at which being passes into existence—the point where a work appears or disappears, between the existence and the inexistence of art. I contend that it is at this moment that a consciousness of the Real is at its most intense. And while this moment of appearance is not an event in the sense Badiou reserves for the term, it is the moment at which an entity is able to demonstrate most clearly its relation to an event.

3.2 Art that attempts to trace itself back from this moment of appearance (where the Real and the event seem most closely tied to existence) to the event itself is art that exhibits the temporal operation I term *evental distension*. Directly defined, evental distension refers to entities in existence that attempt to reproduce the eventality of the event upon which they are predicated, or to which they have become bound. I use the term distension to indicate the various processes and objectal[12] qualities that reside *within* the art object, so that the attempt to recreate the event in this sense might be thought of in terms of an internal expansion proper to the Real object, and not an equivalent subjective constitution of the object. To rephrase this definition, art objects are

[12] As opposed to objective, *objectal* refers to the properties of the object itself.

not attempting to trace themselves in accordance with the subject or truth of an event, to once again borrow Badiou's terminology, and, to return specifically to the aesthetic case, it is beyond the interests of such works to engage directly in questions of mimesis, meaning, or content. Rather, such works attest to their raw facticity, their status as objects independent of subjects. In so doing they are directed towards generating the conditions in which the Real appears immanently distanced from itself, so that it might be regarded as a describable entity. If this were genuinely possible, then art might quite easily be associated with the preeminently divine power of being able to create *ex nihilo*, since what is the power of the *eventality of the event* if not the pure immanence of the Real? Claiming this power through some sort of aesthetic meaning or intention is too easily refuted. If art is capable of this type of *evental distension*, it needs to manifest in a zone of undecidability, an exemplary zone in which process and object overlap to the extent that what is actually an object, or some dominant part of an object, takes on the appearance of an event.

Primary amongst such examples is Steve Reich's *Piano Phase*, but explorations of this type of evental distension might also incorporate many other types of process-oriented (principally abstract) art that operate by exposing questions of temporality as their principal substance, or, alternately, assert that their substance has an inalienably temporal dimension. Such works invariably fail in the task of recreating the event, but it is not always possible to *know* that they have failed. In fact, the best of the works that appear to re-effect the event, that manifest within an evental distension, present very precisely that aspect of the consequences of an event that seems *more knowable*. This is one of the definitions that Giorgio Agamben offers as the true example—that which is more knowable.[13] So we proceed by suggesting that in the best examples of evental distension—examples that I believe are exposed in minimalism—the illusion of the work's ability to reproduce or gain access to the eventality of the event is genuinely remarkable.

It is this experience that might be identified in the unsettling ambiguity that is identifiable with the phase-shift in Reich's composition and that equally characterizes other examples of process-oriented

[13] Giorgio Agamben, *What is a Paradigm?: A Lecture by Giorgio Agamben*. European Graduate School, August 2002. Stable URL: http://www.egs.edu/faculty/agamben/agamben-what-is-a-paradigm-2002.html.

minimalism. The work itself might thus be designated by the somewhat paradoxical name of instantaneous process art, and it is this instantaneous character that lends considerable weight to its association with a sudden emergence, or an event.

4. Phasing in: Three Modalities of Time

4.1 Neither unique nor limited to minimalism,[14] evental distension nonetheless finds a particularly forceful exemplification in many of the techniques employed by the best works of processual objecthood. The following presents a necessary, if somewhat heavy-handed, exegesis of some of the central technical concerns of the piece. Without these, the reader who is unfamiliar with the composition might be expecting a transcendent burst of epiphanic bliss. In reality, there is always an element of risk involved in compositions capable of inducing a genuinely sublime experience—a very banal risk, one should emphasize: that the delight and satisfaction of one member of an audience might be balanced by disgust and contempt in another. Others might find the piece utterly banal, and, to be fair, approached with certain less flexible formal expectations, such repetitive music may easily be charged thus.

4.2 As one of the defining compositions of musical minimalism, it goes without saying that much has been written on the technicalities of *Piano Phase* and what these accomplish aesthetically. Firstly, in typically minimalist terms, the composition is an exposition of the most fundamental musical matter—the combination of pitch and rhythm, a very simple melody constructed from a consistent flow of equal-lengthened notes, eighth notes and eighth note rests. Secondly, it presents a complex array of effects produced when this very simple melody, played simultaneously on two pianos in unison, is gradually shifted out of phase.

Piano Phase was the composer's first purely instrumental attempt at phase-shifting or phasing, a process he had discovered in his earlier tape compositions.[15] Phasing, as it emerges in Reich's early *oeuvre*, refers to an essentially temporal process involving an acceleration and/

[14] I use the term here not as a strict historical or stylistic designator.

[15] See Keith Potter, *Four Musical Minimalists: La Monte Young, Terry Riley, Steve Reich, Philip Glass* (Cambridge: Cambridge University Press, 2000), 167–8; Edward

or deceleration between two or more relatively short musical fragments that initially sound together. All phasing combines temporal linearity (the identitarian affirmation of Meillassoux's ancestral time) with cyclical time. To the extent that any phasing composition begins and ends, it reinstigates linearity, but constitutively, phasing involves a cyclical progression. To be more precise, during the phasing process, generally short fragments move progressively apart until, at the point of greatest temporal distance,[16] they once more begin to converge until they have returned to their original relative positions.[17] Thus a cycle is completed, which may, theoretically, be repeated without disrupting the essential unity guaranteed by the process itself.

From a conceptual perspective what takes place is essentially a demonstration of the necessity of temporal consistency. Once this consistency is interrupted by acceleration or deceleration, the relationship between identical melodies breaks down as one is progressively displaced in relation to the other, and with it the very linearity of time. Regular pulses become inconsistent, indiscernible, a fact heightened by the difficulties of executing the composition, and, as already noted, the progress of time from essential linearity—a composition moves steadily from beginning to end—to a combination of linear and cyclical time. In the case of *Piano Phase*, phasing happens between two melodic fragments with a consistent pulse. This pulse is interrupted and destabilized in the process of acceleration or phase-shifting.[18] The shift is undertaken by one pianist accelerating in relation to the other. Both play an identical melody, and the initial melody is held, unvaried in pitch and rhythm, by the second pianist throughout a cycle of phasing.[19] Partly because the melody consists of regular and easily discernible sixteenth notes, it is fairly straightforward if not immediate for the listener to discern when singularities are reached. Thus, within a full cycle of phasing, particular singularities are reached whenever

---

Strickland, *Minimalism: Origins* (Bloomington and Indianapolis: Indiana University Press, 1993), 186–7.

[16] Ibid., 184.

[17] This is the case in a fully cyclical process of phasing. See Paul Epstein. "Pattern Structure and Process in Steve Reich's *Piano Phase*." *The Musical Quarterly* 72: 4 (1986): 495. JSTOR Stable URL: http://links.jstor.org/sici?sici=0027-4631%281986%2972%3A4%3C494%3APSAPIS%3E2.0.CO%3B2-A, and Potter, op. cit., 184.

[18] Keith Potter refers to these as "fuzzy transitions," ibid., 180.

[19] The anchoring role is taken by the first pianist in the first two sections of the composition and exchanged in the third, with the second pianist holding the rhythmically stable melody.

the beats of the first and second piano coincide after a period of uncertainty, which emerges as the second pianist moves out of phase with the first. Structurally, *Piano Phase* consists of three full cycles, also called sections. The first and third consist of a single melodic fragment held by both pianists,[20] while the second consists of two distinct melodies: that of the first pianist reminiscent of the melodic material of the first cycle and that of the second foreshadowing the material of the third cycle.

Each cycle begins with a single piano repeating a particular melodic fragment a number of times (only guidelines are provided as to the specific number of repetitions). At an agreed point, the second player joins the first, either in unison (cycles 1 and 3) or counterpoint (cycle 2), and this new singularity is then repeated, followed by the first phase-shift. The process of acceleration by one piano continues in the phase-shift until the first beat of the stable line is coincidental with the second beat of the moving line, at which point the acceleration stops and this new singularity is repeated a number of times, once again providing the listener and performer with a stable point of reference, before the phasing process continues. Various points of contingent stability are reached in this way and phasing proceeds until the first beats of both lines are once again coincidental, at which point an initial phasing cycle is complete.

4.3 The psycho-acoustic aspects of the composition have not remained unexplored. Indeed, Epstein's rigorous formal analysis includes significant connections in this regard and might furthermore be used to ground important argumentation as to the progress from form to effect, and from effect to affect.[21] Perceptual ambiguity—a compelling field—arises when temporal fluctuation is juxtaposed with contingent stability, phase shifts or fuzzy transitions with sections of rhythmic coincidence and melodic stability. Somatic experience and psychological response overlap as the maximally ordered system, which is exposed with an unremitting procedural clarity that marks minimalism as a whole, is plunged into disorder. Listeners frequently attempt to re-establish order within the disordered periods of flux (phasing or

[20] For a detailed discussion of the melodic structure and its effects, see Epstein, *op. cit.*, 495–8, and Potter, *op. cit.*, 183–5, 187.

[21] *Op. cit.*

phase shifts) or anticipate disorder even during the sustained periods of melodic order that punctuate the phasing process as a whole. The restlessness produced by the simultaneous play of anticipation and resolution, invokes a specific tension that is clarified only when we consider more closely the temporal complexities at work in the composition and the way in which the evental distension they underpin functions as a sublime aesthetic.

4.4 Epstein's penetrative analysis is particularly revealing of the technicalities of this spatio-temporal ambiguity.[22] That this ambiguity resides in Reich's specific appeal to temporality above space (real or conceptual) is evidenced in the composer's emphasis on process as a means of providing him "direct contact with the impersonal and also a kind of complete control."[23] Freed from the responsibility of having to represent anything except the objecthood of the work itself, the composer is now able to explore the situation in which Reich affirms "compositional process and a sounding music...are one and the same thing,"[24] the essential unity of form and content Mertens identifies as characteristic of minimalism in general.[25] In this complex relation between listener, performer and self-generation, time finds three principal expressions.

The indifferent, ancestral time of the Real is the first of these, and it is this time of absolute becoming that the present piece ultimately affirms as the ground for other temporal explorations. Second, one must consider the multiple individual temporalities of individual entities, literally finite marks of appearance and disappearance within ancestral time that indicate existence. Every entity has a unique temporality in this sense, but, most important, such times are subtractions from ancestral time, which in no sense diminish the status of the Real. The third proposition is cyclical time, which offers itself as an alternative to ancestral time. Through a paradoxical split between what can be perceived as progress and change, and what is always in the process of eternally recurring—the famous proposition Nietzsche develops from

[22] Epstein, op. cit., 497.
[23] Steve Reich, "Music as a Gradual Process," in *Writings on Music: 1965–2000* (Oxford: Oxford University Press, 2002), 35.
[24] Ibid.
[25] Wim Mertens, *American Minimal Music: La Monte Young, Terry Riley, Steve Reich, Philip Glass*, Trans. J. Hautekiet (London: Kahn and Averill, 1983), 89.

Stoic *cairos* "infinite time...at once delimited and made present"[26]—cyclical time offers a unique model for the totalization of infinity, in the notion of the eternal which is both transcendental and imbued with a present transformative energy.

On closer inspection, Reich's temporal exposition follows quite closely the suggestion made earlier regarding the persistence of temporality within the Real. As far as ancestral time is concerned, any performance of *Piano Phase* begins and ends. In so doing, it occurs within what philosopher of time Steven Savitt might call a temporality of *absolute becoming*. Absolute becoming reaches precisely the same conclusion from the perspective of metaphysics as Meillassoux's ancestral time does from a strictly non-metaphysical perspective. Temporal passage is, quite simply, "the ordered occurrence of events."[27] Savitt asserts that there is no *intrinsic* connection "between this sort of passage and either freedom, spontaneity, and emergence on the one hand, or determinism, necessity, and reductionism on the other."[28] I would agree with all except necessity, which I assert is always implicit in such passage, although admittedly what I refer to here in echoing Meillassoux's proof (which ties contingency to necessity) emerges in a very different philosophical register from that of Savitt. Meillassoux's thought is unapologetically speculative, whereas for Savitt there is nothing self-consciously profound about absolute becoming,[29] And upon this continuum, for genuine events and singularities to take place, being must operate within the often banalizing, but for that no less significant, ambience of the Real.

This is the first indifferent and unswerving modality of time in *Piano Phase*. If we are to understand in *Piano Phase* anything of significance—particularly with respect to temporal irregularity or aesthetic undecidability, here termed simultaneity—such significance is always situated in the infinite unfolding of the Real, indifferent to any particular reality as such. Irregularities define themselves within or

[26] Giorgio Agamben, "Critique of the Instant and the Continuum," in *Infancy and History: The Destruction of Experience*, trans. Liz Heron (London: Verso, 2007), 111.

[27] Steven F. Savitt, "On Absolute Becoming and the Myth of Passage," in *Time, Reality and Experience*, ed. Craig Callender (Cambridge: Cambridge University Press, 2002), 160. Obviously the ontological significance assigned by Badiou and Savitt to the event differs.

[28] Ibid., 165.

[29] Ibid., 165–6. There is a striking resemblance between this proposition and Lyotard's statements on the sublime quoted below.

against a finite entity—the composition—that appears or begins, and disappears or ends (there can be an actual desistance or a projected ending of the imagination). That these irregular entities (or sub-sets) might institute radical doubts as to this appearance or disappearance, might seem to manipulate the steady flow of chronological time—all this occurs within the indifference of the Real which is essentially infinitely divisible or reinterpretable, irreducible to any entity or number of entities, precisely because such entities can only be contingencies that take place, and hence are inscribed under the law of the Real.

In this light it is possible to interpret Reich's claim that the listening process "always extends farther than…can [be] hear[d]" in two senses.[30] First, as the affirmation of the genuine infinity that inhabits the subject when, as Badiou suggests is the case in true subjects, it is tied to the infinity of thought.[31] But equally, we discover individual human subjects whose judgments are always relative and, inasmuch as they might be seen as perturbances in an uninterruptable path of the Real, are in excess of the neutrality of any information offered up for interpretation. The second interpretation of Reich's claim regarding extension resonates through Stoianova's claim that minimalism "generat[es] the present at each moment. Aimless wandering without beginning, multi-directional motion without cause or effect."[32] It invokes simultaneously the inherent directionality of the work, as well as the multiplicity in the experience of the listener. This is then the second, *cairological* temporal modality of *Piano Phase*—the multiple lines followed by individual listeners that the aural ambiguity of the phase shift offers up for decision as restless simultaneity.

Mertens is correct to identify in such process music a conflict between so-called clock time, which he associates with the dialectic progression of history, and macro-time, which he claims is a "higher level…beyond history…which has been called *now* or *stasis* or *eternity*,"[33] and he concludes that minimalism "attempts to unite the historical subject with non-historical time."[34] The problem with

---

[30] Steve Reich, "Music as a Gradual Process," op. cit., 35.

[31] Bracketing here an apparent tension between Meillassoux's thought on correlationism, which must in any case be understood in reference to determinate rather than indefinite entities.

[32] Ivanka Stoianova, qtd. in Mertens, op. cit., 89.

[33] Mertens, op. cit., 92.

[34] Ibid., 92.

Mertens' assertion is less with regard to his separation of two types of time than in the failure to apprehend that these are not in any sense contradictory. This failure rests on a perplexing inability to distinguish the indifferent passage of clock time from the Real, and on an incorrect conflation of eternity (that which is transcendental, immanent) with infinity. There exists (and has for quite some time existed) an exact mathematical demonstration of actual infinity that is precisely unconflatable with the ineffable endlessness that has been the dominant metaphysical understanding of infinity. It is upon these that Badiou's ontology rests. Infinity is hence no longer a surreal category forwarded to take account of an inexpressible divine spark but becomes, in the hands of Badiou, the indifferent condition along which any being *qua* existence must pledge itself as difference, in radical solidarity, in order to revive meaning and truth for the subject. This conceded, historical time is necessarily contingent and characterized by multiplicity—multiple and limited paths of individual entities that become more and more specific as they emerge in affirmation of specific singularities or truths within the Real.

It was suggested above that cyclicality—which might appeal to the dominant Greco-Roman "circular and continuous time"[35] or to the later Stoic *cairos*—presents a third temporal modality to the ambience of the ancestral, or of multiple individual historical times that are exposed at the moment of the phase shift. Mertens emphasizes that "the piece is built up cyclically; the second player starts playing a [... beat] further on in relation to the first player after each acceleration so that, eventually, after a certain number of repetitions, both players reach unison again."[36] That the same melodic material is phased against itself in the first and third cycles accounts for the composition's simultaneously symmetrical and cyclical structure. A singularity can be structured in such a way that a continuous displacement of its elements will amount to a return to the original material. It is all too easy here to overlook the most obvious aspect of cyclicality: that the determination of a cycle rests entirely on its being in relation to a stable, or relatively stable, point of reference. The progress of the second pianist—moving in and out of phase (a shared pulse) with the unchanging stability of the first pianist—is only cyclical inasmuch as

[35] Agamben, op. cit., 100.
[36] Ibid., 49.

it discovers, in acceleration, the temporal technique of forcing open an otherwise closed system of a common pulse, at its most strict: of unison and repetition *ad infinitum*. Cyclicality only presents the illusion of an eternal recurrence of the same, the *experience* of cyclical time that effectively emphasizes the significance of evental distension. In fact, the material it presents in each piano is non-identical. Distinct temporal paths continue to be followed by each performer, reasserting the steady progress of ancestral time (always underpinned by the steadiness of one piano in relation to the other) and the Real. The start and end of the three internal cycles that constitute *Piano Phase*, and of the composition as a whole, while emphasizing the radically unsettling potential of restless simultaneity and the phase shift, do not bend time back on itself in any ontologically satisfying way.

4.6 In a suggestive patterning of the convergence of the human body with the physical nature of the environment, Michel Serres suggests that

> [t]he more the human body is young and the more it is possible, the more it is capable of multiplicity, and the more time it has: not time in its length and duration, but the more kinds of time, the more varieties of river beds it has to flow down, the more valleys it has before it. The more undetermined it is.[37]

The youth of the body we might here associate with the experiential novelty that inheres in the unsettling listening experience we encounter in the case of *Piano Phase*, an experience I call *simultaneity.* Restless simultaneity is a condition of temporal multiplicity that ties together two modes of undecidability, the first of perception and (sensory) experience in relation to an entity, the second on the actual constitutive undecidability of the entity itself. Entities that operate according to such restless simultaneity subtract themselves from the Real as the singular multiplicity (the composition's existential aspect) that appears or is experienced as multiple singularities (the aesthetic or sensory experience of simultaneity in relation to the composition). But, to the extent that its primary characteristic is its undecidability—its restlessness—simultaneity makes no claims to be able to translate or alter the Real as such.

---

[37] Michel Serres, *Genesis*, trans. Geneviève James and James Nielson (Ann Arbor, University of Michigan Press, 1995), 32–3.

The promise of a speculative realism with regard to the experience of simultaneity is that multiplicity and infinity are no longer markers of some deferred utopianism but quite simply the immanent properties of being, retained in existence through the emergence of events and their institution as genuine novelty. In light of Meillassoux's convincing argument regarding the Real, I now offer that, if the condition of the Real is genuinely multiple, this does not in any significant sense contradict the linear, irreversibility of time. As Badiou suggests, events that institute genuine novelty occur as a subtraction from this multiplicity. These subtractions institute the possibility of that which is strictly infinite—a generic truth procedure, a true subject-thought[38] unlimited by the finitude of human existence. In this case, they do not reduce what is in any case the strictly irreducible quality of multiplicity, but rather they exist as an ontological subtraction that is also a supplement.

A similar phenomenon is indicated above in terms of the confluence of linear and cyclical time in *Piano Phase*, their coexistence being realized by the smallest subtraction or addition (here, in terms of acceleration or deceleration). In this case, the simultaneity of individual temporal trajectories followed by individual subjects in listening to the composition, can be seen as the play of finitude—literally the finitude and limitedness of individual perception of our bodies—along the genuinely infinite consequences of an event (truth, subject-thought): a play that itself takes place in the overarching multiplicity of the Real, without reducing the fact that ancestral time is an essential quality of all that exists.

What, then, of the situations in which subjective time delineated according to individual will is asserted over any of the considerations mentioned above—in which a listener just ignores the complexities unfolding in the composition, hearing a single if undifferentiated wall of sounds? In such cases, the positing subject seems to forcefully effect a present tense—a compulsory, if often unintentional, consensus of sorts—often at the expense both of Real time and of the temporal play

---

[38] See Alain Badiou, "Truth: Forcing and the Unnameable," *Theoretical Writings*, ed. and trans. Ray Brassier and Alberto Toscano (London: Continuum, 2006), 121–136. I discuss aesthetic aspects of this truth in Marc Botha, "The Right to Theory As The Right to Truth: Resisting Resistance and Para-Ontology," *TkH: Teorija koja Hoda / Walking Theory*, 6, "Pravo na teoriju/Right to Theory" (Belgrade, Serbia, 2008): 31–9.

that marks existence. This positing subject, a subject that exists towards finitude and destruction and not infinity, is placed in the terrifyingly threatening position of being able genuinely to extinguish potentiality itself—the possibility that being takes place. In addressing this threat, I suggest that the experience of restless simultaneity erodes the sort of positing subjectivity whose relation to truth and time is habitually one of structural violence and assertion and that forces itself blindly in relation to other entities.

Despite the tension that may result from Reich's phasing—a tension I associate here with the sublime—the temporal trajectories of *Piano Phase* exist in non-coercive relation to each other and the listening subject. They expose a vulnerability at the heart of contemporary experience that seems intimately tied to the constitutive complexity of our human organismic experience.[39] This experience, which habitually manifests as "an indefinitely multiple memory of all the pasts we have inherited and to which we are sensitive,"[40] provokes an oblique but promising form of community. The phase shift, undecidable and multiple, unites seemingly opposing temporal modalities in their restlessness precisely at this threshold of undecidability, an empty anonymous space where the operation of indecision is itself a type of decision, an example of that which Agamben terms "[w]hatever singularity," the very process of demonstrating oblique belonging, of exemplarity, "which wants to appropriate belonging…and thus reject all identity and condition of belonging."[41] Inasmuch as restless simultaneity is the experience of this exemplary operation, Steve Reich's *Piano Phase* provokes such a contradictory community. Perhaps this is community in the Real.

5. Phasing out: Simultaneity and the Sublime

5.1 It is the sensory and conceptual ambiguity most often named sublime that is perhaps most instructive in holding together what I here call simultaneity. The sublime manifests as a question of alterna-

[39] Isabelle Stengers, "Complexity: A Fad?," in *Power and Invention: Situating Science*, trans. Paul Bains (Minneapolis: University of Minnesota Press, 1997), 13–6.

[40] Ibid., 17.

[41] Giorgio Agamben, *The Coming Community*, trans. Michael Hardt (Minneapolis: University of Minnesota Press, 1993), 87.

tion—a play of oppositions, prompted aesthetically, between form and formlessness, pleasure and pain, control and powerlessness. Is this not what we encounter in any performance of *Piano Phase*?[42] We saw quite precisely how, despite the ease with which certain formalist models map onto the composition, something remained under-articulated through such models. This apparent failure of form occurs on account of the migration of form into the process of formation—evidenced clearly in the multiple temporal trajectories determinable in relation to the phase shift. Formlessness is thus the prevailing order to the extent that in the process of formation no appropriate form can be aesthetically determined, except by projecting onto the composition a completion—an ending. But, although the composition does contain certain suggestive structures, this most basic element of its start and its completion is the very thing that cannot be deduced from any cue in the composition itself (intervention of instructions aside), since any of the cycles might legitimately be repeated. Time is embodied, limited, and formed only retrospectively to the extent that the piece begins and ends, and contingent stabilities are maintained in the sections that punctuate the phasing process in general.

5.2 Here it is possible to seek clarification in a celebrated essay of Jean-François Lyotard, *The Sublime and the Avant-Garde*, which, in presenting the event in terms of the interrogative mood, well characterizes the problematic yet stimulating undecidability that invades the restless simultaneity evident in *Piano Phase* and that is associated with the sublime:

> The event happens as a question mark 'before' happening as a question. It happens is rather 'in the first place' *is it happening, is this it, is it possible?* Only 'then' is any mark determined by the questioning: is this or that happening, is it this or something else, is it possible that this or that.[43]

---

[42] A potentially significant distinction is retained here between the immediacy of live performance and its technological mediation as recording.

[43] Jean-Francoise Lyotard, "The Sublime and the Avant-Garde," in *The Inhuman: Reflections on Time*, trans. Geoffrey Bennington and Rachel Bowlby (Cambridge: Polity, 1991), 90.

For Lyotard, the marker of the sublime is very precisely in the presentation of the unpresentable.[44] There can be little doubt that existing in the conditions of the Real, something is presented in *Piano Phase*. We have also deduced that there is something unpresentable that resides in the process of shifting phase. The question that concerns this discussion most directly, however, is whether or not it is possible to conflate this unpresentable with the event. Clearly, Lyotard would affirm this. Following Kant, the unpresentable aspect in the sublime emerges from a failure of form to negotiate content and concept, but this Lyotard extends to a properly ontological level. The interrogative mood, the mark of the event as a question—as a radical and sublime uncertainty—occurs in relation to existence. Properly epiphanic, Lyotard's event opens up the metaphysical problem of potentiality—"is it possible?".

5.3 The charge of the future thus resides not *in* the present but *as* the very presentation of the present. Returning to Lyotard, the interrogative is the mark of presentation of the present, the way in which the present obliquely enters presentation. The problem for Lyotard is that real presence always seems to imply a decision as regards the sublime question—that the question moves towards its predication in a sublime object, entity, or situation.[45] But does this decision imply an exhaustion of potentiality? Lyotard tells us that the sublime question as event is paradoxically always withheld and announced: "*Is it happening?*... [T]he mark of the question is 'now', *now* like the feeling that nothing might happen: the nothingness now."[46] Lyotard's move from the indefinite, sublime, suspension—"nothing might happen now"—to the predicative "the nothingness now," figures for the presentation of impotentiality in the act.

Let us be clear. Lyotard suggests that the mark of the event is its sublime uncertainty—will it happen, will it not? As such, it evades any normal presentation because it precedes manifestation. Offered to the senses, taking place as its own uncertainty, there is something in the sublime event that remains unpresentable. For this reason, any possible resolution to the question—entity, manifestation, presentation—

[44] Ibid. Also, Jean-Francois Lyotard, *The Postmodern Condition* (Manchester: Manchester University Press, 1984), 78, 81.
[45] Lyotard, "Sublime," op. cit., 93.
[46] Ibid., 92.

must first be suspended as the force of the question itself. The force of the question is thus the presentation of the unpresentable part of the event—of pure ontology. For Lyotard, the sublime question is what allows the unpresentable (the positivity of the event) to offer itself to presentation as negative presentation—the future guaranteed by the event *now*.

The point of disagreement, from the perspective of the sort of realism I expose above, occurs in the gap between the eventality of an event, the moment of its pure happening as a rupture, as trans-being, and its appearance. Following Badiou, the present suggestion is that an event cuts across being and non-being and in so doing produces a situation that is absolutely new. This Badiou calls the Void of a situation. There exists in relation to the Void of an event the following paradoxical condition: it is a genuine rupture within pure being that is still part of pure being. It comes unexpectedly from inconsistency, and as such it carries with it this inconsistency. But it cannot be self-reflexive. The event does not know itself as such, and consequently there is nothing in the event *qua* event that can be considered genuinely epiphanic. It does not *reveal itself* as the source of a startling new discovery or process. Any such revelation, if it occurs, takes place retrospectively: an event is decided upon, named outside of the evental site itself.[47] Were this not the situation, were it possible to bring into presentation the eventality of the event, what we would really be presenting is the act of a rupture of being and the Void that this rupture exposes—in other words, inconsistency and destruction. It would be tantamount to claiming that in order to affirm the existence of death, one first needs oneself to die. So there is a gap between an event and the decision that subsequently directs this event towards appearance, if this decision takes place at all.

An event can only appear in existence at a certain point after it has been decided. The very fact that it appears, attests to its not being a presentation of the event. To elucidate: if one accepts Badiou's argument that the presentation of an event amounts to a presentation of inconsistency or of the Void (since inconsistency is at the very heart of an event to the extent that it emerges unexpectedly from the inconsistent multiplicity of being *qua* being), it is impossible that this presentation can take place in existence without simultaneously destroying the

[47] The evental site alludes to the space or topography Badiou claims is proper to an event.

existence of that which it seeks to present. So, when an event appears in existence, it is not the event itself but a trace of the event. This trace, what Badiou calls a subject, is a testament to the decision that was taken as regards an event. It maps the consequences of the event under the name of the event.

It is precisely on this count that Lyotard and Badiou diverge as regards the event.[48] In asserting that the sublime is the presentation of the event—the manifestation of an effective coincidence of the answer, "[h]ere and now," fully manifest in the sublime aesthetic object, with the question that is coincidental with the immediacy of the event, "is it happening?"[49]—Lyotard asserts that what is taking place is nothing other than the presentation of the unpresentable. What is asserted, in other words, is the presentation of the event, which to Badiou is strictly impossible. In so doing, Lyotard effectively forces the appearance[50] of the event in the sublime work to coincide with its presentation. Henceforth, the event is located in being *qua* existence, having been presented in the aesthetic work in the ontologized form of a sublime feeling. The unpresentable which needed to remain outside of presentation[51] attested to only as an interrogative feeling of the intensity of the unpresentability of the event, has thus been forced into presentation.

The error lies precisely in situating the site of the event in appearance and being *qua* existence, rather than in the trans-being of the evental site, which is characterized by pure multiplicity, before it is by the finite markers of existence. If we reconsider Lyotard's fundamental formulation of the event through the sublime—the presentation of the unpresentable—are we not returned precisely to what I

[48] The key differences between Badiou and Lyotard might be situated in their apparent divergence regarding, with Badiou insisting that being *qua* being is thinkable and presentable in mathematics, while Lyotard follows in the wake of Heideggerian ontology, in which being is always submitted to the conditions of a certain givenness. If Badiou relies firstly on axiomatics, Lyotard refuses to subordinate experience to thought.

[49] Lyotard, "Sublime," op. cit., 90.

[50] The point of appearance might be described as the moment at which an entity enters into a situation or a world. From this moment on it can be relationally affirmed as something that exists. See Alain Badiou, "Being and Appearance," *Theoretical Writings*, op. cit., 175. Appearance of temporal processual entities in *Piano Phase*, and indeed of the sublime, is precisely what is presently in question.

[51] Which, Badiou claims, can only be presented indirectly in the re-presentation of the metastructure which guarantees the consistency of the elements of any situation.

propose above as an evental distension? The unpresentable, whether from the perspective of Badiou or Lyotard, marks the substance of the event. Accepting this, presentation clearly involves an attempt to find appropriate objects or processes to approximate the event. This approximation does not take the form of a mapping of the consequences of the event. Rather, recalling the argument earlier, evental distension involves the presentation of entities—objects or processes—that attempt to reproduce the *eventality of the event*. For Lyotard, such entities are sublime. Matching the tensions between presentation and the unrepresentable, the sublime steps in to answer *here and now* to the sublime question: *Is it happening?* Sublime feeling was advanced as the primary experience (the others being formal and theoretico-ontological) of the restless simultaneity instigated by the phase shifting in *Piano Phase*. One might claim that, through an examination of the sublime, the phase shift reveals itself to be exemplary of what I refer to above as evental distension. We must remember, first, that evental distension is not some mystical property but is subject to the same condition of the Real as all entities. Phasing demonstrates that the coincidence of a steady pulse (temporal) and melodic (abstract spatial) consistency in the various stable sections of the composition are contingent on very specific conditions. The gradual dissolution of these conditions, as these stabilities dissolve into instability and undecidability, seem to point to the pervasiveness of the condition of contingency in the Real.

A more general structural problem enters the fray in that it becomes difficult to pin down the location of an evental distension in *Piano Phase*. Despite its formal alternations between disunity and uninterruption, its fragmentary cyclicality and processual disruptions, there can be little doubt that *Piano Phase* is still a single work. Where might evental distension be located then: in some specific moment within the phasing process, in the first phase shift, in all the phase shifts, in the composition as a whole? To address this problem, one must recall that in being directed to the eventality of the event, evental distension seeks to demonstrate *that* the event happens. As such, it presents a less localized and more processual aspect of what Lyotard claims in relation to the sublime and the event. To be clear, the situation of restlessness is no less under the all-inclusive sway of the Real and offers no possibility of transcendence in this regard. To this extent it is necessary to note a definite disagreement with and divergence from Lyotard's exact nomenclature. But to the extent that he realizes that one of the primary opportunities for approaching the eventality of the event—the sublime, the question in its absolutely non-predicative state—his proposition seems exemplary of the same ontological desire expressed through evental distension. And what is at stake in evental distension? I would suggest it is nothing less than an attestation to the Real. In a world where the virtual is increasingly all-pervasive, there seem few more urgent vocations.

References

Agamben, Giorgio. *The Coming Community*, translated by Michael Hardt. Minneapolis: University of Minnesota Press, 1993.

———. "Critique of the Instant and the Continuum." In *Infancy and History: The Destruction of Experience*, translated by Liz Heron. 99–115. London: Verso, 2008.

———. *What is a Paradigm?: A Lecture by Giorgio Agamben*. European Graduate School, August 2002. Stable URL: http://www.egs.edu/faculty/agamben/agamben-what-is-a-paradigm-2002.html.

Badiou, Alain, "Being and Appearance." In *Theoretical Writings*, edited and translated by Ray Brassier and Alberto Toscano. 167–181. London: Continuum, 2006.

———. *Being and Event*. Continuum: London, 2005.

———. "The Event as Trans-Being." In *Theoretical Writings*, edited by Ray Brassier and Alberto Toscano. 99–104. London: Continuum, 2006.

———. "Truth: Forcing and the Unnameable." In *Theoretical Writings*, edited and and translated by Ray Brassier and Alberto Toscano. 121–36. London: Continuum, 2006.

Botha, Marc. "The Right to Theory As The Right to Truth: Resisting Resistance and Para-Ontology." Translated and published in Serbo-Croat by Irena Sentevska. *TkH: Teorija koja Hoda / Walking Theory* 16, "Pravo na teoriju/Right to Theory." Belgrade, Serbia, 2008. 31–39.

Derrida, Jacques. "Ellipsis." In *Writing and Difference*, translated by Alan Bass. 371–78. London: Routledge, 1978.

Epstein, Paul. "Pattern Structure and Process in Steve Reich's *Piano Phase*." *The Musical Quarterly* 72: 4 (1986): 494–502. http://links.jstor.org/sici?sici=0027-4631%281986%2972%3A4%3C494%3APSAPIS%3E2.0.CO%3B2-A,

Hallward, Peter. *Badiou: A Subject to Truth*. Minneapolis: University of Minnesota Press, 2003.

Kant, Immanuel. *Critique of Judgment*. Trans. Werner S. Pluhar. Indianapolis: Hackett, 1987.

Lyotard, Jean-François. "The Sublime and the Avant-Garde." In *The Inhuman: Reflections on Time*, translated by Geoffrey Bennington and Rachel Bowlby. 89–107. Cambridge: Polity, 1991.

———. *The Postmodern Condition*. Manchester: Manchester University Press, 1984.

McTaggart, J. Ellis. "The Unreality of Time," *Mind*, New Series, 17: 68 (Oct., 1908): 457–74.

Mertens, Wim. *American Minimal Music: La Monte Young, Terry Riley, Steve Reich, Philip Glass*. London: Kahn and Averill, 1983.

Meillassoux, Quentin. *After Finitude: An Essay on the Necessity of Contingency*. Translated by Ray Brassier. London: Continuum, 2008.

Potter, Keith. *Four Musical Minimalists: La Monte Young, Terry Riley, Steve Reich and Philip Glass*. Cambridge: Cambridge University Press, 2000.

Reich, Steve. "Come Out—Melodica—Piano Phase." In *Writings on Music: 1965–2000*. 22–25. New York: Oxford University Press, 2002.

———. "Music as a Gradual Process." In *Writings on Music: 1965–2000*. 34–36. New York: Oxford University Press, 2002.

———. *Piano Phase (for two pianos or two marimbas)*. London, Universal Edition, 1980.

———. "Piano Phase." *Early Works*. Perf. Steve Reich, Nurit Tilles, Edmund Niemann, Russell Hartenberger. Nonesuch, 1990.

Savitt, Steven F. "On Absolute Becoming and the Myth of Passage." In *Time, Reality and Experience*, ed. by Craig Callender. 153–67. Cambridge: Cambridge University Press, 2002.

Serres, Michel. *Genesis*. Translated by Geneviève James and James Nielson. Ann Arbor: University of Michigan Press, 1995.

Stengers, Isabelle, "Complexity: A Fad?." *Power and Invention: Situating Science*. Translated by Paul Bains. 3–19. Minneapolis: University of Minnesota Press, 1997.

CHAPTER TWELVE

MUSIC-MAKING TIME?

Helen Sills

Summary

Our sense of time and temporality is largely founded upon our sense of hearing. The potential of the auditory system to experience an expanding range of temporalities in the course of a human lifetime is vast. This paper traces the limits and constraints that are vital to the formation and development of our senses of hearing and time and, in so doing, finds that each level of limits and constraints gives rise in turn to a qualitatively new level of temporal experience, ranging from linear clock-time to timelessness. Limits and constraints govern the physical formation of the auditory system from the earliest weeks of our existence, as cells are precisely selected and neuronal connections are finely timed. As these physical developments are completed in childhood and early adulthood, a qualitatively new set of limits and constraints establish a sense of linear time and a new cognitive awareness of movement in time as having *meaning*. The culmination of this process may be seen in a composition such as Stravinsky's *Requiem Canticles*, where the relationship between sound, movement, and time is very highly developed. Stravinsky limits and constrains his sound materials to create immense time worlds that stretch into timelessness, despite their brief performance time.

In the beginning is…SOUND. Our sense of hearing is the first of our five senses to develop. The first sounds that we hear in the womb introduce us to the rhythms and patterns of human activity and prepare us for the physical world of linear time into which we will soon be born. This paper explores some of the vital limits and constraints that organize our processing of sound and the ways in which they shape our developing sense of time. Our natural development in the dimension of time is determined by a continuing process of limits and constraints.

Before we are born, the growth of the auditory system follows a precise timetable, stimulated by a range of sounds that is limited but already registering the basic rhythms, patterns, and pitch contours of human life. In childhood, an innate, *qualitatively new* set of organizing constraints emerges to promote the development of our cognitive skills and the bases for our understanding of *meaning*. These limits and constraints help us to orient ourselves within linear time, to order

Jo Alyson Parker, Paul A. Harris and Christian Steineck (Eds), Time: Limits and Constraints, pp. 289–304

ourselves within it, and to live more easily with its patterns and cycles. If we live in a musical environment, or undergo even basic musical training, our understanding of sound patterns and their meaningful movement may continue to develop, so that we expand our experience of linear time into the greater variety of temporal qualities that lie in the world of sound. If we continue to develop our listening skills in adult life, we find that yet another *qualitatively new* level of limits and constraints can release the mind and spirit into an immense variety of temporal experience and even timelessness. Lastly, this paper explores a high point of musical creativity with sound and time: the organizing constraints employed by Igor Stravinsky in *Requiem Canticles*, his final masterpiece of movement and meaning in temporal qualities.

This paper not only selects examples of limits and constraints that organize the human life span but also shows how the emergence of *qualitatively new* levels of limits and constraints bring about a development from the general, the cellular, microsecond time-world in which every human being is formed before birth, to the very particular example of musical genius and the very specialized musical creativity in many qualities of time and even timelessness, which becomes possible through the continued constructive power of limits and constraints.

1. Limits and Constraints on the Formation of the Auditory System

Let us look briefly at some of the limits on movement and timing that constrain and focus the formation of the auditory system before birth.

(1) Neurons must be produced in the right numbers and, from the eighth week, *constrained to follow a structure of highways to their correct locations.*[1] These are determined by biochemical signaling

[1] Protoneurons, such as those destined for the auditory system in the temporal lobes are produced between the third and about the eighteenth week after gestation, their rate of production peaking at the seventh week. Production averages perhaps 500,000 neurons *per minute* during this period. By the end of 7–8 weeks (the end of the embryo period), the main areas of the brain are in place, the cells of the future neo-cortex are beginning to appear and the brainstem has developed an elaborate system of projections to influence the migration and differentiation of cells into neural and glial cells (non-neural support cells).

and by glial cells which act as guides and even provide nourishing "rest stops." The migration is orderly: the first protoneurons to leave stop in the closest layer of their destination, the next make a second layer, and so on, until all six cortical layers are populated.

(2) The neuronal population is *pruned* by "programmed cell death" ("apoptosis"). Every neuron has to compete for its place in the general scheme; otherwise it is eliminated.[2] A rapid expansion of connections between cells occurs in spite of the fact that different brain structures emerge at different times, and there is much sorting, editing and correction of initial errors to focus the system's functions.

(3) Once targets have been achieved, different types of cells begin to contribute their individual patterns of electrical and chemical activity to the growing neural networks.[3] Complex signal transduction schemes between molecules permit very precise *temporal* control of cell behavior over a wide range of times, some allowing information to be transferred rapidly, others more slowly and with longer-lasting effect. To encode information that varies so widely over time, the concentration of relevant signaling molecules must be carefully controlled: having multiple levels of molecular interaction facilitates the intricate timing of diverse events.

(4) The developing brain is also *shaped by action potential activity.*[4] Patterns of electrical discharges trim axon branches, adjust intercellular chemical messages, guide the formation of cell layers and determine whether connections are to be strengthened or eliminated. *Timing* is crucial: ganglion cells generate bursts of firing lasting a few seconds, after silences of between 30 seconds and 2 minutes. These predictable and rhythmic patterns sort and correlate neighboring cells and connect them into groups according to function. Asynchronous waves in seemingly random directions

[2] Those neurons that survive begin to sprout axons and dendrites. As they travel and compete to connect, growth cones on the tips of their axons grow towards molecules of specific trophic factors that they find along the route, guiding them to the 'correct' address. (This is an astonishingly precise journey that has been compared to that of a baby crawling from New York to a specific street and house in Seattle!)

[3] The amplification of molecular signals, which may occur at several stages along transduction pathways, has the advantage that a small number of signal molecules can ultimately activate a large number of target molecules, guaranteeing a physiological response.

[4] Action potential activity is necessary to ensure the withdrawal of side branches from the axons and the elaboration of the terminal branches.

ensure that cells that must *not* 'fire and wire' together are properly segregated.

(5) Between 15 and 25 weeks cells are *constrained into orderly layers* in the neo-cortex, so that, by 6 months, even the foundation for the higher psychological functions are in place. Cells in the corpus callosum, carrying information between the two hemispheres, appear at 10 weeks and during the next 3 months undergo vast *selective pruning*, confining contacts between the hemispheres to certain cortical zones and separating groups of *inter*hemispheric cells from *intra*hemispheric cells.

By the time we are born the—as yet—immature wiring of the auditory system has already been constrained and shaped, to a precise timetable by genetic instructions, bio-chemical signaling, programmed cell death, signal amplification, waves of electrical firing patterns, and the crucial stimulation of environmental sounds.[5]

2. The Initial Processing of Sound and Time

Like sheep, goats and guinea pigs, we process sound before we are born. Auditory centers develop in the brain stem at 13 weeks, and the auditory system begins to process sounds between 16 and 20 weeks before it is anatomically mature. The temporal lobes begin to take over the processing of sound as their connections with the cochlea start to mature from 25 weeks, but even until four months after birth, functional activity is restricted largely to the top outer layer of grey matter.

Our first sensations are somewhat muffled, of limited pitch range, and chiefly derived from the rhythms of the maternal environment, such as heartbeat, breathing, digestion, and footsteps. This muffling does not appear to prevent the hearing of prosody, rhythm, accentuation, and even the difference between vowels and consonants and male and female voices. But the sensation of timbre *is* affected because all but a few lower partials are rendered inaudible.[6]

---

[5] See Schatz 1993, 15–26; Purves et al. (eds.) 2004, 165–7, 309–11, 516–25; and Gregory (ed.) 1991, 102, for the material of this section.

[6] Pitch processing cells arrange themselves tonotopically along the length of the inner ear, with high frequencies at the base of the basilar membrane and low frequencies at the apex. At first the response range of the cochlea to pitch frequency is limited to between 200–1000 Hz, but as this broadens, so does the ability to separate

The heartbeat of the fetus aligns itself with that of the mother from about 32 weeks. The fetus not only registers the memory of internal and external patterns of sound *but also the emotional connotations that accompany them.* Movements—allied to emotions such as joy, fear, or anger, which are transmitted biochemically—are the first movements to be associated with *meaning*. Evidence suggests that we develop a sophisticated pre-natal ability to process and assimilate sound patterns at complex levels. Very young babies can distinguish their mother's voice, prefer to hear people speaking their mother's language, and are able to "recognize" a story heard prenatally. These experiences also mark the beginning of the human ability to construct cognitive representations. The fetus bonds with the mother as signals from sounds, movements, and the biochemical expression of emotional states come together to form a representation of her. Just as in later life, the attentive listener finds that musical sounds can synthesize into a sound 'object', so for the unborn child environmental influences come together to convey the presence of an Other, which has spatial extent and is filled with rhythms associated with body movement (Parncutt 2006, 15).

3. New Constraints Bring Meaning to Movement

Soon after birth a *qualitatively new* set of organizing constraints and predispositions appear, to structure our hearing as a means to bring coherence and intelligibility to the mass of aural information that we receive.[7] They continue to develop through childhood.

simultaneous frequencies, discriminate rapid sequences of sounds (as in speech) and hear quieter sounds (Parncutt 2006, 3).

[7] At birth, however, the visual system comes into play and usually becomes the more relied-upon source of information. Unlike the ear, the eye *immobilizes and isolates* objects in order to recognize them. The ear by contrast has the ability to process several streams of sound at once. The sense of hearing connects us to the flexibility and flow of changing patterns in time, especially several patterns moving at different speeds simultaneously. It also bridges the qualitatively different domains of moving sound and the motion of physical gesture, enabling us to 'listen' with attention, to connect with the living essence of things, and to be moved by them with our whole being. Watching *silent* movements in space—however loving, tragic, or war-like—does not move us to nearly the same degree as hearing the sound shapes that correlate to their motion and *e*-motion, as the makers of early silent films soon discovered. Even the great Diaghilev, once planned a ballet that would be silent, without music. Deeply unhappy at the effect of emphasizing spatial movement without sound, he later wrote to Stravinsky: "After having 32 rehearsals of the *Liturgie* we have concluded that absolute silence is death.... Therefore dance action must be supported not [necessarily]

Newborn babies (1) are predisposed to respond to music-like qualities of rhythm, variations in pitch inflection and dynamics, especially from the care-giver, and love to be sung to by a high-pitched, slow, emotional voice; (2) prefer the simple ratio intervals of 4ths, 5ths and 8ves; (3) prefer unequal step scales to equal divisions of the octave (Trehub 2003, 7).

During childhood:

(1) We find it easier to process *regular* sound events and tend to regularize those that are slightly irregular. Almost everyone, from about the age of four, can tap in time, and babies have even been known to adapt the rate at which they suck to a regular pulse.
(2) We prefer *simple duration ratios*, finding it easier to produce rhythms in a 1:2, rather than a 1:3 ratio, and we tend to categorize rather irregular events as "half as long" or "twice as long."
(3) From the age of about two months we acquire a *temporal zone of optimal processing* within a range of 300–800ms, and we are most sensitive to change if events occur about every 600ms.
(4) We are predisposed to *segment and group* events into perceptual units if they share similar physical characteristics or occur close in time. Infants as young as 6–8 months have the same ability as adults to group sounds according to timbre and pitch (although larger pitch changes are necessary for infants), intensity, tone duration, and pause duration (Drake and Bertrand 2003, 24–29).
(5) Remarkably, babies are readily aware of *relationships between sound stimuli* in which one musical element has been changed: they can, for example recognize a tune that is performed in a different rhythm or at a different pitch from usual. As long as the salient features are preserved, they can discern that the 'speed' at which sounds are presented to them can vary: until about the age of six months they will respond to a familiar tune if it is presented at a different tempo and even notice changes of tempo within a familiar tune, as long as the pitch relations are preserved. After the age of six months, infants remember the specific tempo and timbre of music with which they are familiar and fail to respond if these elements are changed (Saffran 2003, 38).

---

by music but by sounds, id est, by filling the ear harmonically... (Stravinsky and Craft, 1960).

(6) Babies prefer *consonance* and quickly learn to discriminate consonance from dissonance: unlike monkeys, babies will turn away from dissonant sounds from the age of four months!
(7) Our prenatal ability to discern *separate streams* of sound is enhanced within a few days of birth by the onset of *timbre-processing* cells.

The auditory system finally reaches maturity at about the age of 16. Some abilities to distinguish sounds that we had at birth will not be developed by our particular cultural environment, and some, like perfect pitch (an absolute memory for pitch frequency) will fade away unless they receive support from early music training. But generally speaking, these constraints on our aural processing bring universal benefits: they facilitate group dynamics, forge group identities, and allow us to enjoy and share with others all kinds of group musical activities and emotional experiences.

These constraints come into operation as the developing brain seeks to interact with its cultural environment. Innate predispositions and exploratory activity work together to gather intelligible patterns of information and meaning. From birth we respond to sound with *movement*, communicating with our caregivers by showing awareness of emotional signals, and turning our heads, for example, when attracted by change and novelty. If we watch children of all ages engaging in spontaneous musical play, unsupervised by adults, the organizing constraints of regularity, simple duration ratios, optimum pace of change, segmenting, grouping, relating, separation of streams of sound, and preference for consonance can be seen to shape their learning in two particular aspects: the experience of form and structure through their own physical movements and the exploration of emotions through relating musical movements to the movements and moods of their environment, such as trickling brooks, buzzing bees, frightening ghosts, wicked villains, crowing cockerels, hidden cuckoos, and the like.

Universally, sung and chanted games, passed on orally from one generation of children to the next, combine music, text, and physical movements such as hand-clapping, counting, skipping, imitative movements, and ball-bouncing, and they are performed in pairs, rings, or line formations. Children actively create their own opportunities for ordered musical play and constantly repeat their favorite experiences in order to deepen and develop them. Formally, children's games are frequently repetitive and cyclical, becoming more complex as repetitions accrue variations and changes. Ostinati are found in many movement patterns, which can create substantial polymetric complexity against the text and which are often interrupted by imitative movements or rhythmic changes at points of rest or emphasis in the text. Children naturally seek to participate, as a group, in formulaic yet imaginatively changing musical play in order to develop their memory for recurring events within a formal structure (Marsh and Young 2006, 290). At first, familiarity with patterns of repetition and recurrence is built up through events that occur *chrono*logically, that is, with events that always repeat in the *same* order (Bamberger 2003, 87).

At the same time, a child's emotional development guides the ways in which the movement of musical sounds become endowed with *meaning*. As we have seen, new-born babies are innately predisposed to detect emotional information from speech and singing: connecting emotional meaning and musical information would appear to be primarily an innate and schematic process. By the age of three years, children are able to display nearly the entire range of adult emotions, and, by the age of four, they can identify happiness, sadness, anger and fear in music. Children as young as four can manipulate the tempo, dynamics, and pitch of a familiar song if asked to present it in a happy or sad way, and they tend to use physical gestures to facilitate their communication of emotion. While schematic connections guide the first three years of life, growth in emotional-musical connections from the ages of three to seven years are predominantly veridical, built up from specific representations of emotion in music that arise out of their particular culture. By the age of about eight, children have acquired a significant collection of musical-emotional experiences, and many will have gathered a wide repertoire of musical styles, such that the veridical connections made between three and seven years become highly internalized and integrated with the innate predispositions of infancy. At first, meaning comes from the direct association of musical movements with external situations or moods. As these veridical connections accrue, they form a schematic collection of musical and extra-musical rules that subconsciously shape our perception of emotional meaning in music (Schubert and McPherson 2006, 202–3).

These childhood experiences of emotion and meaning in music are of immense value in developing our cognitive and imaginative faculties, our relational skills, and metaphorical thinking. For music can embody, through what Ian Cross has described as its "floating intentionality" and its "transposable aboutness," one or many meanings. In contrast to the more focused referentiality of language, it can explore and express abstraction and ambiguity in all its qualities, not in terms of objects *but in terms of events, ongoing musical structures in time*:

> And it is conceivable that music's "transposable aboutness" is exploited in infancy and childhood as a means of forming connections and interrelations between different domains of infant and childhood competence such as the social, biological and mechanical.... The floating intentionality of the music can provide for the child a space within which she can explore the possible bindings of.... multidomain representations. Hence one and the same musical activity might, at one and the same time be about the trajectory of a body in space, the dynamic emergence or signification of an affective state, the achievement of a goal and the unfolding of an embodied perspective. (Cross 2003, 51–52)

From the very first, sound has stimulated our growth, and it is vital that we continue to develop our innate aural predispositions through music-making, for they will play a vital part in our general cognitive development.

Specialized musical training during childhood builds on the twin paths of form building and the acquisition of meaning to train the mind in two ways that have important implications for our understanding of time. The first is the development of holistic memory; the second is the refined perception of expressive movement.

Holistic memory is developed by aural training and by the study of the structure of a musical composition. To hold onto a musical motif or theme throughout a composition and to visualize mentally its changing character, relationship to other materials, and structural context, is to develop the facility for relating events across time and gathering them into a unified whole. (What was the purpose of a change in pitch, or rhythm, or mood? How did a motif defy expectation or alter the character of the piece?) The recognition of repetition and recurrence develops from a perspective that is linear and chronological to one that is multi-dimensional and holistic. To follow the working out of the materials is not to compare sounds that were heard, say, one, two or seventeen minutes ago, but to relate them as a part of the whole *beyond* clock time in the newly created time world of the music.

Specialized music training not only develops an appreciation of meaning in the movement of sound, but also of *changing* meaning. Just as in the physical world, living things have their own unique way of being and moving in time, the character and meaning of a musical work is largely determined by its tempo, pacing of change, and rhythmic behavior. Rhythmic motifs convey meaning by imitating physical gestures, quasi-familiar speech patterns, the trajectories of emotional experience, or common life events, such as goals reached or matters left with open-ended question marks. Harmonic relationships can convey emotional color, a change of timbre can transport us to a different location, and a change in the intensity, texture, or volume can change the mood. Musical phrases may complement, contrast, or combine, communicating meaning by their connective 'talk'. Moments of significance create their own qualities of time, achieving depth or breadth in a different temporal 'umwelt' from that of normal everyday life. Our attention is drawn away from clock-time to the degree that we become absorbed in meaning, and to develop our sense of hearing is to continue our natural development into a wider range of temporal experience.

4. Constraints Synthesize Sounds into a Temporal Construct

Limits and constraints have so far shaped the early stages of our life. Before birth, constraints operate on the growth of the auditory system, limiting the sound-scape and its associated emotions to the physical time world of the mother. In childhood, we are predisposed to organize sound events into small related units that enable us to develop a more holistic memory for different kinds of movement that carry meaning. Finally, in adult life, limits and constraints on the form and materials of musical constructions can expand our temporal experience still further. Stravinsky's approach to composition illustrates this well. It was Stravinsky's life's work to seek meaningful "rhythmic manners" to create strong temporal qualities, and to heighten the *temporal* shape of a musical form to give it an architectural quality. He was uniquely suited to this task by having a "wonderful ear" and being nurtured in a professional musical environment and the richly evocative sounds of life in St. Petersburg.[8]

---

[8] He liked to distinguish the vibrations of notes played on the piano, and later in life analyze, for instance, the noise of an aircraft as containing seven pitches—but it also made him uncomfortably aware of radio waves everywhere.

In searching for that "strength which is born in constraint and dies in freedom," Stravinsky found that a yet another *qualitatively new* level of organizing constraints emerges. As a composer who had a strong belief in God, he felt strongly that the task of communicating meaning clearly and coherently through form and content was akin to continuing God's work in the world as a co-creator. Stravinsky identified with this strong spiritual purpose from his early works and found that, to communicate *spiritual* truth in sound, it was first necessary to make a careful selection from the infinity of possible materials, and then to impose severe limitations on them. As he remarked,

> one can build only on a resisting foundation: whatever constantly gives way to pressure, constantly renders movement impossible…my freedom will be so much the greater and more meaningful the more narrowly I limit my field of action and the more I surround myself with obstacles…. The more constraints one imposes, the more one frees one's self of the chains that shackle the spirit. (Stravinsky 1946, 65)

Moreover,

> it is more satisfactory to proceed by similarity rather than contrast. Music thus gains strength in the measure that it does not succumb to the seductions of variety. What it loses in questionable riches it gains in true solidity…. Similarity is born of a striving for unity. Contrast is everywhere. One only has to take note of it. Similarity is hidden: it must be sought out, and it is found only after the most exhaustive efforts. (Stravinsky 1946, 32)

5. Stravinsky's Constraints on Musical Materials

Stravinsky composed *Requiem Canticles* in 1965–66, calling it his "Pocket Requiem" and regarding it as a brief summary of his life's work. He imposes limits and constraints on both form and expression of meaning and in so doing releases a new level of organizing constraints, which spring from the fundamental properties of movement in time and space and engage the deepest, most spiritual aspects of ourselves. It is as though nature, given appropriate nurture, has been directing our growth towards this spiritual outcome from our earliest weeks of life.

The formal events of *Requiem Canticles* are succinct and segmented, and they relate to each other as a pattern of temporalities. The essential meaning of the spiritual text is embodied, not through expressive melody or harmony but through qualities of movement conducive to timeless meditation upon the end of time, both personal and cosmic. The text concerning petition, judgment, awe and lamentation is from the Introit and Sequence of the Mass for the Dead, and it concludes with the entire responsary "Libera Me" from the Burial Service.

(1) Stravinsky's first constraint is on *form*: he avoids the *linear* development of materials, and severely curtails the length of performance time. With the utmost economy of musical movement, its nine brief and uncompromisingly stark movements last just 15 minutes:

1' 5"	PRELUDE	
	Exaudi	1' 38"
	Dies Irae	0' 50"
	Tuba Mirum	1' 03"
3' 5"	INTERLUDE	
	Rex Tremendae	1' 10"
	Lacrimosa	1' 45"
	Libera Me	1' 00"
2' 17"	POSTLUDE	

Like icons set within a symmetrical retablo, each succinct movement is created from the pacing of rhythmic combinations that express his *spiritual* response to the text. Each employs a seminal temporal technique from an earlier work to create its own uniquely paced temporality as part of the whole construct.

(2) Stravinsky's second constraint is upon the emotions and moods that are associated with particular arrangements of pitch, both horizontal and vertical. Stravinsky limits the arrangement of the 12 pitches of the octave to *two serial row forms,* directing our attention instead to the quality of their *motion* and *rhythmic behavior* as a spiritual response to the text. He generates variety from similarity by using inverted and retrograde forms and also vertical and diagonal lines from their rotated hexachords. Stravinsky's one concession to this severe discipline is that intervals may be repeated, either to create a softer, more lyrical quality of movement or to distinguish lines from each other.[9] Sculptured from

[9] *Requiem Canticles* is created from two complementary row forms, emanating from the central pitch 'F', each having its own intervallic character, the second

> many forms of a proscribed pitch "palette," the nine brief and contrasting constructs synthesize in the memory into a uniquely shaped temporal form whose dimensions have an architectural quality.

By avoiding the "development" of materials, curtailing the performance time, and confining relationships of pitch to a multitude of ordered row forms, Stravinsky directs our attention to the differing temporality-creating movements of each of the nine short texts. He transforms the formal relationships of a work from those of rhythm and pitch, etc., into temporal relationships arising from *variable pacing* and *density.* As a result of these constraints, the form becomes predominantly temporal *and a new level of organizing features emerges.*

6. Stravinsky's Techniques for Shaping a Temporal Form

(1) The combination of regularity and irregularity:
Stravinsky juxtaposes lines of regular and irregular movement so that our attention is engaged in following their separate and differing activities. This is the case in the Prelude, where our attention is gradually deepened and drawn into the following structure by a regularly increasing number of solo instruments that repeat their lines, against the irregularly moving row form in the tutti parts.

(2) The exploration of variable density:
The second movement, "Exaudi," is built on immense contrasts of density. Slow, transparent lines of changing timbres transport us to a much lighter and larger spatial context. The great density produced by the rocking movement of the choir's repeating intervals not only expresses the intensity of their prayer but also emphasizes the vast distance that the petition—*hear* my prayer—has to penetrate. The wide variety of density evokes a vast time-scale, from the cosmic and eternal to the intimate physical gestures of human grief that are heard in the central movement—"Interlude." These physical movements anchor the work to everyday reality and enable us to relate to it.

---

containing two perfect 5ths and being more tonal than the first. Each of the nine movements is built upon one of the rows with the exception of the Interlude and the Postlude which employ both.

There are also immense contrasts of density *between* movements. The monolithic gravity of "Rex Tremendae" is created by the constant overlapping of large intervals in the choral lines, while repeated staccato chords in the flutes and lower strings beat against their opacity as if upon "the cloud of unknowing." It is followed by the wide spatial effect of "Lacrimosa," in which four contrasting timbral groups of differing densities maintain separate accompanying roles throughout, and act as a four-dimensional frame for the very expressive vocal part.

(3) Pitched row forms against human speech:
This combination promotes high activity on both sides of the brain: in the speech processing centers at the same time as in the melodic contour/timbre centers. In "Dies Irae," we move from the immediacy of high speed, pitched movement, as though traversing the universe, to the historic prophecy of King David and the Sybil regarding the end of the world. Our sense of time is dramatically deepened as the brain works intensely to process the low-pitched spoken prophecy held in tension against high-pitched, kaleidoscopically changing row forms of contrasting timbres. Stravinsky returns to this technique in "Libera Me," where he evokes the atmosphere of the Russian Orthodox Church in its ancient practice of "tuilage"—the reciting of prayers and singing of chants at differing speeds. Time is stretched and deepened as we hear the text in two pitch ranges (sometimes different parts of it simultaneously), and work to relate the two speeds of their unfolding.

(4) Varied pacing of repeating intervals:
Our temporal experience is stretched and contracted by the pace at which the dyads (repeating intervals) undulate, like sound waves through time and space. The brisk dyads of "Prelude," for example, change to the slow, beseeching dyads of "Exaudi," and the commanding, then slowly penetrating dyads of "Tuba Mirum" change to the rapid and highly charged expressive figures of "Lacrimosa."

(5) Resonance stretching into eternity to seal the form:
Densely compacted row forms are heard on very resonant instruments, imitating the strike of a fundamental note on a bell, with its partials (here the horn, with its 'partials' on flutes and harp) followed by the ringing of changes. Not only do we become absorbed in the chang-

ing resonances, but both the extended *temporal spacing* of the 'bell' strikes and the *slightly irregular number* of changing partials seem to lie slightly beyond our ability to regularize them into a conscious pattern, and lift us beyond any focused sense of time.

7. Conclusion

And so, it can be seen that throughout our lifetime, limits and constraints shape the processes by which we mature to full awareness of the temporal dimension of life. Before birth, many constraints on the prolific production of neurons shape the functioning of our most time-sensitive sensory system, the auditory system, to accustom us to the world of linear time. In childhood our hearing is subject to a *qualitatively new* set of organizing constraints that predispose us to develop the flexibility of our cognitive skills, to become aware of structure and meaning in our environment, and to deepen our experience of time beyond that of linear clock time. Ultimately, by continuing to develop our listening skills, we find that carefully selected, meaningful movements of sound, arising from *qualitatively new* constraints on linear development and expressive content associated with emotions and moods, can synthesize into a holistic structure that appeals to that part of ourselves that is spiritual. Becoming absorbed in a unified sound object of diverse musical perspectives and temporal dimensions can free us, temporarily at least, from the experience of linear time, and allow us our first intimations of the nature of eternity.

References

Bamberger, Jeanne. 2006. "What develops in musical development?" Pp. 69–91 in *The Child as Musician*, ed. Gary E. McPherson. Oxford: Oxford University Press.

Cross, Ian. 2003. "Music, cognition, culture, and evolution." Pp. 42–56 in *The Cognitive Neuroscience of Music*, ed. Isabelle Peretz and Robert J. Zatorre. Oxford: Oxford University Press.

Drake, Carolyn and Daisy Bertrand. 2003. "The quest for universals in temporal processing in music." Pp. 21–31 in *The Cognitive Neuroscience of Music*, ed. Isabelle Peretz and Robert J. Zatorre. Oxford: Oxford University Press.

Fischbach, Gerald D. 1993. "Mind and brain." Pp. 1–14 in *Mind and Brain*. Freeman and Co.

Gregory, Richard L., ed. 1991. "Brain development." Pp. 101–10 in *The Oxford Companion to the Mind*. Oxford: Oxford University Press.

Marsh, Kathryn and Susan Young. 2006. "Musical play." Pp. 289–310 in *The Child as Musician*, ed. Gary E. McPherson. Oxford: Oxford University Press.

Parncutt, Richard. 2006. "Prenatal development." Pp. 1–31 in *The Child as Musician*, ed. Gary E. McPherson. Oxford: Oxford University Press.
Purves, Dale, et al., ed. 2004. *Neuroscience*. Sunderland, MA: Sinauer Associates.
Saffron, Jenny R. 2003. "Mechanisms of musical memory in infancy." Pp. 32–39 in *The Cognitive Neuroscience of Music*, ed. Isabelle Peretz and Robert J. Zatorre. Oxford: Oxford University Press.
Schatz, Carla. J. 1993. "The developing brain." Pp. 15–26 in *Mind and Brain. Freeman* and Co., 1993, 15–26.
Schubert, Emery and Gary E. McPherson. 2006. "The perception of emotion in music." Pp. 193–212 in *The Child as Musician*, ed. Gary E. McPherson. Oxford: Oxford University Press.
Stravinsky, Igor. 1946. *The Poetics of Music*. Oxford: Oxford University Press.
Stravinsky, Igor and Craft, Robert. *Memories and Commentaries*. Faber 1960, 101.
Trehub, Sandra E. 2003. "Musical predispositions in infancy." Pp. 1–20 in *The Cognitive Neuroscience of Music*, ed. Isabelle Peretz and Robert J. Zatorre. Oxford: Oxford University Press.
———. 2006. "Infants as musical connoisseurs." Pp. 33–49 in *The Child as Musician*, ed. Gary E. McPherson. Oxford: Oxford University Press.

SECTION IV

FINAL QUESTIONS

CHAPTER THIRTEEN

PAULINE ESCHATOLOGY: THINKING AND ACTING IN THE TIME THAT REMAINS

Steven T. Ostovich

Summary

The letters of Paul are becoming newly legible two thousand years after he wrote them. This is evident in the recent interest in Paul from philosophers and cultural critics reviewed below. This essay focuses on particular passages from Paul's letters to uncover his understanding of time, that is, his eschatology, and what this eschatology entails for understanding reason and the reasonableness of pursuing justice in a perennially unjust world. Paul was convinced that the resurrection of Jesus marked the rupture of time in promising the temporal presence of the Messiah at the *parousia*. This messianic time is limited time wherein "thinking" is a historical rationality of promise, hope and remembrance rather than an escape into the timeless, and "acting" is a form of political theology. Eschatology has been a difficult topic for Western thinking especially in modernity. It has either been secularized as philosophy of history or dismissed as politically dangerous. This essay uses Paul to claim that now is the time for this time. Paul's eschatology also provides a way to think about J. T. Fraser's "sociotemporality."

Let me begin with a three-part preamble. First, the central term in what follows is "eschatology," by which I mean thinking about/understanding of/doctrine concerning the end times where "end" (eschaton) means rupture or interruption rather than goal or purpose (telos). The originary context for this term is the apocalyptic literature of ancient Israel and the Jewish and Christian religious traditions (although this kind of literature also might be found in other traditions including Islam). The literature begins with Daniel (167 BCE) and extends through the New Testament letters of Paul, the gospels, and the Apocalypse to John, but most of the literature remains extra-canonical. Reading this literature is a challenge in part because the symbolically rich language that characterizes it always remains ambiguous.

Second, eschatology as interruption is hard to take seriously as a concept. It might even be characterized as philosophically "silly": can one really live such beliefs? Can one establish rational expectations and a sense of responsibility for the future if one really believes the narrative arc of life is always and at every moment subject to rupture?

Jo Alyson Parker, Paul A. Harris and Christian Steineck (Eds), Time: Limits and Constraints, pp. 307–327

We can respond to this difficulty by trying to bring eschatology under some kind of what we consider critical control. We can read eschatology as concerned with predicting the future and collapse the eschaton into linear teleological time as the end of history or the end of the world. Eschatology then is reduced to determining when this will happen. In some respects this collapse is encouraged by apocalyptic literature itself as it seeks to give hope to oppressed believers by communicating God's plan for the world in which the end of the suffering caused by evil forces is at hand. On the other hand, experience counts against this reading as, time and again, attempts to pin down the exact time of the end have led to folly. And the literature also repeatedly points out that we do not and can not know the date of the end.

An alternative attempt to bring eschatology under control and make it less disruptive is to privatize it as a matter of the end and destiny of the individual believer, that is, of the believer's death and expectations for the afterlife. This has long been the approach in Christian theology wherein eschatology as dealing with the "last things" typically is put at the end of catechisms and rendered politically impotent through privatization. Joseph Ratzinger sees this approach as more than simply a reaction to disappointment with the delay of the eschaton (connected by Christians to the parousia of the Risen Lord, the promised "second coming"). Writing in the context of Christian-Marxist dialogue and its concern with history and politics and the development of liberation and political theologies with their attempts to recover the political core of Christianity, Ratzinger defends reading eschatology in terms of death and personal immortality as part of the original and ongoing focus in Christianity on the individual and her relationship to God also evident in the language of prayer. This is a way to avoid the excesses characteristic of apocalyptic movements such as those associated with the end of the first millennium of the Christian calendar. And while Ratzinger recognizes the danger of reducing Christianity to the religion of bourgeois individualism criticized in liberation and political theologies, he does not want to sell the nourishment of tradition for the pottage of contemporary politics. He would rather "strike a balance and come to understand the real promise of faith more deeply."[1] Still, it is hard not to feel here that the essential flavor of eschatology,

[1] Joseph Ratzinger, *Eschatology: Death and Eternal Life*, trans. Michael Waldstein, ed. Aidan Nichols (Washington: Catholic University of America Press, 1988), 15.

with its message of God's coming reign in love and justice, all political terms, has been lost.

At the same time, it remains hard to work out an understanding of eschatology as interruption and what this means for how we live and act. Walter Benjamin, in the last of his theses on the philosophy of history, writes of each moment being "the small gateway in time through which the Messiah might enter."[2] What does it mean to live like this, with this hope? In what follows I will try to develop an understanding of time as eschatological and the implications of this understanding for our existence in the context of the New Testament writings of Paul. For Paul, eschatology is tied to justice, that is, it entails concrete actions and commitments even though these fail: we are called to live "as if not," that is, as if this were not an unjust world, and to live justly despite our lack of success in achieving justice. This means Paul is framing a political theology. It also means that, in terms of J. T. Fraser's theory of time's conflicts, eschatology entails sociotemporality.

Which leads to the third point of my preamble: we must learn to take theology seriously if not literally. The larger project of which this paper is a part might be entitled "thinking theologically." By this I mean turning to traditional theological modes of discourse as providing rational alternatives to modern categories of reason as well as ways to tie thought and action intimately together. With specific reference to eschatology, this work might be described as developing what Jacob Taubes calls a "messianic logic."[3] As you can tell already, the particular direction this kind of theological thinking takes us is towards political life.

This should not be construed as an apology for religious faith or as making claims about God's existence; neither is it anti-religious. It is perhaps closer to trying to do theology without God, something recent work in philosophy and cultural criticism also attempts (as will be explored briefly below). Even so, it remains hard to think in, about, and through religion in modern culture where religious belief

---

[2] Walter Benjamin, "On the Concept of History," *Walter Benjamin: Selected Writings Volume 4, 1938–1940*, trans. Edmund Jephcott and Others, ed. Howard Eiland and Michael W. Jennings (Cambridge, MA/London: Belknap Press of Harvard University Press, 2003), 397.

[3] Jacob Taubes, *The Political Theology of Paul*, trans. Dana Hollander, ed. Aleida Assmann, Jan Assmann, Horst Folkers, Wolf-Daniel Hartwich, Christoph Schulte (Stanford: Stanford University Press, 2004), 10.

is defended as a matter of private feeling or dismissed as irrational because unscientific (except perhaps as a phenomenon for social scientific investigation).

Why bother with theology? A response might be in terms of experienced crises of modern reason, but to be more positive, there are opportunities provided by theological thinking that otherwise might fail us. One such opportunity is relevant to the work of the International Society for the Study of Time: to think about time as "eschatological" and to do so in "secular" terms, that is, in Carl Schmitt's formula, to "secularize" eschatology.[4] What I am thinking of here might be compared to how other religious understandings of time have been secularized in the past, for example, linear as opposed to cyclical time as a product of prophetic reflection in Ancient Israel, a linear time that as under the hand of God, has a direction and goal, an understanding of time that informs scientific thinking and is secularized as scientific progress, as pointed out by H. R. Butterfield and others.[5]

You might want to point out (with Karl Löwith) that eschatology already has been secularized as philosophy of history.[6] But turning to philosophy of history is not an adequate embodiment of the sense of temporal rupture that is an essential part of eschatological thinking; it is more an attempt to avoid or tame eschatology by reducing it to the goal (telos) of historical processes (whether these processes are understood in Hegelian or Marxist, Romàntic or natural terms). There is a fear of eschatology and its disruptive potential in Western culture that must be overcome if eschatology is to prove useful as a temporal and rational model. This fear is in part based on historical experience of how radical apocalyptic sects enact their understandings of eschatology in politically destructive fashion; and it is in part an academic response to the apparent irrationality of eschatological hope in a culture that understands truth in transcendental, abstract categories that

---

[4] Schmitt, in writing about legal theory and political theology, claimed, "All significant concepts of the modern theory of the state are secularized theological concepts"—Carl Schmitt, *Political Theology: Four Chapters on the Concept of Sovereignty*, trans. George Schwab (Chicago and London: University of Chicago Press, 1985), 36.

[5] H. R. Butterfield, *The Origins of Modern Science: 1300–1800* (New York: Free Press, 1957), 222–229.

[6] Karl Löwith, *Meaning in History: The Theological Implications of the Philosophy of History* (Chicago: University of Chicago Press, 1949).

allow for control, a control the interruption of eschatology resists.[7] The key task is to understand eschatology critically.

1. Paul among the Philosophers

In what follows the New Testament letters of Paul (or at least passages from the undisputed letters) will provide the texts for trying to think about time as eschatological and what this means for thinking and acting. Paul's writings are informed by the memory of the resurrection of the Messiah as a sign that Paul was living in the time of the end and as ground for the hope in the promised new community of justice. This eschatological understanding guided Paul's advice to the communities with whom he corresponded. Eschatology is the basis for Paul's "political theology" and hints at what J. T. Fraser calls "sociotemporality."

Paul's letters may seem an odd choice for describing a *political* theology (a propaedutic task to framing "sociotemporality"). Isn't Paul after all the champion of existential faith centered on what makes one "righteous" (*dikaiosyne*)? Could there be anyone more concerned with personal, private, individual salvation? And doesn't this make Paul more "Christian" than Jesus (that is, Christianity originates in Paul's mission, not with Jesus)? This view of Paul was the one developed by (Protestant) biblical scholarship at the close of the nineteenth century and remains favored today. And yet recent developments in reading Paul have challenged these assumptions and shown Paul's letters to be newly legible two thousand years after they were written. These developments come from two areas: in New Testament scholarship, a "New School" of Pauline studies has developed and is marked by a keen awareness of Paul's Judaism and the fact seemingly obvious from his letters that Paul always remained a Jew in his thought and practice even as he wrote in Greek;[8] and interest in Paul has developed

---

[7] See, for example, Anson Rabinbach, *In the Shadow of Catastrophe: German Intellectuals between Apocalypse and Enlightenment* (Berkeley/Los Angeles/London: University of California Press, 1997). Rabinbach examines the work of German intellectuals both between the world wars (Walter Benjamin, Ernst Bloch, Hugo Ball) and later (Martin Heidegger, Karl Jaspers, Max Horkheimer and Theodor Adorno) using apocalyptic leanings in some of these writers as (negative) critical markers.

[8] This position remains somewhat controversial among biblical scholars, but the volume of work done from this perspective has steadily increased. In fact, more has been done than can be summarized adequately here. One example will have to suffice, an example particularly relevant to the concern of this essay: Neil Elliott, *Liberating*

in philosophy and cultural criticism to the extent that in 2005 Syracuse University (USA) hosted a conference entitled "Saint Paul among the Philosophers" where the presenters were themselves philosophers and not biblical scholars. In both contexts Paul's thinking like a first-century Jew, his concern with justice and community, and at least in some cases his understanding of time and living in the eschaton come to the fore.

Let me sketch these new developments through brief reviews of four recent works on Paul:

- Daniel Boyarin, *A Radical Jew: Paul and the Politics of Identity.*[9] Boyarin, a Talmudic scholar at Berkeley, describes Paul as a "Jewish cultural critic" (106) who "mapped" a biblical ethical dualism onto Platonic hermeneutical-anthropological-ontological dualism to produce a "political ontology" based on the crucifixion as promising a new creation.
- Alain Badiou, *Saint Paul: The Foundation of Universalism.*[10] The French philosopher Alain Badiou describes himself as an atheist and claims, "Basically, I have never connected Paul with religion." (1) Paul is all about the resurrection as "event" inaugurating a new subjectivity. Badiou intentionally focuses on the resurrection as an event in the realm of fable or myth, not history, and therefore capable of a universality particular historical events cannot achieve. Badiou marks Paul's "provocative centering of thought upon its fabulous element" (5) as a new way of thinking: "*Paul is an antiphilosophical theoretician of universality*" (108). The subjective universal is the "universal singularity" that can move us beyond the contemporary impasse of a political life framed by the "abstract homogeneity of capital and identitarian protest [against this homogeneity]" (13). Paul provides a model for this in a politics of love (more than

Paul: The Justice of God and the Politics of the Apostle (Maryknoll, NY: Orbis Books, 1994). Elliott advances a reading of Paul as a Jew as a way to overcome what he describes as "the Theological Marginalization of Paul's Politics" (72).

[9] Daniel Boyarin, *A Radical Jew: Paul and the Politics of Identity* (Berkeley/Los Angeles/London: University of California Press, 1994).

[10] Alain Badiou, *Saint Paul: The Foundation of Universalism*, trans. Ray Brassier (Stanford: Stanford University Press, 2003).

faith or hope—or justice, according to Badiou) as practical, concrete action.[11]

- Theodore W. Jennings, Jr., *Reading Derrida/Thinking Paul.*[12] Jennings is located in the Christian theological tradition as a seminary professor, but he too is concerned with saving Paul from the "theological ghetto" (1, 157) to which Paul has been consigned. As the book's title suggests, Jennings turns to Derrida's work as the context for reading Paul, and he respects (and cites more than once) Derrida's description of "a *thinking* that 'repeats' the possibility of religion without religion" (*Gift of Death*, 49). Jennings pushes "justice" rather than "righteousness" as Paul's concern (and the proper derivation of *dikaiosyne*) against Rudolph Bultmann and an existentialist reading of Paul—Paul was no Stoic: "What is crucial here is that Paul supposes that justice must have a corporate or social character... where many of his Stoic contemporaries [and later Existentialist readers] resigned themselves to an individual instantiation of justice." (162) Following Derrida, Jennings pushes Paul to a concern with "law as such" (and not just Mosaic law) as it relates to justice and explores the "heterogeneity of justice and law" (see below).
- Jacob Taubes, *The Political Theology of Paul.* The title of this book by the late professor of hermeneutics (Berlin) states his point: Paul is a political theologian. Taubes reads Paul's letter to the Romans as "a *political* declaration of war on Caesar" (16) and describes Paul's task as the legitimation of a new people of God. Taubes reads Paul as a Jew ("I regard him... as more Jewish than any Reform rabbi, or any Liberal rabbi"—11) who challenges the domination of Western thought by Greek and who would (proleptically) agree with Taubes's claim that, "a Bible lesson is more important than a lesson on Hegel." (5) The "messianic logic" referred to above is non-Greek: to arrive at this logic one must "try to think from the center in a Jewish way" (10) because it is a logic that responds to events paradoxically. (This paradoxical logic will be developed below.)

[11] This turn to a "politics of love" is something Badiou shares with several other contemporary theorists outside Pauline studies. It seems to be an attempt to avoid the pitfalls of political realism by engaging in a new kind of political philosophical discourse—see David Nirenberg, "The Politics of Love and Its Enemies," *Critical Inquiry* 33:3 (Spring 2007): 573–605.

[12] Theodore W. Jennings, Jr., *Reading Derrida/Thinking Paul* (Stanford: Stanford University Press, 2006).

Two things are important to take from these (and other similar) works: Paul thought as a Jew and somehow this is different than the Greek thinking still dominant today (it is worth noting how often the names of Kierkegaard, Pascal, and Barth show up in these books); and Paul was a political theologian. Now let's substantiate these claims regarding Paul in more specifically eschatological and temporal terms

2. Letters of Paul

What did Paul think regarding the eschaton, that is, the interruption of time and history? I want to respond to this question by turning to two passages from Paul's undisputed letters. I realize that this can amount to an exercise in proof-texting, but I believe these passages to be representative of Paul's (richly complicated) thought and highly suggestive regarding Pauline eschatology and political theology.

1 Thessalonians 4:13–5:11

> (13) But we do not want you to be uninformed, brothers and sisters, about those who have fallen asleep, so that you may not grieve as others do who have no hope. (14) For since we believe that Jesus died and rose again, even so, through Jesus, God will bring with him those who have fallen asleep. (15) For this we declare to you by the word of the Lord, that we who are alive, who are left until the coming of the Lord, will by no means precede those who have fallen asleep. (16) For the Lord himself, with a cry of command, with the archangel's call and with the sound of God's trumpet, will descend from heaven, and the dead in Christ will rise first. (17) Then we who are alive, who are left, will be caught up in the clouds together with them to meet the Lord in the air; and so we will be with the Lord forever. (18) Therefore encourage one another with these words. (5) (1) Now concerning the times and the seasons, brothers and sisters, you have no need to have anything written to you. (2) For you yourselves know very well that the day of the Lord will come like a thief in the night. (3) When they say, "There is peace and security," then suddenly destruction will come upon them, as labor pains come upon a pregnant woman, and there will be no escape! (4) But you, beloved, are not in darkness, for that day to surprise you like a thief; (5) for you are all children of light and children of the day; we are not of the night or of darkness. (6) So then let us not fall asleep as others do, but let us keep awake and be sober; (7) for those who sleep sleep at night, and those who are drunk get drunk at night. (8) But since we belong to the day, let us be sober, and put on the breastplate of faith and love, and for a helmet the hope of salvation. (9) For God has destined us not for

wrath but for obtaining salvation through our Lord Jesus Christ, (10) who died for us, so that whether we are awake or asleep we may live with him. (11) Therefore encourage one another and build up each other, as indeed you are doing.[13]

Elements of this passage to consider:

- The passage is full of images typical of apocalyptic literature (see 4:16 and 5:8), and we do well to remember that Paul writes as part of a literary tradition already three hundred years old in his day. One new element introduced here: what later millenarians, post-dispensational millenarians, etc., call the "rapture" (see 4:17). This is the only place in the Bible such a notion is mentioned.
- A particular "memory" is at work here, the memory of the resurrection. Many other passages testify to this importance of the resurrection in Paul's reasoning (his account in Galatians and the accounts in Acts of Paul's experience on the road to Damascus; the summary of his gospel in 1 Corinthians 15, etc.). Resurrection as a concept and possibility originates in Daniel 12 where the raising of the dead pertains to martyrdom in the final conflict between good and evil, light and darkness, and it is latched onto by those who preach Jesus of Nazareth is the Messiah as a sign that Jesus' resurrection is the "first fruit" of the eschatological harvest, that is, that we are living in the time of the end.[14]
- If there is a memory at work in this passage, there also is a hope, a hope given content by Paul. It is a hope for the "death of death" (Jennings) or, in Pauline terms, the death of sin, the bridging of the gap (sin) between God and creation. It is the "hope of salvation" that serves as a "helmet" for the believer in the conflict based on loving service of one another. This aspect of Paul's hope may only

[13] The translation is the New Revised Standard Version with one change: the NSRV has "died" where Paul wrote "fallen asleep." I'm not sure why this is done: "fallen asleep" clearly means "died," and the NSRV may be trying to face death squarely rather than use Paul's seeming euphemism, but this occludes the clever parallelism Paul establishes between death expectation and moral conduct in what follows.

[14] Boyarin makes the crucifixion more central to Paul's worldview than the resurrection and does so in part to retrieve Paul's hope for a new creation. This is wrong: resurrection is the starting point of Paul's faith (see 1 Corinthians 15), and new creation is addressed in eschatological terms in Romans 8. Badiou, on the other hand, focuses on the resurrection, but he mistakenly removes it from history; Paul (and Daniel as well as other apocalyptic texts) looks to a historical event even as this event has mythic significance and remains inaccessible to historical proof.

be intimated here, but it is part of his source in Daniel—Daniel 12 is an affirmation of the meaning of individual life in the context of persecution and death, but it affirms as well that the source of that meaning is the hope for living in the new community of God's justice. So, too, this community aspect of hope explains why Paul addresses his audience as "saints" no matter where (or how!) they live. And Paul offers a picture of the justice to be hoped for in this new community in the famous passage in Galatians 3:28: "There is no longer Jew or Greek, there is no longer slave or free, there is no longer male or female; for all of you are one in Christ Jesus."

- Despite the language of fear in 1 Thessalonians ("no escape!"; the connection of the day of the Lord and wrath), Paul typically (and also here) does not seem to equate justice with retribution in the manner of some apocalyptic literature—see, for example, the story of Daniel in the Lions' Den where not only is Daniel saved from certain death but his accusers and their wives and children are thrown to the lions after Daniel is rescued (Daniel 6:25). (The element of revenge in eschatological justice remains even after eschatology is privatized in Christianity: one of the joys of heaven is seeing the suffering of those in hell, an idea mocked by Nietzsche in *On the Genealogy of Morals*.)[15] While it is as difficult to define justice in Paul as it is in general, it is clear that for Paul justice is liberating.[16]
- Paul plays with the temporal categories typical of apocalyptic literature. Often what is revealed in this literature is God's plan for the world and how recent events indicate that plan is not only being followed but is approaching its end. Paul points to the ambiguity of this plan with images one finds elsewhere in the New Testament too (perhaps as one way of dealing with delayed expectations): the day of the Lord will come "like a thief in the night" (5:2) or even more vividly like labor pains upon a pregnant woman (5:3—we kind of know when, but not exactly, so be prepared; and while there is much suffering associated with the birth pangs of the eschaton, something good is expected to come out of it). And yet Paul tells the

[15] Nietzsche quotes both Thomas Aquinas and Tertullian—the latter at length and in Latin—on judgment as retribution and the joys of watching those in hell suffer. See the first essay, section 15. Friedrich Nietzsche, *On the Genealogy of Morals and Ecce Homo*, trans. Walter Kaufman and R. J. Hollingdale (New York: Vintage Press, 1967), 48–52.

[16] See Elliott, *Liberating Paul*.

Thessalonians they do know: they are of the light and not in darkness, but this knowledge is practical rather than theoretical—that is, it is manifest in concrete actions. To know means to live a certain way, the way of love in hope of justice.

- Finally, the meaning of living eschatologically is key to responding to one other issue raised by this passage from 1 Thessalonians: Paul seems to have been wrong, that is, he expected the Risen Lord to return in his lifetime (see 4:17)—something that still has not happened two thousand years later (and a delay that other New Testament texts—all written after Paul died—tried to address). What claim can Paul make on our thinking given this fundamental error? And yet to call Paul "wrong" here is to repeat the mistake of reading eschatology in terms of teleology, to see time only as linear. The real question is what it means to live in another time, an alternative time now.

This practical knowledge is the described in the next passage.

1 Corinthians 7:29–31

> (29) I mean, brothers and sisters, the appointed time has grown short; from now on, let even those who have wives be as if they had not wives, (30) and those who mourn as if they were not mourning, and those who rejoice as if they were not rejoicing, and those who buy as if they had no possessions, (31) and those who deal with the world as if they had no dealings with it. For the present form of this world is passing away. (Author's translation)

This passage makes even more explicit what 1 Thessalonians suggests about the implications of memory and hope and what it means to live in eschatological terms. In part this is because Paul is responding to very specific problems in Corinth about which he has heard in a letter from the Corinthian community (7:1)—never forget these are letters. What about sex and marriage as we wait for the kingdom? How should we deal with what seems like a case of sexual immorality? Paul's advice reflects pastoral concern. And yet more is going on here.

This "more" can be summarized in a very short phrase that Paul uses repeatedly in 1 Corinthians 7, *hos me*, "as if not" ("married as if not married, mourning as if not mourning, rejoicing as if not rejoicing," etc., to paraphrase the passage). A new community has been called to live according to this *hos me* principle as it seeks justice. The contemporary Italian philosopher Giorgio Agamben makes the connection between *hos me* and calling: "*Hos me*, 'as not': this is the formula concerning messianic life and is the ultimate meaning of *klesis* [calling]." And this vocation to messianic life is both radical and universal: "Vocation calls for nothing and to no place. For this reason it may coincide with the factical condition in which each person finds himself called, but for this reason, it also revokes the condition from top to bottom. *The messianic vocation is the revocation of every vocation*."[17] Paul himself describes justice in this community in, for example, the passage referred to above, Galatians 3:28, a description that remains critically negative according to this principle "as if not": none of the oppressive distinctions obtaining now will remain "in Christ," neither Jew nor Greek, slave nor free, male nor female.

Letty Russell, a feminist liberation theologian, makes this *hos me* principle politically explicit in ways that are instructive for both what she gets right and how she is wrong. For Russell, this Pauline principle establishes a "horizon of freedom" that makes possible a sense of living proleptically in a community of justice and to "begin to build a world solidarity now" in the "journey on the road to freedom."[18] Pauline eschatology makes possible proleptic temporal mistakes, that is, forward-looking mistakes regarding time (as opposed to anachronisms) in which living "as if not" really is about living "as if" the eschatological reality of God's reign were already present now. Russell captures the political effectiveness of the *hos me* principle but is still trapped in the progressive mythology of dialectical history and thus misses the radical negativity of living according to "as if not." Agamben's insight is worth summarizing extensively in this regard.

> The Pauline *hos me* seems to be a special type of tensor, for it does not push a concept's semantic field toward that of another concept. Instead, it sets it against itself in the form of the *as not*: weeping as not weeping. The messianic tension thus does not tend toward an elsewhere, nor does it exhaust itself in the indifference between one thing and its opposite. The apostle does not say: "weeping *as* rejoicing" nor "weeping as [meaning=] not weeping," but "weeping *as not* weeping." According to the principle of messianic *klesis*, one determinate factical condition is set in

[17] Giorgio Agamben, *The Time That Remains: A Commentary on the Letter to the Romans*, trans. Patricia Dailey (Stanford: Stanford University Press, 2005), 23. Elliott reads this passage as describing an "apostolic praxis" of "challenging the ideology of privilege" in "holy solidarity with those not yet free."—*Liberating Paul*, 204–214.

[18] Letty M. Russell, *Human Liberation in a Feminist Perspective: A Theology* (Philadelphia: Westminster Press, 1974), 45–49.

> relation to itself—the weeping is pushed toward the weeping, the rejoicing toward the rejoicing. In this manner, it revokes the factical condition and undermines it without altering its form.... In pushing each thing toward itself through the *as not*, the messianic does not simply cancel out this figure, but makes it pass, it prepares its end. This is not another figure or another world: it is the passing of the figure of this world.[19]

The vision of justice operative in Paul works according to the *hos me* principle. It cannot be brought under control in the categories of modern reason because it cannot be given positive conceptual content. Still, it remains critical in its aleatory quality demanding a new understanding of reason as responsibility, that is, as responding to a calling (*klesis*). Justice is a promise—a promise that assumes the form of an announcement in messianic time according to Agamben (91)—and remains open and at issue. And as rooted in the critical negativity of "as if not," justice is universal in the manner stated above: not as an abstract concept removed from particular conditions but as for everyone.

The danger lurking in Russell's reading of *hos me* is the aestheticization of politics: "as if not" is a critically negative principle, but "as if" is aesthetic. In some respects this is the difference between Pauline eschatology and the "realized" eschatology of the Gospel of John. The Gospel of John makes stark distinctions between "light and darkness," "spirit and flesh," "life and death," while stressing the life-giving power of the former half of each pair of terms (see, for example, the discourse with Nicodemus in John 3:1–21). Such positive description of what is hoped for risks the deformation of hope we experienced too often in the twentieth century as visions of utopian community have been used to justify injustice, oppression, apartheid, even genocide.

In part this deformation stems from our desire to escape responsibility for making decisions, for judgment: if we can define our hope in terms of positive content, it becomes a goal we no longer have to

[19] Agamben, *The Time That Remains*, 24–25. This kind of thinking leads Agamben to create a misleading and unnecessary distinction between the "messianic time" he sees here and eschatology where the latter is about another world, the world of eternity in his view (see pp. 62–78). Perhaps he is concerned about the philosophical principle of eternity or about the conflicting understandings of time in apocalyptic literature, but his coinage of a separate messianic time is not helpful and not supported by the relevant texts. When Agamben describes messianic time as a "particular conjunction of memory and hope, past and present, plenitude and lack, origin and end" (1–2), he also is describing Pauline eschatology. And as Agamben claims, this "time that remains" is "the only real time." (6, 68)

think about, and we can shift our attention to the more manageable task of devising strategies to achieve our goal instrumentally. As Jennings points out, Paul confronts this desire:

> Precisely because Paul's argument is in a certain way double-edged, it cannot offer a program that is merely to be put into effect in any and all situations. That is, it is not a text that abrogates the concrete responsibility in any situation to come to terms with a double and contradictory imperative, one that is not accidental or the result of unclear analysis, but one that is ineluctable and thus one that makes possible because necessary the decision that cannot exonerate itself of responsibility by reference to a mechanism that makes decision unnecessary.[20]

Judgment cannot be avoided but in its eschatological critical negativity must be exercised in changing situations according to principles that themselves change with the situations. That is, we can never know in advance what justice is, but we can experience concretely what it is not, its absence and the presence of injustice.[21] The result: what Taubes describes as a creative form of "nihilism." (72)

But is it nihilism or "quietism" that is on offer in Paul? Doesn't Paul encourage a kind of eschatological quietism with regard to worldly conditions that, although unjust, are passing away? Isn't this precisely the point of a passage like Romans 13 with its opening injunction, "Let every person be subject to the governing authorities," even though those authorities in this case represent the Empire? How, then, can Taubes claim he finds in "Romans 13, nihilism as world politics" (74)? Hasn't Romans 13 been used repeatedly in Christian history as an excuse for

---

[20] Jennings, *Reading Derrida/Thinking Paul*, 75. This issue harkens back to the distinction between the critical concept of *anamnesis* (remembrance) in Walter Benjamin and Theodor Adorno and the aesthetic concept of *anagnorisis* (recognition) in Ernst Bloch—see Taubes, *The Political Theology of Paul*, 70–76, who confuses this issue by putting Adorno on the side of Bloch; Agamben corrects this in *The Time That Remains*, 35–43; see also, David C. Durst, *Weimar Modernism: Philosophy, Politics, and Culture in Germany 1918–1933* (Lanham, MD: Lexington Books, 2004), 110–111; and Vincent Geoghegan, "Remembering the Future," *Not Yet: Reconsidering Ernst Bloch*, ed. Jamie David Owen and Tom Moylan (London/New York: Verso, 1997), 15–32.

[21] The political theologian Johann Baptist Metz captures the temporal dimensions of this critically negative justice in the future oriented concept of the eschatological proviso—how all social structures are provisional insofar as they necessarily fall short of the promised reign of God—and anamnestic solidarity with the dead and their demand for justice. See: Johann Baptist Metz, *Zur Theologie der Welt* (Mainz: Topos Taschenbücher, 1968); *Glaube in Geschichte und Gesellschaft* (Mainz: Matthias-Grünewald, 1977); and *Memoria Passionis: Ein provizierendes Gedächtnis in pluralistischer Gesellschaft* (Freiburg: Verlag Herder, 2006).

tyranny? Reactions to this problem have been many and varied, but Jennings provides an interesting and helpful recent response through his following the principle of the "heterogeneity of justice and law" referred to above. Jennings points out that Paul's readers in Rome have direct experience of the difference between law and justice in the persecution they face from imperial authorities—thus the admonition at the end of Romans 12, "Do not be conquered by evil but conquer evil with good" (12:21), is not an abstract principle but represents a response to conditions in Rome and, as Karl Barth already maintained in his now classic commentary of Romans, serves as the introduction to the discussion of authority in Romans 13.[22] The (Roman) law is not just. But at the same time, Jennings maintains, "Even if it is necessary to recognize the heterogeneity of law and justice, it is also necessary that justice come to expression, come into being, as law." (74) Here again, Paul's argument is "double-edged," so that it cannot simply "be put into effect in any and all situations." Historically this can be seen from the experience that, as Jennings points out regarding Romans 13, "Although this passage from Paul has often been invoked to justify tyranny, it also has been read as providing the basis for a resistance to tyranny." (75) Judgment and criticism remain necessary—and risky.

Eschatology has ontological consequences and therefore consequences for our thinking as well as our acting. Or perhaps it would be more accurate to say eschatology undermines our distinction between thinking and acting and offers a political form of reason and a model of knowledge as transformative. Eschatological communities are living in another world. This other world is not eternity; neither is it the aesthetically constructed proleptic embodiment of hope. It is rather a different temporal order, what J. T. Fraser describes as a temporal umwelt.

Fraser borrows this term from Jakob von Uexküll, a theoretical biologist who uses "Umwelt" to describe the relationship between a species and its environment and how members of a species can only "know" what they have the receptors to know. Fraser takes this principle and "extends" it (the "extended umwelt principle") on the basis of the human ability to relate to its environment through instruments and mathematics as well as physiologically. He uses this extended principle

[22] Karl Barth, *The Epistle to the Romans*, trans. from the 6th edition by Edwyn C. Hoskyns (London: Oxford University Press, H. Milford, 1957).

as a way to circumvent the traditional philosophical problem of the relationship of the knower to the known or the relationship between appearance and reality: "the Umwelt of a species appears as a complete world which contains everything that can be known and hence, everything there is."[23] Or more directly, use of "the *extended umwelt principle*...amounts to equating epistemology with ontology: the world is the way we find it to be through the many forms of human knowledge, even if some of its features appear to be counterintuitive."[24]

The last phrase is important when considering eschatology. There is a temporal dimension to umwelts for Fraser, a temporal dimension that underlies his hierarchical theory of time and time's conflicts (see below). What I am maintaining here is that apocalyptic texts and eschatology as a concept put us in a particular umwelt, a particular temporal world, a strange and counterintuitive reality. As Agamben puts it, this world is

> Not a world of substance and qualities, not a world in which the grass is green, the sun is warm, the snow is white. No, it is not a world of predicates, of existences and essences, but a world of indivisible events, in which I do not judge, nor do I believe the snow is white and the sun is warm, but I am transported and displaced in the snow's-being-white and in the sun's-being-warm. (129)

This temporal topos is the promised new creation.

Going beyond Paul's letters might help clarify what this new creation is about. Elisabeth Schüssler-Fiorenza describes the last book in the New Testament, the Apocalypse to John, as a rhetorical exercise in world-shattering and new creation. This is a helpful way to read a text that is even crazier and subject to more misreading than Paul's letters. Responses to the Apocalypse mirror and sum up responses to most apocalyptic literature. One option has been to read the text (supposedly) literally, the fundamentalist's reading in temporally linear terms; one gains control of the text here by deciphering its strange images to fix the (correct—this time!) date of the eschaton. The other option is the biblical critic's where the text is deciphered in terms of its historical context of origin and its significance is fixed in the (over and done

[23] J. T. Fraser, *Time and Time Again: Reports from a Boundary of the Universe* (Leiden/Boston: Brill, 2007), 39.

[24] J. T. Fraser, *Time, Conflict, and Human Values* (Urbana and Chicago: University of Illinois Press, 1999), 25.

with) past; this too is a linear temporal reading even if done in the opposite direction. Neither is taking eschatology seriously. Schüssler-Fiorenza instead believes the text is an exercise in rhetorical world construction creating a new place, or perhaps better, a new temporal order or umwelt in which the community can live. This is a world where lambs become lions and exercise power in their powerlessness as in Revelation 5.[25] And the eschatological community is called to live in this world.

The temporal umwelt of eschatological communities involves an understanding of time as disruptive, that is, as non-linear. The logic of this umwelt also is non-linear, that is, non-Greek, not even the modified line of dialectical progress whether spiritual or material. It constitutes what has been referred to above as Taubes's "messianic logic," a logic characteristic of the faith of Paul's communities as it was later of Jewish ghettoes "among whom Greek and Greek philosophy were a non-issue" and for whom "the internal logic of events demanded a faith that is paradoxical, that is contradicted by the evidence." (10) This is a logic capable of thinking the paradoxes contained in Paul's eschatological principle *hos me*, as if not. One of these paradoxes is that we are called to live justly knowing that this will remain an unjust world. A paraphrase of Paul in this relation might run as follows:

> We live in the promise of eschatological justice even as the world around us is unjust. Our hope is not that at some point in the future the world will be made just; neither is it that we can live in the world "as if" it were just. No, it makes sense for us to live as if the world were not unjust, that is, to act justly not as a way to realize God's kingdom but as the only action that makes sense in this world. Justice is not a goal but a rational principle that gives direction to our actions without thinking of coming to an end.

This understanding of justice and action also makes possible taking eschatology seriously in its full disruptive power. Time is not a line: it is not teleological, about ends and goals; time is eschatological, about interruption. And interruption is not a future possibility but a constant reality that also concerns the past, that is, the past is not over and done with but can disrupt the present. What this means can be

---

[25] Elisabeth Schüssler-Fiorenza, *Revelation: Vision of a Just World* (Minneapolis: Fortress Press, 1991), 1–37, 58–62. David L. Barr gives a liturgical rather than rhetorical reading of this text but to the same effect: David L. Barr, "The Apocalypse as Symbolic Transformation of the World," *Interpretation* 38/1 (January 1984), 39–50.

explained most directly in terms of justice: *justice is due the dead*. If, as maintained above, justice as an eschatological promise working on the *hos me* principle is universal because it extends to everyone (rather than as an abstract concept), it also extends to the dead who have been denied justice.

Can any sense be made of this demand for justice to the dead? After all, the dead are past, and the past is a given beyond our capacity to influence through action. Our temptation has been to read the promise of justice solely in future terms because justice is absent from the present and because the dead are dead. But no amount of future progress and glory can make sense of the suffering the dead, which means either that the demand for justice for the dead is nonsense or what is claimed here: the demand for justice for the dead breaks the hold of teleology and linear logic and puts us in the eschatological umwelt. A new way of thinking is needed, a "messianic logic" that takes the form of what might be called anamnestic (remembrancing) reason, thinking based on anamnestic solidarity with the dead in their demand for justice. In other words, eschatological interruption is not just about hope for the future but about memory and memories that become dangerous in their disruptive power.[26]

3. Sociotemporality

These texts from Paul are offering us a new way to think and act in time, a time understood eschatologically as subject to rupture, a knowledge that makes it possible for us to live in hope according to the principle "as if not" while at the same time responding to the remembrance of the dead and their call for justice. Eschatology is hard to understand and/or hard to take seriously: again, we have a tendency to read texts like these in terms of an escape into eternity (personal immortality); or we continue to take time as linear and simply plot the eschaton as

[26] Exploring dangerous memories would take us too far beyond Paul's texts and will not be done here except to mention one particular dangerous memory in the life of Paul and his communities: the memory of the cross and resurrection of Jesus, a memory that defines Paul's eschatological communities already liturgically according to 1 Corinthians 11: 23–32. Dangerous memories play a central role in Metz's political theology—see the books in note 19, above; see also Bruce T. Morrill, *Anamnesis as Dangerous Memory: Political and Liturgical Theologies in Dialogue* (Collegeville, MN: Liturgical Press, 2000); and Steven T. Ostovich, "Dangerous Memories and Reason in History," *KronoScope* 5, no. 1 (2005): 41–57.

the end point (near or distant—or failed) on that line (the tendency to search the texts for the exact time of the end). Something more is going on here, however, something Paul describes in terms of how an awareness of time's constraint moves us to live in memory and hope. And this something more also can be described in terms of what J. T. Fraser calls sociotemporality.

Sociotemporality, the most complex temporality in his hierarchical theory of time, is "the postulated level-specific temporality of a society" (*Time, Conflict, and Human Values*, 37). There is an essentially social or more properly political element in eschatological thinking; this is why thinking about the eschaton leads to a political theology, as seen above. Eschatology depends on memory and hope for understanding; memory and hope in turn demand a communal context—an individual's personal memory and hope in isolation have no effect before birth or after death; and so someone like Paul is very much concerned with the promise of a new community, a community of justice.[27] Fraser develops a hierarchical theory of time based on the extended umwelt principle, a theory of "nested" canonical levels of time. The first four canonical levels of time in Fraser's theory all have some form of physical [not material; not natural] embodiment—nootemporality/mind, biotemporality/living organisms, eotemporality/solid objects, prototemporality/particle-waves. Political reality is the physical manifestation of sociotemporality.

Eschatological thinking might help us to understand sociotemporality more fully. Fraser describes the "unresolvable creative conflict" between temporal levels in the hierarchical theory of time. He identifies the conflict relevant here (between sociotemporality and nootemporality) as the conflict between the individual and the group, between "conduct that supports the common good and conduct that supports the personal good" (*Time, Conflict, and Human* Values, 39). This seemingly locates the conflict in the will. Eschatological thinking like Paul's takes us further by helping us see that while this conflict may be experienced first in the will, it also is a matter of reason and the intellect in its demanding an alternative understanding and response to the temporal worlds in which we live (umwelts). Eschatology involves making sense of the demand for justice, even the neo-sensical demand

[27] See W. James Booth, *Communities of Memory: On Witness, Identity, and Justice* (Ithaca, NY, and London: Cornell University Press, 2006).

to render justice to the dead: "as if not" is the principle by which to pursue the good of justice which is the aim of rational activity in the political world.

We seem to be condemned to a life of conflict, but as Fraser points out, it is conflict that characterizes the various canonical levels of time. For sociotemporality, the passage quoted above continues,

> the conflicts themselves, working in the social present, may be seen as constitutive of society. A society is a group of people with a family of conflicts that defines them and distinguishes them from other societies. (39)

Even revising the nature of these conflicts by extending them from the will to the mind as well does not make them go away. For Fraser, "resolution" of temporal-level conflicts is achieved by movement to the next higher level of temporal integration. In this case, might that be eternity? I don't think so based on how we have responded to the creative conflicts of eschatology: as stated above, one way we have tried to escape from the demands of eschatology is by privatizing it, by making it a matter of my end, that is, my flight into eternity. The task is to take eschatology seriously as a category of temporal existence, thinking it through in terms of anamnestic reason, the promise of justice, and the call to live "as if not."

References

Agamben, Giorgio. *The Time that Remains: A Commentary on the Letter to the Romans*. Trans. by Patricia Dailey. Stanford: Stanford University Press, 2005.

Badiou, Alain. *Saint Paul: The Foundation of Universalism*. Trans. by Ray Brassier. Stanford: Stanford University Press, 2003.

Barr, David L. "The Apocalypse as Symbolic Transformation of the World." *Interpretation* 38.1 (January 1984): 39–50.

Barth, Karl. *The Epistle to the Romans*. Trans. from the 6th edition by Edwyn C. Hoskins. London: Oxford University Press, 1957.

Benjamin, Walter. "On the Concept of History." In *Walter Benjamin: Selected Writings Volume 4, 1938–1940*, trans. by Edmund Jephcott and Others, ed. by Howard Eiland and Michael W. Jennings. Cambridge, MA, and London: Belknap Press of Harvard University Press, 2003.

Booth, W. James. *Communities of Memory: On Witness, Identity, and Justice*. Ithaca, NY and London: Cornell University Press, 2006.

Boyarin, Daniel. *A Radical Jew: Paul and the Politics of Identity*. Berkeley, Los Angeles, and London: University of California Press, 1994.

Butterfield, H. R. *The Origins of Modern Science: 1300–1800*. New York: Free Press, 1957.

Durst, David C. *Weimar Modernism: Philosophy, Politics, and Culture in Germany, 1918–1933*. Lanham, MD: Lexington Books, 2004.

Elliott, Neil. *Liberating Paul: The Justice of God and the Politics of the Apostle*. Maryknoll, NY: Orbis Books, 1994.
Fraser, J. T. *Time and Time Again: Reports from a Boundary of the Universe*. Ledien/Boston: Brill, 2007.
———. *Time, Conflict, and Human Values*. Urbana/Chicago: University of Illinois Press, 1999.
Geoghegan, Vincent. "Remembering the Future." In *Not Yet: Reconsidering Ernst Bloch*, ed. by Jamie David Owen and Tom Moylan. London and New York: Verso, 1997.
Jennings, Theodore W. *Reading Derrida/Thinking Paul*. Stanford: Stanford University Press, 2006.
Löwith, Karl. *Meaning in History: The Theological Implications of the Philosophy of History*. Chicago: University of Chicago Press, 1949.
Metz, Johann Baptist. *Glaube in Geschichte und Gesellschaft*. Mainz: Matthias-Grünewald, 1977.
———. *Memoria Passionis: Ein provizierendes Gedächtnis in pluralistischer Gesellschaft*. Freiburg: Verlag Herder, 2006.
———. *Zur Theologie der Welt*. Mainz: Topos Taschenbücher, 1968.
Morrill, Bruce T. *Anamnesis as Dangerous Memory: Political and Liturgical Theologies in Dialogue*. Collegeville, MN: Liturgical Press, 2000.
Nietzsche, Friedrich. *On the Genealogy of Morals and Ecce Homo*. Trans. by Walter Kaufman and R. J. Hollingdale. New York: Vintage Press, 1967.
Nirenberg, David. "The Politics of Love and Its Enemies." *Critical Inquiry* 33.3 (Spring 2007): 573–605.
Ostovich, Steven T. "Dangerous memories and Reason in History." *KronoScope* 5.1 (2005): 41–57.
Rabinbach, Anson. *In the Shadow of Catastrophe: German Intellectuals between Apocalypse and Enlightenment*. Berkeley, Los Angeles, and London: University of California Press, 1997.
Ratzinger, Joseph. *Eschatology: Death and Eternal Life*. Trans. by Michael Waldstein. Ed. by Aidan Nichols. Washington: Catholic University of America Press, 1988.
Russell, Letty M. *Human Liberation in a Feminist Perspective: A Theology*. Philadelphia: Westminster Press, 1974.
Schmitt, Carl. *Political Theology: Four Chapters on the Concept of Sovereignty*. Trans. by George Schwab. Chicago and London: University of Chicago Press, 1985.
Schüssler-Fiorenza, Elisabeth. *Revelation: Vision of a Just World*. Minneapolis: Fortress Press, 1991.
Taubes, Jacob. *The Political Theology of Paul*. Trans. by Dana Hollander. Ed. by Aleida Assmann, Jan Assmann, Horst Folkers, Wolf-Daniel Hartwich, Christoph Schulte. Stanford: Stanford University Press, 2004.

CHAPTER FOURTEEN

HOW DEATH WAS INVENTED AND WHAT IT IS FOR

Frederick Turner

Summary

Death is a comparatively recent arrival in the universe. One doesn't require the term "death" to describe the metamorphoses of matter and energy. Oscillators, counters, alarm-clocks (counters with thresholds), and the chemistry of cyclic catalytic reactions can exist in the inanimate world and can trigger radical changes, precursors of death's emergence. But only with living organisms does death begin, at first as the destruction of the organism but later as a design feature of cells and multicelled animals and plants. Death answers the need to balance reproductive genetic stability against real world experience. Death is life's way of distinguishing target information from "jottings" acquired during the computation of an algorithm—and limiting the thermodynamic cost of discarding the latter. Death is a tool in constructing adaptive futures. In human cultures death is an editor of information and an incentive for post-biological forms of survival.

For human beings the most pressing and important constraint or limit is clearly death.

But what is death? Matter and energy change their state and their arrangement, but we do not require the term "death" to describe their metamorphoses. Only with living organisms does death make its appearance, at first as the destruction of a specific ordered organism, but later, more strangely, as a wired in and designed feature of cells and multicelled animals and plants. Only living things can die—the first of many paradoxes we will meet in this investigation. Death emerged at a certain point, comparatively recent in the history of the cosmos; and it did so in four stages.

In the first stage, death was simply the cessation of life: the negation of life's attempt to survive and be immortal. It was the worst thing that could happen to an individual organism. Collectively, however, death was not such a disaster, for the earliest organisms were clones and thus expendable. Since life was a pattern of information coded into the genes (including instructions for the construction of an information processor and an information storage device) the destruction of excess copies of it did not matter. Indeed, it is this feature, life's

Jo Alyson Parker, Paul A. Harris and Christian Steineck (Eds), Time: Limits and Constraints, pp. 329–338

independence of the fate of its particular constituent atoms and molecules, that defines it as life. A sonnet survives the paper shredder as long as it's still on the hard disk and survives a crash if there's a zip drive backup. Death was a necessary cost of doing business, like taxes—for a lineage of asexually-reproducing bacteria it was the attrition rate that had to be beaten by the reproduction rate.

The emergence of death in the history of the universe, though late, was not an altogether unprecedented event. Several of its ingredients already existed in primitive forms, which are worth looking at more closely since they will help us understand the strategies of later adaptations. Obviously, though particles, atoms, molecules and chunks of matter do not die, they can be destroyed in various ways—broken, melted, turned into energy or transmuted. Some chunks of matter, such as crystals and polymers, can grow and resist destruction. Physicochemical processes such as cyclic catalytic chemical reactions, hurricanes, fires, rivers, stars, and suchlike can establish a rhythm and renew themselves. They can come to an end either by the exhaustion of their feedstocks or by the accumulation of a product that either smothers them or turns out to have a quantitative threshold at which some other emergent property becomes fatal to its parent. Death is a convenient metaphor for their cessation—stars and rivers "die," hurricanes and fires "die out"—though it is only a metaphor.

Clepsydras and scaled candles based on such processes were used by early humans to measure time: today we use radioactive decay, in the same paradoxical faith that the intervals between the cycles are equal. J. T. Fraser has illuminatingly characterized physical structures as clocks, whose range of possible oscillations constitutes an archive of preservative adaptations to enable their continuity. Those clocks can even be selected for, in that brittle ones with few viable frequencies tend to be more easily destroyed than multimode flexible ones.

The physicochemical world can even give us counters, that accumulate a measurable product (we use one for carbon dating, for instance), and alarm clocks, where the counter material crosses a natural threshold, reaches some critical mass and "goes off" (as when an orbital dust-cloud generates planetesimals that can rapidly snowball into a planet).

Life needed all these features, and more, to work its marvelous bootstrap operation: in a sense bioclocks are not just a feature of biology, they *are* biology. But with these uses came a new, terrible, emergent property including all of those features: death. Catabolism, the process by which we break down our food for our own uses, is a fire that consumes us too. We digest ourselves and burn ourselves with the oxygen we breathe. In internalizing for its own purposes the external vicissitudes that previously smashed structures of inanimate matter, life may indeed have discovered the secret of preserving its code independently of a particular physical embodiment; but it also created an internal enemy to join the old external threats.

But life is ingenious. The second stage in the development of death came at the collective level of asexually reproducing clone groups and lineages. The ability to die, when combined with the fact that the source code—the RNA and DNA—can be altered by random mutation or genetic grafts from other species and still sometimes produce viable individuals, had a useful emergent property of its own. In effect, a field of cloned individuals with small random differences could act as a vast distributed sense organ, a kind of tongue to taste the environment. The death of one member of the group was a bit of data, and the overall pattern of deaths constituted information on the collective level. Death was nature's way of saying "don't live there unless you have different genes,"\and differential rates of reproduction constituted a crude response to this advice. When we look at the patterns formed in a petri dish of bacterial lineages and spots of toxin, we see the result. Death has been pressed into service—as a very uncomfortable found object, so to speak, but one that could be put to use.

The third stage in the evolution of death coincided with the origins of sexual reproduction. Sex systematically shuffles the genes of the male and female partners to create unique individuals. Sex would, in a way, be an abomination to an asexually reproducing organism, which would sensibly regard survival as being survival of its own specific DNA sequences. By shuffling the chromosomal deck, sex trades off exact preservation of the genetic message against a riskier but far more rewarding chance at huge adaptive success, and a far swifter accumulation of the species' collective knowledge of the world. Here death changes from an ad hoc borrowing, so to speak, to a shaped and crafted tool.

To be fair, asexual organisms had already capitalized on the tendency for snippets of RNA and DNA to be exchanged among individuals by viral transfer and other random events. This exchange accelerated the rate and limited the dangers of mutation, and provided a crude cross-reference system that could connect different strains of the species. It thus increased the scope, sensitivity, and interpretive depth of the clonal "tongue." With true sexual reproduction, however, this process is changed from being an ingenious capture of necessity to a distinct survival policy, with a sophisticated competitive system for analyzing and using the results. Mate choice puts the species, at the social level, in charge of its own evolutionary development. Mock battles, sexual display and signaling, rank and kinship systems, and the beginnings of politics and instinctive game-strategizing elaborated and hugely accelerated the accumulation of the information archive of a species, which in turn served as a collective memory and a set of epigenetic plans for future contingencies.

But at once a major problem presented itself, known in economics circles as the Betamax problem. (The VCR video recording system was inferior, it was claimed, to Sony's Betamax, but had got there first and had monopolized the invested capital technology, so there was no room for the better system.) If sex was extremely efficient at producing a wide variety of viable individuals, without further modification that advantage would be lost—those individuals would face a world whose ecological niches were already dominated by their own well-adapted parents. Favorable mutations would never get a chance to show their superiority and never get reproductively scaled up so as to be able to take charge of the niche themselves and exploit it better—or create a new niche. In terms of informational phase-space, a more efficient basin of attraction would never be found because the lip of the existing basin was too high.

Some method of lowering that lip, leveling the playing field, so to speak—some means of clearing out the older generation to make room for the new—was required. Enter death in a new role, as the essential partner of sex, as aging, life's programmed-in and quite elaborate system of self-destruction. Current longevity research has found a whole battery of life-shortening mechanisms, including counters to limit the number of cell-divisions, biological thresholds that set off lethal changes, alarm clocks set to go off after a given number of cycles, chromosome-ends (known as telomeres) designed to fray until essential information begins to be lost, apoptosis or cell-death triggers, programmed decreases in the ability of cells to resist prions and oxidants, and the like. Death by aging is enforced with multiple backup systems. Death is no longer something that *happens to* living things; it is something that they *do*.

Cancers are especially interesting in this context, as they reveal clearly the deeply paradoxical nature of the death-life *entente*. The essential feature of a cancer cell is that it breaks the contract, so to speak, that healthy cells subscribe to—to die when their time comes. Cancer cells are immortal in a petri dish, and in a living body they will grow and divide and survive at the cost of the whole organism. So their bid for immortality results in the shortening of the life of their host and thus in their own untimely death. But the paradox has yet another iteration. For the body's multiply redundant aging system actually uses cancers as one of its backups. Once the useful breeding life of an organism is over, the immune response and the DNA repair mechanism is allowed to degrade, thus permitting cancer to do the job of finishing off the aged organism if the other backups have not yet done their work. The gene pool thus votes, as it were, to privilege sexual reproduction—which creates a new and different generation of individuals—over mere ontological survival of the individual (or, what is the same thing, its clone). In another sense, it "votes" for its own collective existence as an interbreeding set of unspecified individuals, rather than for some pure and specified lineage. There is even a sort of political economy in this trade-off that might illuminate the macro-politics of the human world.

Another disciplinary perspective may also offer insight into the peculiar ingenuity of the sex-death system. In the bizarre world of computational physics, where information science abuts upon thermodynamics, the search for a perfectly "frictionless" computation engine is the core mental experiment, and the idea that the universe is essentially a computational system computing its own existence and future is the core assumption. These abstract considerations actually have very practical and lucrative applications. Perhaps the largest fundamental problem the computer industry faces is simple heat. That is, computation is the more efficient, the faster it is. Speed requires the shrinking of the components so that the electricity or light that runs and connects them has less distance to travel. But computation is also a thermodynamic process; and, if computational circuitry and transistors are sufficiently miniaturized, the waste heat given off by their work will not have time or space to escape and will melt the components and destroy the processor. Thus the search for computational systems that finesse the Second Law of Thermodynamics. One promising approach is to reverse each computation after its result has been collected, thus leaving the system in its initial state.

But the point here is that the analysis of exactly *where* the pesky waste heat was being generated in the computational process turned up something extremely interesting. It was not in the input of the data. Nor was it in the actual computation. Nor was it in the announcement of its results. Where the heat came from was in the destruction of the data that the process had generated on its way to the conclusion—the jottings, so to speak, on the electronic slate, the chalk on the blackboard that the puffing math student must erase. The heat was like the bill that arrives at the accounting department each month after the shredding van has departed. In our information-multiplying world we are already beginning to see the macro version of the overheating laptop in the form of pollution credits, vast home storage lots, garage sales, contested landfill sites, Yucca Mountain, the Superfund, and, of course, global warming. Our wealth, it turns out, was not so much in what we possessed, but in what we could afford to throw away.

Life itself is, among other things, an immensely powerful computational system. The brain burns a large proportion of the body's intake of sugar and oxygen, and regulates its temperature meticulously with sweat, hair insulation, internal liquid cooling, hats, and other devices. Ingeniously the waste heat from its massively parallel and massively nonlinear computational activity is put to good use in maintaining the body's temperature and metabolism. But there is a kind of "waste heat" that is not so easily managed: if an individual's whole life—from the species gene pool point of view—is a "jotting," an experimental setup that has proved its point and that now clutters the store-room, how does nature dispose of it without destroying the valuable information itself?

Death is the evident answer. The information gained from the individual's life is at first preserved only in terms of its differential reproductive success; but in advanced social species that useful information, purged by death of its jottings, gets passed down in the form of social learning, imitation, and ritual. Later, with human beings, speech, writing, the arts, and history enormously widen the bandwidth of what can be passed down. Significantly, it is when humans enter the scene that death becomes an object of consciousness—the central issue, the wages of sin, the fruit of the tree of knowledge, the monstrous specter, the mystical act of heroism or sanctity that it is. Death assumes its most perfect form exactly when its information-purging function becomes most urgent. It is the ultimate cooling device, the graphite rod in the reactor, the fin on the refrigeration system, the fan in the laptop, the inhibitory synapse that stops the brain of the species from breaking into an epileptic seizure.

So it is here that we enter the fourth stage in the evolution of death. "After life's fitful fever, he sleeps well," says Macbeth of the murdered Duncan. The irony is that timely death is indeed the breaking of the generational fever, but Duncan does not die of natural old age. Death takes on a richer and different meaning again in the human context of culture and language and self-consciousness. The old meanings of death still apply—it is still often an accidental and wasteful thing, as it was for the most primitive forms of life. It is still a demographic sensor, indicating what kinds of ecologies and regimes are good to live in, as it was for bacterial lineages. It is still, as aging, the essential partner with sex in an accelerated process of evolution, clearing out the dead wood of the past and leaving proven innovations that are now better represented in younger versions. It is still the biologically-programmed editor, excising the notes and workings by which solutions were found, dismantling the scaffold that once supported the early stages of the logic. It is still the refrigeration system of the overheating cultural information-generating machine.

But death now took on yet another identity, as a cultural instrument of great power. The consciousness of death became a huge spur to cultural work. When that work had technological results, it could ensure the survival of whole tribes and nations of people and condemn others to extinction. As Dr. Johnson said, death marvelously concentrates the mind. By setting a definite, though unpredictable, end to the continuity of our individual existence, it defined a realm of collective cultural achievement, commemoration, and quasi-immortality. The superiority of a combined cultural/genetic information transmission system over a purely genetic one is elegantly signaled in the phenomenon of menopause when coupled with female longevity. The reason why one half of the species should consistently tend to survive its reproductive function—at great metabolic and ecological cost—must surely be that its members act as culture-bearers, as tribal memory, as the carriers of the "old wives' tales" that embody the collective wisdom of the group.

By setting an end to one kind of future, conscious anticipated death brought into being another kind of future that could take the form of either posterity or eternity. To envisage that other world, discontinuous with our present consciousness, was to be released into a vast world of possibilities, counterfactuals, spiritual imaginings—the world of what is there when we're not there, or not there in our present form. Death is a great corner—and behind that corner who knows what is waiting? "The undiscovered country, from whose bourn / No traveler returns"? The human religions intuited from this that there was another dimension of existence, and this realm then became a huge space for new "jottings," for mental experiments and hypothetical constructs. If one could glimpse its inner order, it could perhaps explain the rich and infuriatingly suggestive order of this world—and so science was born in the form of "timeless" mathematical concepts and "timeless" physical laws. In a word, one could say that death, as a reflexive concept half-grasped by the human mind, gave us the future. Paradox again.

The acceptance of death became one of humanity's noblest virtues, the gateway to a realm of exploration and discovery, the proof of our moral superiority to our own mortality, the central creative moment and fulcrum of all fictions and works of art. It was deeply involved with the ritual—and then the ethical practice—of sacrifice. In our willing death we became the scapegoat that carries the sins of the village away with it, but also the priest of that sacrifice and thus the agent and friend of a merciful god. What was a way of clearing away the dead wood of waste information became the flames of the sacrificial pyre; what was the chilling cure of life's fever becomes the portal to a new world of warmth and light; what was a biosexual imperative became a means of communication with our nascent conception of the divine.

But now even death cannot keep pace with the relentless exponential results of Moore's law, the decrease in the doubling-time of the body of human information. How do we distinguish good information from waste information when information, freed from the human brain, has become immortal? We can no longer rely on forgetting, the dark side of the hippocampus's marvelous work of memory-condensation and memory-potentiation. In the past, you could kill the poet who knew the old stories, and they would die. You could kill the people who made the civilization that preserved the writings, and most of those writings would be lost. Now even the most determined efforts to kill the Internet are futile. So the undiscovered country is increasingly not behind the corner or beyond the "bourn" of death, but here around us all the time—it is the incomprehensible wealth of our own information. Religious or secular-ideological regimes that try to crush the Internet in their countries are quite consistent in their attempt to stamp out a plausible rival to the mysteries of paradise.

At the same time our desire to live forever has never been stronger. If the undiscovered country is here, it would be a pity to leave before we had a chance to explore it. We are less and less content with the religious recommendation to accept death and see in it a great change into some mystical state of felicity. In our stories—recent movies like *Wings of Desire* and *City of Angels*, for instance—it is the undying spirit-beings that envy the fleshly mortals, not the reverse. We don't want immortality in heaven—we want it here on earth, as David Brooks wittily points out in his wise book *Bobos in Paradise.* So just as death begins to lose its function as a killer of unnecessary information, it also loses its spiritual appeal as the portal to a better world.

What is the new role of death? Should we, as Dylan Thomas puts it, "go gentle into that good night"? Or rather "rage, rage against the dying of the light"? Suppose our medical knowledge of aging became so perfect that we could halt or even reverse it? What of Hans Moravec's dream of uploading the constituents of our consciousness and selfhood into an immortal machine? How would we deal with the smothering overload of memory, obligation, irrelevant information?

If there is no line of demarcation between the now and the future, no separation of today from tomorrow, then the future itself becomes stunted and misshapen. We feel this when we are deprived of sleep and cannot dream: dream begins to invade waking life, and we are left in a state of nightmare that is nicely characterized by Macbeth's famous lines, where the very word "tomorrow," the symbol of hope, is rendered meaningless by repetition:

> Tomorrow, and tomorrow, and tomorrow
> Creeps in this petty pace from day to day,
> To the last syllable of recorded time;
> And all our yesterdays have lighted fools
> The way to dusty death. Out, out, brief candle!
> Life's but a walking shadow; a poor player,
> That struts and frets his hour upon the stage,
> And then is heard no more: it is a tale
> Told by an idiot, full of sound and fury,
> Signifying nothing.

An unrelieved and unpunctuated continuity of life might well begin to feel like this. Death might indeed be a good sleep after such a fitful fever.

Some religions have already anticipated these problems. The Greeks and Romans believed that the soul would be washed in the forgetful waters of the Lethe before its metempsychosis into a new life. The reincarnation system of the Hindus and some Buddhist traditions contains the same information-purge. The Christian doctrine of the resurrection of the body and paradise also includes some sort of purgation process, whether in the form of purgatory or being "born again."

I have argued that the prolongation of human life would require institutions and technologies to periodically reboot the human personality. The rejuvenated individual would be purged of his past legal, social, and economic holdings and be freed of his past obligations of memory. And the burdens of possession and status would be passed on, lest the rich and powerful old overwhelm and dominate the eager and ambitious young, and the transformative benefits of our rebellious adolescence be lost. This would be a sort of nominal death without its literal and biological accompaniment; perhaps it would be an opportunity, as some science-fiction authors have suggested, to perform a neuropsychological edit of one's own memories.

Perhaps we see an anticipation of this adjustment in the new sociological phenomenon of vigorous retirement, the "snowbirds" who in all previous eras would already be dead of disease and poor hygiene and who form a second society all over the warm temperate regions of the planet.

If we do manage to abolish death as a terminal event, I do not believe death will simply go away. It is too valuable and proven a device to abandon. I suspect that if death does not take place at the end of a life, its aura, terror, awe and darkness will simply migrate to all the shadows of life itself; it will become synonymous with time itself (as it is already in the Sanskrit language). That is, our life is full of little deaths: the hiatus of sleep, the darkness of inattention, the absence of distant dear ones and estranged friends, forgetfulness, the experiences elsewhere on the globe that we cannot share because of present activities and obligations, the inability to relive a first experience of some lovely thing, the alienation of one's past self, all that happens behind one's back, the manifold ignorances of any human life, the opportunity costs of any economic transaction. Those shadows will take on the numinous force of death and provide the necessary spur to work and the necessary altar of sacrifice. Hello, darkness, my old friend.

RESPONSE
NEED DEATH HAVE A PURPOSE?

William R. LaFleur

It's difficult, especially in modern times, to find someone who finds something *good* in death and is willing to say so in public. Ever since Descartes surmised that at some future date our medical technologies might make us physically immortal, modern men and women have been waiting for that day. In our own time that old hope easily mutates into hype—one that is hardly disinterested. Turner writes: "our desire to live forever has never been stronger. If the undiscovered country is here, it would be a pity to leave before we had a chance to explore it. We are less and less content with the religious recommendation to accept death and see in it a great change into some mystical state of felicity." And even the old religions, ones that once had insisted that biological life surely does not exhaust what "life" might mean, usually find it prudent in our time to stick to the ordinary biological life of humans (and often fetuses as well) as what needs protection at all costs.[1] The person who today stands up to say that there may be *good* to be found in the end of individuated biological life-forms will run the risk of being tagged a nihilist—ideationally a "Goth" but without the expected chains, skulls, and black garb.

Frederick Turner's essay bravely assumes that risk and does so in an unusually interesting and provocative fashion. As in all his writing, he here too provides an extraordinary inter- and cross-disciplinary essay—as exemplified by the fascinating section that correlates findings in the computational sciences with thermodynamics. And his stereoscopic approach to domains within the sciences is outstripped by a readiness to move with ease between the sciences and attention to metaphor as usually housed within the humanities. Bold weddings of usually disparate parties make for fascinating reading. Turner borrows from the hard sciences to give us some new metaphors for use

---

[1] Kenji Doi, a Protestant theologian and bioethicist in Japan, suggests that so much weight placed now on *biological* life within the Christianity of the West misconstrues what the terms for "life" meant within the early Church.

Jo Alyson Parker, Paul A. Harris and Christian Steineck (Eds), Time: Limits and Constraints, pp. 339–341

in talking about death—the "ultimate cooling devices," the "graphite rod," etc.

His outline is basically evolutionary. At its second stage, for instance, "Death was nature's way of saying 'don't live there unless you have different genes,' and differential rates of reproduction constituted a crude response to this advice." Later it became a "means of clearing out the older generation to make room for the new." When humans enter the scene "The consciousness of death became a huge spur to cultural work," and, quoting Dr. Johnson, Turner notes that "death marvelously concentrates the mind."

I have no quibble with the claim that death functions in all these ways...and more. What I find—and find problematic—in all these is an unbroken depiction of death as *instrumental*. Here is where what in me has been shaped by Daoism and Dôgen surfaces to prompt me to ask: If, as Turner claims, "the acceptance of death became one of humanity's noblest virtues," does that acceptance as acceptance not get chipped away as soon as it become conditional? Once we begin to describe death as having a *role* to play we are trying to fit it in as a bit-player in a production in which someone/something else—most likely "Life"—has top billing and will in the end be standing up alone while the others (including death), having functioned as fated adversaries, are themselves dead on the stage. Death is made to serve a *telos* not only outside itself but in many ways its own denial. It serves a far stronger master. This is what worries me in Turner's language about death as being "pressed into service" and "a shaped and crafted tool."

The problem is how to let death be death and at the same time avoid the at-the-ready charge that just saying so is to embrace nihilism. We in the West are still in the process of unpacking why Daoists and many Asian Buddhist insisted that letting death be death without instrumentalizing it was anything but nihilistic. Even a whiff of instrumentalism seems absent in the statement by Dôgen (1200–53) that: "When there is life, there is nothing at all apart from life. When there is death, there is nothing at all apart from death. Therefore, when life comes, you should just give yourself to life; when death comes, you should give yourself to death. You should neither desire them nor hate them."[2]

---

[2] Translation of "Shôji," in Norman Waddell and Masao Abe, trans. *The Heart of Dôgen's Shôbôgenzô* (Albany NY: State University of New York Press, 2002), 106.

An eagerness to instrumentalize death certainly did not first arise with the articulation of the Anglo-American philosophy we call "utilitarianism." But, given that this philosophy has such large role in the values that get configured in contemporary society, there is, I suggest, value in questioning it own utility in some domains. Because it points to the uncritical dominance of such values floating around in the heads of many people within at least contemporary Anglo-American societies, I like a cartoon that showed up in *The New Yorker* a few years ago. Because copyright laws prohibit it replication here, I ask the reader to try to imagine it through my words here. The cartoon shows two couples meeting on a city sidewalk. One of these couples is, however, accompanied by its two young children, and these are being introduced for the first time to the other pair, a couple that obviously is childless. The male of the latter is bending down to scrutinize the little ones he has just met. With a dumbfounded look on his face he says: "They are cute. But what are they *for*?"

I know I would be unable to state precisely what children—especially my own—are *for*. Goblins of reduction lurk within the question itself. Some things in our lives and our world are better left, I suggest, never described in terms of their ultimate *teloi*. Might there be something instructive in the fact that Daoists and many Asian Buddhist thinkers clearly recognized this problem and simply refused to go on an Aristotelian hunt for ultimate purposes? We may, I assume, hazard some guesses at more proximate values, and Turner's essay is unusually helpful in recognizing some of them. But, sensing that he is really eager to find the ultimate reason why death exists, I would prefer not to join in such a hunt. If the question "what is it for?" seems strange and out-of-place concerning the existence of a child, I think we are approaching a time when it will seem to be so concerning death as well.

CHAPTER FIFTEEN

PREFATORY REMARKS TO CHAPTER FIFTEEN REFLECTIONS: LET A DIALOGUE BEGIN

J. T. Fraser

I am grateful to Professor Steineck for his neo-Kantian critique of the generalized umwelt principle, a tool of the hierarchical theory of time. That theory is a framework for an integrated, interdisciplinary study of time.

The hierarchical theory of time recognizes six stable integrative levels in nature. They are (1) the world of particles with zero restmass that are always on the move at the speed of light and in whose proper frame time vanishes; (2) the world of particle-waves with non-zero restmass always on the move at speeds below that of light; (3) the world of massive, ponderable masses gathered into stars, galaxies and groups of galaxies; (4) the world of living organisms; (5) man as a species and as an individual of that species; and (6) the collective institutions of human societies to the extent that they function as semi-autonomous systems. These integrative levels form a hierarchically nested relationship. This means that the structures and processes of each level subsume those of the level(s) beneath them, are restrained by the lower level regularities, while also possessing new degrees of freedom of their own.[1]

Kantian thought emphasizes the intuitive and the spiritual over the empirical and material. The hierarchical theory of time is one in natural philosophy in the tradition of Newton and Whitrow. It builds its reasoning on whatever is known about the idea and experience of time through the sciences of matter, life, the human mind and society, and through the many faculties of the humanities.

---

[1] For an early formulation of the hierarchical theory of time see the chapter "Time as Conflict" in J. T. Fraser, *Of Time, Passion, and Knowledge*, 2nd ed. (1975; Princeton: University Press, 1990), 435–446. For an up-to-date summary of that theory, in the form of a collection twenty-two published papers see *Time and Time Again: Reports from a Boundary of the Universe*, Leiden: Brill, 2007.

Jo Alyson Parker, Paul A. Harris and Christian Steineck (Eds), Time: Limits and Constraints, pp. 343–349

Scholarly and scientific work, favoring interdisciplinary and integrated views, has been going on within and without the International Society for the Study of Time and has led to novel ideas, such as the identification of qualitatively different causations, temporalities, and their evolution. The focus of the present paper is the generalized or extended umwelt principle, a tool of the hierarchical theory of time.

1. Of the Heavens and of Human Values

In the conclusion of his *Critique of Practical Reason*, Immanuel Kant wrote that two things filled his mind with ever-increasing wonder and awe, the more often and the more intensely his thoughts are drawn to them: the starry heavens above him and the moral law within him. His remarks are broadly valid. I believe that the sense of awe, especially if placed in the service of survival, does indeed inspire and drive the human search for knowledge of many kinds.

Beholding the heavens and human values together in the context of the universe carries a question: by what path, by what means did the inanimate, dumb matter of the starry heavens—in its primeval form—bring about life, man, and the minding faculties of the human brain that make it possible for that brain to formulate moral laws? What was the path from cosmogenesis to biogenesis, to the human intellect and culture?

Kant himself initiated the tracing of such a path by inquiring into the history of the cosmos in his 1755 *Universal Natural History and Theory of the Heavens or an Essay on the Constitution and Mechanical Origin of the Whole Universe Treated According to Newton's Principles*.[2] It is a brilliant work and a joy for twenty-first century readers.[3] Also, Christianity is never far from the arguments. "I have tried to remove the objections which seemed to threaten my position from the side of religion."[4]

[2] Immanuel Kant, *Universal Natural History and Theory of the Heavens or an Essay on the Constitution and Mechanical Origin of the Whole Universe Treated According to Newton's Principles*, trans. by W. Hastie, revised and edited with an introduction and appendix by Willy Ley (1755; New York: Greenwood Publishing Corporation, 1968).

[3] It anticipates the likelihood of a plurality of Milky Ways. See G. J. Whitrow, "Kant and the Extragalactic Nebulae." *Quarterly Journal of the Royal Astronomical Society* 8 (1967): 48–56. We can say "with intelligent certainty and without audacity *Give me matter and I will construct a world out of it*" (17).

[4] Ibid., 15.

The Newtonian, mechanical universe is the simplest modern view of its subject and the easiest to relate to human experience. But the genius of the work suggests that he would have been equally successful—in a later epoch—to tackle problems of the primeval atom of general relativistic cosmology, of biogenesis and organic evolution, of the depth psychologies of our epoch and even of the universe of particle-waves. But not necessarily those of time.

Kantian philosophy appeals to the spiritual and the intuitive (derived from unspecified sources) and eschews the empirical and the material. Those who are set to trace a continuity between "the starry heavens" of inanimate matter and "the moral law within," a creation of the human mind in service of the body and society, cannot be excused from seeking to understand the origin and evolution of the cosmos, the origin and evolution of life, of man, and of society.[5] There are many ways of tracing such a path. One such journey, following the evolution of causation and time, is the subject of the chapter on "Constraining Chaos" in this volume.

2. Reality in the Worlds of Man, Animal, and Matter

Let us return to the schools of neo-Kantianism. There are very many of them. As a group, they make for a large and rich forest. So as not to get lost in it, I follow Prof. Steineck's guidance and deal, first, with the notion of reality, in some detail, in the hierarchical theory of time and, then, and only passing, with human values: the true, the good, and the beautiful.

We begin with a functional definition or test of reality, formulated in the tradition of the natural sciences. It is called the umwelt principle. In its generalized or extended form—coming up later—it is a tool in the integrated study of time. One that makes possible the coordination of insights concerning the nature of time, derived from the many ways of human knowledge.[6]

---

[5] On these issues, see: "The Secular Mystery of the First Day," (cosmogenesis); "Time and the Origin of Life" (biogenesis); "Temporal levels and reality testing" (depth psychology); "From chaos to conflict" (the universe of particle-waves). All of these are reprinted in *Time and Time Again, op. cit.*

[6] An entry may be had through J. T. Fraser, "The Extended Umwelt Principle: von Uexküll and the Nature of Time," *Semiotica* 134 (2001): 263–74. Reprinted in J. T. Fraser, *Time and Time Again: Reports from a Boundary of the Universe* (Leiden: Brill, 2007): 39–49. The generalized umwelt principle has been discussed and employed in the following of my books: *Of Time, Passion, and Knowledge*, 75 (1975); *Time as*

Early in the twentieth century, the German biologist Jakob von Uexküll became interested in the question of what constitutes reality for animals.[7] He noted that "an animal's receptors determine its world of possible stimuli, its effectors its world of possible actions." He called the combination of world-as-perceived and world-as-acted-upon the animal's *umwelt.* The word *Umwelt,* in this context, has been translated as "self-world," "phenomenal world" "perceptive universe," and "species-specific universe." It is defined as "the circumscribed portion of the environment that is meaningful and effective for a given species."[8] Note that here the "environment" is our human umwelt.

Uexküll's idea of "umwelt" may be easily formulated for the sensory systems of humans. We may think of auditory, tactile, and olfactory worlds, distinct but coordinated. From there it is a small step to the recognition of scientific instruments as exosomatic extensions of endosomatic systems.[9] With their help we can explore animal umwelts, such as the ultraviolet world of butterflies, available to them but not to us, because vertebrates have no sense organs through which they could see in the ultraviolet. Or, if we extend our senses through the use of radio telescopes, we may enter the world of electromagnetic radiation and explore the universe of a selected part of the spectrum—while remaining deaf to bird songs.

But we should go even beyond technical devices. For instance, if after thorough inquiry, a domain of nature is revealed to us consistently and exclusively through stable mathematical relations, then whatever we can learn through those mathematical relations, will become parts of our reality. For instance, the formulas of quantum theory make it possible for us to enter the umwelt of particle-waves where instants, precisely located in time and points, precisely located in space, do not

Conflict, 19–21 (1978); *The Genesis and Evolution of Time*, 19–23 (1982); *Time the Familiar Stranger*, 108–110, 368 (1987); *Time, Conflict, and Human Values*, 23–25 (1999); and in many of my published papers.

[7] Uexküll, Jakob von and Georg Kriszat, *Streifzüge durch die Umwelten von Tieren und Menschen, Bedeutungslehre* (Frankfurt am Main: Fischer Verlag, 1970). The *Steifzuge*...part was translated and published as "A Stroll Through the Worlds of Animals and Men," in *Instinctive Behavior: the Development of the Modern Concept*, ed. And trans. Claire H. Schiller (1934; New York: International Universities Press, 1957), 5–80.

[8] English and English, *Comprehensive Dictionary of Psychological and Psychoanalytic Terms* (New York: David McKay, 1958).

[9] "The Extended Umwelt Principle," in J. T. Fraser, *Time as Conflict: a Scientific and Humanistic Study* (Basel and Stuttgart: Birkhäuser Verlag, 1978), 19–21. See also, "Temporal Levels: a Fundamental Synthesis "in *J. Social and Biological Structures*, vol. 1 (1978), 339–355. Reprinted in *Time and Time Again, op. cit.*, pp. 153–174.

exist, where instants of time and locations in space may only be probabilistic. Or, we may enter the umwelt of photons in whose proper frame everything happens at once, where time does not pass.

Such conditions are alien to our experience and hence counterintuitive. For instance, Einstein never felt at ease with the probabilistic world of quantum theory. He often said that God does not play dice with the world. He presumed that in the physical world all causes are deterministic, and times and distances are well defined. It took the Copenhagen interpretation of quantum theory to reveal that the probabilistic nature of time in the world of particle-waves is not so because we are unable to demonstrate the underlying determinism. Rather, it is so because, on that evolutionary level, precise locations in time and space do not yet exist.[10] The times and distances in the laws of particle-waves are not statistical so as to hide precise, deterministic relationships. They are probabilistic because on that organizational level the cosmos is not evolved enough to have well-defined instants and locations. The cry of an infant is not a way of hiding his well-articulated thoughts. He does not yet have such thoughts to hide.

Since 1966, by meticulous attention to the temporal aspects of whatever we know of the laws and regularities of matter, life, and the human mind, the integrated study of time revealed in nature an open-ended, nested hierarchy of qualitatively different causations and qualitatively different temporalities. For a sampling of some of a few of the details, see "Constraining Chaos" and its references in this volume.

It is obvious that the hierarchical theory of time, employing the generalized umwelt principle, does not interpret reality as an immense store of well-defined, pre-existing details about the world and time, from which people may select. Rather, it sees the world as comprising

> acts of creation whose uniformity for each species is guaranteed by the psychobiological uniformity of members of that species. [....] For Kant the Ding-an-sich, though not knowable existed, nevertheless, independently of the perceiver. In the view held here, final reality is relative to the perceiver. Yet, for the reader and this writer it is not ambiguous because [they] share the psychobiological unity of their species.[11]

---

[10] See the chapter on quantum theory in J. T. Fraser, *The Genesis and Evolution of Time: a Critique of Interpretation in Physics* (Amherst: University of Massachusetts Press, 1982), 62–98.

[11] J. T. Fraser, *Of Time, Passion, and Knowledge*, 2nd ed. (1975; Princeton: Princeton University Press, 1990), 75–76.

This makes the generalized umwelt principle itself solidly Kantian. This idea must have been in the mind of von Uexküll when he remarked that, through his work, "biology has ultimately established its connection with the doctrine of Kant, which it intends to exploit in the Umwelt theory, by stressing the decisive role of the subject."[12] The integrated, interdisciplinary examination of time also directed attention to some other, interesting matters, such as the likelihood that mathematics is an empirical science whose laws and regularities became programmed into all species through natural selection but has been abstracted only by humans. This particular idea was born from noting certain striking similarities between the structure and properties of the axiomatic system of mathematics on the one hand and, on the other hand, the nested hierarchical structuring of the laws of nature revealed by the hierarchical theory of time.[13] Another suggestion that surfaced was a definition of truth as a recognition of permanence in reality.[14]

Consistently, the world at large is then best understood as a dynamic model, created by the knower and the known together.[15] "External reality does not appear as an immense store of well-defined information from which an individual may select some, as a student of a language selects words from a dictionary. Instead, it will comprise acts of creation whose uniformity for each species is guaranteed by the psychobiological uniformity of the members of that species."

The extended umwelt principle allows us to identify all the qualitatively different temporalities present in human reality: those of inanimate matter, of living organisms, those of distinct minding faculties, and that of a person as a member of a society.[16]

The scientific and humanistic details of the hierarchical theory of time have been set forth in several books and many paper.[17] Each such detail abides by the modes of inquiry and self-checking of the field of knowing where it is located. The extended umwelt principle, to be examined below, is the theory's particular interpretation of reality,

---

12 Jakob von Uexküll, "A Stroll Through the Worlds of Animals and Men," *op. cit.*, 13.

13 J. T. Fraser, "Mathematics and Time," *KronoScope*, 3.1 (2003): 152–67. Reprinted in *Time and Time Again*, *op. cit.*, 51–64.

14 J. T. Fraser, "Truth as a Recognition of Permanence: an interdisciplinary critique," in *Zeit und Wahrheit* [Symposium on Time and Truth], ed. Heinrich Pfusterschmidt-Hardtenstein (Vienna: Ibera Verlag, 1995), 120–31. Also in *Time and Time Again*, *op. cit.*, 297–309.

15 *Time as Conflict*, *op. cit.*, 21.

16 J. T. Fraser, *Of Time, Passion, and Knowledge*, 2nd ed. (Princeton: Princeton University Press, 1990), 436–443.

17 See the website TIMEMITH.NET.

so composed as to permit the incorporation into human reality of the many ways of knowing that necessarily enter the hierarchical theory of time as it appeals to the many and different ways that time enters human life and knowledge. The extended umwelt principle thus conjoins into a single system of ideas that have their roots and their functions appropriate for the many compartments of human knowledge. By a poetic image, the extended umwelt principle may be called a theory of the moving boundaries of reality.

How may one sum up these notes? By turning to poetry. To James Russell Lowell (1819–1891):

> New occasions teach new duties
> Time makes ancient good uncouth
> They must upward still and onward
> Who would keep abreast of truth.

In other words: let the dialogue proceed.

Appendix

An exercise, serving as a summary and a review

"[Butterflies] evolved distinct ultraviolet wing and body patterns, visible to one another but not to vertebrates, which serve as the medium of much of their private communication."[18]

> Do butterflies have distinct wing and body patterns?
> In the umwelt (reality) of many invertebrates? Yes.
> In the umwelt (reality) of vertebrates? No.
> In the umwelt (reality) of humans? No.
> In the extended umwelt (reality) of humans? Yes.

"Time that takes survey of all the world...." Is it qualitatively the same kind of time for matter, animals, and man? The answers:

> For human reality, unaided, yes.

For human reality extended by instruments and theories, no. Rather, the universe functions with a nested hierarchy of five qualitatively different temporalities, called the canonical forms of time.[19]

---

[18] Edward O. Wilson, *Sociobiology the New Synthesis* (The Belknap Press of Harvard University, Cambridge Massachusetts, 1975), 241.

[19] J. T. Fraser, *Time and Time Again, op. cit.*, 91.

TRUTH, TIME, AND THE EXTENDED UMWELT PRINCIPLE: CONCEPTUAL LIMITS AND METHODOLOGICAL CONSTRAINTS

Christian Steineck

Summary

This paper seeks to elucidate methodological constraints that apply to a natural philosophy of time as envisioned by J. T. Fraser from the point of view of a critical (Neo-Kantian) theory of knowledge. The seminal concept considered is the "extended Umwelt principle," which in Fraser's theory serves to establish a strong link between ontology and epistemology. It is argued here that this link is problematic. Furthermore, the extended Umwelt principle does not account for the specific characteristics of the human notion of the "world." The human "world" is not just a larger Umwelt, extended by technical instruments, but, to the contrary, a critical tool to reflect on the confinements of any Umwelt. The final section sketches out some revisions to and guiding principles for the further development of the natural philosophy of time on the grounds of the analysis given.

1. Introduction

J. T. Fraser's hierarchical theory of time is arguably the most comprehensive theory on this subject existing to date. As such, it deserves full attention from all disciplines engaging in the exploration of time. This paper approaches the hierarchical theory of time from a philosophical point of view. More precisely, it is based on a critical reading of Fraser's work through Neo-Kantian eyes. Keeping the core insight of Neo-Kantianism—that is, the intimate relation that exists between the concepts of truth, method, and reality in mind—it reflects upon the methodological constraints that apply to a natural philosophy of time. At the same time, it attempts to resolve some tensions between this theory's content and its epistemological and ontological foundations as stated by Fraser himself.

In this introductory section, I will briefly describe the essential characteristics of the Neo-Kantian point of view. The main section of this paper then consists firstly of a summary of Fraser's express statements on knowledge, truth, and reality, and the fundamental epistemological

Jo Alyson Parker, Paul A. Harris and Christian Steineck (Eds), Time: Limits and Constraints, pp. 350–365

concept behind the hierarchical theory of time, the extended (or generalized) umwelt principle. Secondly, I will enumerate the main points of contention between Fraser's and the Neo-Kantian point of view, but also the logical tensions inherent in Fraser's position. This is followed by a critical review of Fraser's "working concept" of truth and of the extended umwelt principle. In the last part of the main section, I propose some consequences for the hierarchical theory of time. Finally, I attempt to anticipate and answer to some questions and criticisms concerning the argument presented.

What is it like to be a Neo-Kantian? Like Kant, Neo-Kantians are mainly interested in the foundations of knowledge, ethics and aesthetics. More precisely: we are interested in the foundations of the *validity* of knowledge etc., its *scope*, its *quality* (valid extension and intension), and its *relation* to other claims to validity. Based on Kantian insights, we think the main task of philosophy is to give a critical assessment of such claims to validity, with the affirmative aim of elucidating their possibility (against skepticism) and the negative aim of curbing excessive aspirations.

Unlike speculative philosophers (metaphysical, existential, positivist or others) we do not think it is our task to produce positive knowledge about the world or human life. However, we see our objective in opening up or clearing possible paths of rational investigation and solving riddles stemming from wrong, confused or excessive application of otherwise valid forms of thought.

In this, we are *empirical* insofar as we relate to given "facts of culture" such as science (Natorp 1910; Cassirer 1956), art (Marx 1981), morality and law (Cohen 1907) instead of deducing them from first principles or producing them by transcendent insight or other forms of intuition. We are therefore essentially opponents of metaphysical idealists and speculative philosophy. But—we are not *empiricists*, because we do not believe that truth etc. is something *received from* experience. We would rather say it is *found in* experience—while going on to say that experience itself is something *created* with the help of organizing principles, which are not given by nature or some celestial spirit, but *invented* by the mind. (On this, see especially Cohen 1914, 31–38; Natorp 1910, 22–26.) Thus, we would insist with Kant that "The conditions of possibility of experience are the conditions of possibility of the objects of experience" (cf. Kant 1787, 197)—and that makes us the natural opponent of all positivist, empiricist or naturalist ideology as well.

Let me sum up this little tour de force of Neo-Kantianism by coming back to the question: what is it like to be a Neo-Kantian?

It means to be open to, even enthusiastic about all forms of cultural achievement while being aware of the limits of human knowledge and existence. It means to be skeptical about all claims to omniscience or omnipotency while securing a place for core human values like truth, freedom, dignity, and beauty. It means to insist on rational analysis while affirming the right of irrational modes of human behavior. Finally: it means to give credit to other philosophical authors by treating their theories seriously, that is, as theories and not as myth, literature or religion. I feel that in these fundamental orientations, Neo-Kantianism is close in spirit to the hierarchical theory of time—and that is why I want to offer Neo-Kantian insights as a, sometimes corrective, philosophical complement to this theory.

2. Fraser's Concepts of Knowledge, Truth, and Reality

In the following, I will briefly sketch out Fraser's explicit statements on knowledge, truth, and reality. My main focus will be on the "Extended Umwelt principle," which forms the centerpiece of Fraser's epistemology and ontology. My description is based on those passages in his works *Of Time, Passion and Knowledge* (Fraser 1990), *Time, Conflict, and Human Values* (Fraser 1999) and his most recent, *Time and Time Again* (Fraser 2007), which expressly reflect on the nature of our subject. It is part of my thesis that Fraser's hierarchical theory of time, in its exploration of the various aspects of temporal realities, goes beyond the confines of Fraser's explicitly stated epistemology and ontology. Readers may therefore find the following description somewhat polemical in the light of the total scope of Fraser's work. This perception is correct to some extent. The description and the ensuing critical analysis are meant to highlight problems of Fraser's explicitly stated epistemological and ontological views, and to contrast them with some important implications of his own theory of time.

Judged by his own general and methodological statements concerning truth, knowledge, and reality, Fraser's standpoint can be described as *empiricist, relativist, conventionalist, scientistic* and *evolutionist*, with each predicate qualifying the other.

To elaborate, it is clear from his description of the mathematical method that Fraser subscribes to an *empiricist* epistemology. Mathematics is arguably the least empirical of all sciences. Still, Fraser describes its method as "the *observation* of mathematical objects, and the abstractions from these observations of such properties as appear to be unchanging" (Fraser 1990, 335; my emphasis). Thus, mathematical truth appears to be ultimately based on *perception*, followed by abstraction and deductive reasoning.

In contrast to positivistic or realistic empiricism, Fraser's empiricism has a surface layer that is *relativistic* and *conventionalistic*, as is obvious from his definition of truth as "that class of knowledge which individual and communal perception judge to be timeless" (Fraser 1990, 321). In his view, truth is a function of human judgement, and it is subject to changes depending on historical and social change: old truths are replaced by new truths which will again be replaced, not necessarily by something more accurate, more complex, or more sophisticated, but simply by something new that is in accord with individual and social predilections of a new era: "If theories accepted at an epoch can appear as fantasies in later epochs, and if the character of fantasies are functions of personalities, then the theories themselves are but reflections of personalities, individual or communal." (Fraser 1990, 335)

As in any conventionalist relativism, the idea of a consistent *theoretical* standard of truth that would allow one to compare the truths of different ages is rejected. However, there is a standard beyond communal human judgement, and that is the evidence provided by natural selection. Thus, behind the surface relativism there is a background realism which is, in its essence, naturalistic: in Fraser's statements, the truth behind the veil of human aspirations is biological evolution, perceived in terms of natural selection. In the light of this background theory, power, the ability to manipulate the environment to the ends of the individual and society, is the ultimate standard of truth:

> Truth may be seen as that knowledge (or, if you wish, those aspects of behavior) which the organism may successfully employ in the control of its destiny. Accordingly, questions of knowledge and truth become functions of the complex implications of many temporalities. This view is to be distinguished from the pragmatism of Peirce and James. For, whereas in pragmatism testing for pragmatic consequences is conceived of as a method of philosophy, *in the relativity of truth to temporal levels the search for "timelessness" is seen as a working method of natural selection.* (Fraser 1990, 330; my emphases)

Although Fraser does explore various fields of humanistic understanding in his works (cf. his engagement with psychoanalysis in Fraser 2007, 177–215), in his epistemological statements he reserves the term knowledge to the realm of the sciences and other fields of factual episteme, which have a direct impact on the manipulation of the environment. In contrast, the humanities, whatever their methodologies, may "pioneer, embrace, tame, protect or nurture these forms of knowledge"—but do not appear to constitute any *knowledge* of their own. (Fraser 2007, 42)

At the heart of this surface relativism and essentialist naturalism and yet somewhat at variance with it, is the centerpiece of Fraser's working concept of reality, that is, the extended Umwelt-principle.

The concept of Umwelt was developed by Jakob von Uexküll (1957) to describe the cognitive and pragmatic spheres of animate beings in general. These spheres are circumscribed by a closed functional loop between the stimuli it may perceive and the responsive actions it is capable of. Umwelten are specific to the sensorical and motorical functions and capacities of a species. Because they are functions of a closed loop, they are also internally complete. The umwelt of a species contains everything that can be acted on or can be known—to the members of this species.

The extended Umwelt principle applies this concept both beyond and below the animal world. Firstly, the content of the human world—human reality—is equated with the human umwelt. In the course of history, this umwelt has vastly expanded beyond the confines of unaided human senses and immediate human action. However, according to Fraser, it is still an *umwelt*, a closed world dependent on human sensori-motor functions and their *extension* through technical instruments and symbolical representation. Although in the human umwelt, the norms of symbolic representation may sometimes counteract primitive biological impulses, ultimately, human reality, like everything else in the evolutionist universe, is shaped and driven by "the needs of an organism" in the struggle for survival: "Human reality…is something that the mind contributes to the body's efforts to survive in its specific environment." (Fraser 1999, 24)

An important part of that contribution is the ability to understand, and thus, manipulate *umwelten* that differ from the familiar human reality. Symbolic representation, technical instruments, and mathematical reasoning allow not only for the integration of the contents of non-human umwelts into human reality. They also enable us to reconstruct and represent their specific structure, even below the level of animate life. As a natural philosophy, the hierarchical theory of time makes use of such reconstructed *umwelten* of photons, massive particles, and inanimate matter in its narrative of the open-ended evolution of time from the atemporal to noo-temporality and beyond. It thus extends the concept of umwelt below the sphere of organic life, into the world of inanimate matter and energy.

Finally, an essential feature of the extended umwelt principle as a theory of reality is its equation of epistemology and ontology: "the world is the way we find it to be through the many forms of human knowledge, even if some of its features appear to be counterintuitive" (Fraser 1999, 25). This statement is essential to the hierarchical theory of time as an ontology of progressively established, increasingly complex levels of temporality.

In spite of this, it is to some extent in conflict with more cautious remarks of Fraser, who is careful to distinguish his theory from a straightforward realism and even professes a proximity to Kantian idealism, in allowing for the constructive powers of the intellect:

> The generalized umwelt principle resembles but is not identical to realism, because it does not maintain that our senses and thoughts present us with details of a world whose furnishings are independent of the knower. [...] It [...] resembles Kantian thought in its insistence that the intellect does not derive its laws from nature but prescribes them to nature. (Fraser 1999, 25)

I surmise that, had Fraser followed up with these remarks, he would have arrived at a revision of some of his bolder onto-epistemological statements much in line with the criticisms elaborated below.

3. Points of Contention and Critical Analysis

As a Neo-Kantian, I disagree with both the surface relativism and the background biologist essentialism of Fraser's concept of truth and knowledge. More importantly, I perceive them to be at odds with his own hierarchical theory of time. This theory argues that each of the subsequent levels of temporalities has its specific norms and structures that cannot be reduced to those of previous layers. This should hold for noo- and socio-temporality as well, which is the level where truth, knowledge, and the notion of reality belong. (For reasons to be explained elsewhere, I treat noo- and socio-temporality as one level.)

One essential characteristic of noo-sociotemporality is that it includes symbolic (or ideal) norms for cognition, behavior and appreciation, which may become constitutive for specific fields of human endeavor and aspiration.

A brief remark on the word "ideal" is in order here: *Ideal* traditionally stands in opposition to *material*. This opposition can be interpreted in various ways. Dogmatic Metaphysics interpreted it in terms of substance, as in Descartes famous distinction between *res cogitans* and *res extensa*. (Descartes 1642, 23–34) However, this substantialist interpretation converts the ideal into yet another kind of the material, and produces an irresolvable dichotomy of two separate realities whose connection cannot be explained. Neo-Kantianists therefore prefer to interpret the opposition in relational terms: the term "ideal" then points to the meaning produced through and communicated by symbolic systems. "Meaning" cannot be presented in space and time (you cannot give space-time coordinates for the definition of a triangle), although it has to be represented by symbols localized in space and time. (On this and the following, see Cassirer 1923, 1–54, esp. 41–54) That is why the sphere of meaning is legitimately differentiated from the realm of physical things, without being separated completely from that realm. Thus, one may conceive of the ideal as part of reality without subscribing to metaphysical idealism: "property," for example, is a norm generated by the symbolic system of law, and thus a typically ideal norm. In contrast to simple possession, which is a relationship of location in space and time, property is not directly demonstrable to the senses. Its whole point is that it may be distinguished from possession. However, if the term "property" is attributable to something, that is, if property is part of a country's system of law, it has become a reality that makes itself felt in many aspects of individual and social life. (On this, see Heinsohn and Steiger, 2002)

The hierarchical theory of time, with its recognition of noo-temporality, allows for the ideal to become a constitutive part of reality. But Fraser's explicit onto-epistemology, the surface conventionalism and background naturalism of Fraser's working concepts of truth and reality, do not account for this. I side with Fraser's hierarchical theory of time against Fraser's relativism and naturalism. I believe that the latter is a received attitude Fraser maintains in his statements on truth and knowledge, partly because his naturalism-conventionalism causes him to disregard the merits of close *philosophical* argumentation. Rhetorical and strategical design may also be involved. Whatever their success, in terms of consistency, the hierarchical theory of time would be stronger as theory without this received attitude. The following argument strives to explain in some detail where and why.

A first problem lies in Fraser's "working concept" of truth and the methodology behind that notion. In his *Time, Conflict, and Human Values*, Fraser takes a conventionalist approach and derives his concept of truth from a survey of "four major families of languages," elucidating the convergent connotations the respective words for truth entail (Fraser 1999, 25–26). This survey leads him to the identification of truth as "the recognition of permanence in reality," more or less equivalent to the above quoted "class of knowledge, which individual and communal perception judge to be timeless" (Fraser 1990, 321).

As a description of commonly shared notions of truth, this definition may be correct. It is, however, an uncritical one (as folk notions are wont to be), and it tends to confound various dimensions of reference pertinent to the assessment of truth.

Philosophers since Plato's famous self-criticism in his "Gigantomachia" (*Sophistes*, 246a–249d) have striven to untangle confusions that arise when "timelessness" is attributed to truth without further qualifications. Most of them nowadays would fully agree with Fraser that "there is nothing in nature to which the idea of truth as eternal verity can correspond" (Fraser 1999, 72; for analogous Neo-Kantian assessments, see, for example, Natorp 1910, 16–22; Cassirer 1923, 41–54). They would even go further to agree that "there is nothing in science to which the idea of truth as eternal verity can correspond" because scientific theories are increasingly put in their historical and cultural context. In both cases, the timelessness of truth would have been attributed to the content of a proposition. Due to the lack of permanence in the natural world (including the human world), the search for timeless objects has come to be commonly understood, at least by historically and scientifically educated philosophers, to be in vain.

However, at least the Neo-Kantians among them would still agree with Plato that the truth of an appropriately formulated statement is *in some respect* independent of time: If it is true that "there was a total solar eclipse over Cornwall on August 11, 1999," this means that this statement will be as true in 500, 5000 and 500,000,000 years as it is now. It also means that predictions uttered in 1700 CE or in 1700 BCE leading to this date of the eclipse would have been true, while others, predicting a total solar eclipse on a date equivalent to August 12, 1998, would have been wrong. This form of timelessness pertains to the *validity* of a truth, but not to its content or its relevance. Furthermore, the actual formulation of this particular truth will have to be adapted to the temporal relation between the time of its being stated and the time of the event it is referring to. The "timelessness" of its validity is thus embedded in a temporal framework. It is not an "absolute" timelessness. This relation has been given particular attention by R. Hönigswald (1875–1947); see, for example, his *Grundfragen der Erkenntnistheorie* ("Fundamental Questions of Epistemology"), first published in 1931 (Hönigswald 1997, 70–81; but see also Cassirer 1990, 410–421).

In a similar vein, many only temporarily valid propositions can be transformed into more "timeless" statements by adding the reference to the framework implied. This is especially important when alternatives have become available. The proposition "now it is night," whose claim to truth was famously ridiculed by Hegel because it turns "stale" when examined in the light of day (Hegel 1986, 84), can reclaim its validity (but not necessarily its relevance) when date and time are added. More significantly, in the sciences, mentioning of specific parameters can save the validity of a proposition after alternatives have become available to a once universal theory: With the invention of non-euclidean geometry, the statement "parallels do not intersect" became wrong, but after reformulation into "in euclidean geometry, parallels do not intersect," it would become true again. In contrast, the Aristotelian assumption that the brain is an organ whose main function consists in cooling the blood (Aristotle 1955, 653b; Gross 1995) turned out to be simply wrong: that is, it was never true, not even in Aristotle's time. There are upper and lower thresholds to the conventionality of truth. Although science as a cultural reality may develop in the leaps and bounces of Kuhnian revolutions or according to the even more anarchic mode described by Feyerabend (1975), the various theories held imply, by their claim to truth and not just to the power of the day, a standard of comparison according to which their respective value may be judged.

Neo-Kantians would not so much emphasize the "timelessness" of this truth-claim, but rather regard it as one element of the critical value of truth: that is, its claim to independence from mere subjective notions or predispositions. "Truth" is related to "method," an established way of verification, which may vary according to the field of investigation. Differentiation of method follows differentiation of problems that are solved by establishing the truths in question (Hönigswald 1997,

132–158). To illustrate, there are many possible truths concerning, for example, Newton's famous apple, to be established by various means of knowledge: Newton's laws of gravity describe certain conditions of its fall. Biology and geology explain why English apples fall in summer and early autumn rather than in December. Psychology and cognitive science strive to explain why the "lightning of truth" strikes often not during a period of concentrated work but in a break after sustained endeavor. Cultural history reveals how the apple fits nicely into the European-Christian tradition as the "fruit of knowledge"—which may have contributed to make the story last in cultural memory. Each kind of truth is coined in different terms and has to be verified by different means. Thus, there is no universal criterion of truth, and cannot be—an idea that fits very neatly into the hierarchical theory of time—but there is also no need to retreat from the challenge inherent to any claim to truth, and to disparage the very idea of theoretical validity.

A truth may be challenged either by (a) the claim that the method established to verify it yields a different result or by (b) a new method pertinent to the relevant field of knowledge. In the second case, the new method has to prove that it is somehow more independent from subjective conditions than its predecessor, and in this sense more comprehensive, although it may explain not everything that its predecessor explained: viz. Galilean physics vs. Aristotelian physics (Cassirer 1990: 407–409). The dynamic of the cultural evolution of knowledge can be accounted for by adequately reflecting upon the relation between truth and method. Once the idea of truth is freed from excessive claims to "eternality of content" or "eternality of relevance," its "timeless" aspects can be understood in their relation to the progressive dialectics of claims to truth, their critical verification, and the establishing of new methods, ideas, and—truths.

With this in mind, let us turn to the extended umwelt principle. The main problem I see in the extended umwelt principle can be stated quite simply: I believe the extended umwelt principle does not account for *reality* as such—neither for the *intension*, nor the *extension* of that concept.

Reality is, first and foremost, a critical concept we use to distinguish adequate ("true") accounts of events, processes, objects, and their aggregates from those that are incomplete, illusory, deceptive, or limited to a specific perspective. The *intension* of *reality* is the qualitative standard we use in measuring accounts of the world. The *extension* of reality is the totality of everything that does exist with all its possible aspects.

Although the word *reality* refers to that which is not a mere figment of the mind, reality itself is a normative concept, an idea. To put it bluntly, in terms of genesis it is a fiction. However, it is a fiction that is productive and even constitutive for the rational appreciation of experience, precisely because it is not something received. The great merit of this fiction is that it allows us to transcend subjective notions, given perceptions and received concepts in search for *ever more adequate and in-depth* accounts of the world. The hierarchical theory of time in its exploration of various layers of temporality is clearly guided by this notion of reality. But the extended umwelt principle does not account for this, for the following logical reasons.

The first reason is that the extended umwelt principle is *intensionally* incomplete. It does not have a conceptual place for the *quality* of reality. Uexküll's concept of *umwelt* was designed to explain the *phenomenal* completeness of the cognitive and pragmatic spheres of each animal. In other words, it refers to an *inside* account of what animate beings perceive of their objective environment. It demonstrates how limited perceptual worlds may appear complete to those living in them. The defining structure of an umwelt is that it consists of a closed loop between sensation and action. Typically, animals are confined to their umwelt because they lack the instruments to break through this closed loop—they cannot conceive of the possibility of an "outside."

In contrast, the world of human beings as they have come to be is, typically, not just a vastly larger *umwelt*, expanded by the addition of other *umwelten* through symbolic representation and technically aided perception, as the extended umwelt principle has it. It is something different in quality. It is not a closed and complete *umwelt*, but a "Welt" (world) that has lost the aspect of closedness, completion and cosiness implied by the German prefix "um-." Human subjective reality is defined by the relationship between the individual's *umwelt* and *the world*. In relating the *umwelt* to the world, individual and species-specific boundaries of perception and conceptualization are put in brackets and assigned question marks. The *umwelt* becomes the "world as it appears to me / us." This is beautifully expressed by Fraser himself in his exposition of the umwelt principle:

> The Umwelt, wrote Uexküll, is like "a soap bubble around each creature" [...] I will consider the content of each "soap bubble" a report about an

> instant in the course of organic evolution, an always transient, always incomplete recognition of the world, whether by animals or man. (Fraser 2007, 39; Uexküll quoted after id. 1957, 11)

As human beings, we still have an *umwelt*, a frame of reference partly defined by the makeup of the human body, which defines the potential and the limits of our immediate perceptual world, and, partly defined by culture, which further guides our cognitive, aesthetic, and ethical intuitions. But we have learned, if to a varying degree, to perceive our *umwelt* as an incomplete representation of the world as such. That is why we ask, is it real? Is it true? And again: Is it really, really true?

The human world is thus characterized by relating to something beyond the individual, communal, and species-specific umwelt. *Reality* is the critical notion that has us systematically put our umwelt into question. *Reality* is not part of the umwelt. Its notion cannot be achieved by a simple expansion or summation of umwelten (which is quite evident from Fraser's own words as quoted above). It involves a transformation, a new quality. This quality is one of reflection: The concept of reality originates when we reflect boundaries that cannot become visible as long as we remain content with, and confined to the purely internal sphere of an *umwelt*.

The second reason the extended umwelt principle is insufficient is that it is extensionally incomplete. In its expansion of the inside account to totality, it systematically eliminates "the outside world" as a frame of reference.

Consequently, those parts of the objective environment not presented in the structure of the umwelt remain outside of the scope of vision. If we equate epistemology with ontology, our ontology lacks a space for that which we do not know, and for those things or events that do not even gleam through the arch of our individual or aggregated experience (cf. Fraser's quote of Tennyson, Fraser 1999, 26).

Consider a palmtree on an oceanic island. It is a living being nicely fit into its ecological niche, with receptors for light, humidity, etc., and effectors for growth, photosynthesis and the like. This umwelt contains all that is necessary and sufficient to "know" for palmtrees in their normal habitat.

The oil industry is obviously not part of this umwelt. The palmtree doesn't know about money. It doesn't have receptors to perceive lobbying in the legislative institutions of the human world. It doesn't have effectors to manipulate human consumption of mineral fuels, let alone political bargaining in international organizations. Still, its very existence may depend on what is decided there.

Reality not represented in the umwelt can have a decisive influence on the existence of an individual being, or even of that of a species. The extension of an umwelt is, by definition, circumscribed by what the receptors of an entity can perceive and what its effectors can reach. The extension of reality comprises things we do not know and things we may be unable to perceive or to affect in any way. Still, these things may prove their reality by having an effect on us or any other living being, thing or particle-wave, for that matter.

Even within the given individual itself, its own reality—the sum total of its existing properties—contains many aspects that are beyond the scope of its umwelt. Proprio-sensation and proprio-regulation are far from being objectively complete. The structure of an umwelt may grossly reduce the number of properties of even the physical appearance of the self-same individual or species that can be accounted for.

The standards of evolution by chance variation and competition for vital ressources may be harsh, but they are also parsimonious. Depending on the environment and the amount of competition, survival may be possible with only a tenacious and fragmentary cognitive link to the internal and external reality—as long as food comes in and enemies stay away. However, if some change in the environment relevant to the existence of some kind of being is caused by something not present in its umwelt, this may spell disaster. Of course, at some point the effects of the cause will take a form perceptible to that being—that is, they will become part of its umwelt—but not the cause itself. The palmtree drowns without knowing what brought about its doom. However, it would be a contradiction to say global warming, the oil industry or the United States' House of Congress are not part of the palm trees *world.* If that were so, they could never affect it. Therefore, there is a difference between what is part of the world (the extension of reality) and what is part of the umwelt. The notions of "world" and "reality" refer not only to everything that is part of the human umwelt at any given point in time, but also to everything that may ever have an effect which could become part of the umwelt of any being that will be known to any being which was, is, or will be present or represented in the human umwelt.

4. Implications for the Integrated Theory of Time

The problems of the extended umwelt principle indicated above imply the following revisions of this theory:

(1) The extended umwelt principle should be understood as a methodological device instead of an epistemological/ontological foundation. Its application leads not to regional ontologies but to the description of minimal temporal conditions for the structural integrity of certain types of entities. It allows us to understand the internal temporal structures operating within different kinds of objects. In addition, it explains why the cosmos itself at early stages of its development does not conform to our usual notion of time and why certain realms of present reality, if isolated, continue to do so. The methodologically constrained extended umwelt principle thus retains its seminal role in exploring the cosmic history of time and the hierarchy of temporal structures realized in the internal organization of various kinds of beings.

(2) The re-interpretation of the umwelt principle as a methodological device leads to a revision of the reductionist regional ontologies of the hierarchical theory of time. Once again, the lower levels, and especially the inanimate physical world come to mind as an example. In Fraser's application of the extended umwelt principle, the reality of this world is adequately and completely represented in the mathematical formulas of physics. Insofar as these do not contain a unidirectional vector of time, there is no directed time in the world of inanimate matter. However, if the umwelt principle is constrained to function as a method of exploring minimal or constitutive temporal conditions, we would instead say that in inanimate matter, unidirectional time is not a constitutive element of the structure of its individual entities. Inanimate physical structures as such may subsist with physical processes being reversed, whereas in organisms, reversion destroys the organic structure itself. This interpretation allows for an enriched understanding of the world of inanimate matter. It is also more consistent with the fact that in the methodology of physics, many aspects of the given object of investigation are systematically excluded from view. Physics is interested in regular relations of quantities, and it treats each object as merely a sample in a series of equivalent events. The specific taste of Newton's apple, though one of its objectifiable properties, is of no concern to the kinetic equations describing its fall, as is its procreational function or its nutritional value. In terms of kinetics (and if we abstract from friction), the apple may be replaced by a piece of marble or clay of equal mass. The specificity of physics' methodology requires complementation by other methods in order to approach a "full" description of the reality of physical objects; and it is not correct to assume that,

for any one given physical object, all that can be meaningfully said about it could be formulated in the terms of physics.

(3) Constraining the extended umwelt principle in this manner makes it consistent with the hierarchical theory of time and in particular with the natural history of time. If the extended umwelt principle were true as an ontology, a natural history of time would and could only commence with the origin of life because inanimate umwelten do not support directed time, and thus, development or evolution. However, in the light of Neo-Kantian methodological idealism, there is room for historical questions pertaining, for example, to the world (not the umwelt) of inanimate matter. The Neo-Kantian would ask: Does the problem make sense? Is it, for example, a meaningful question to inquire about the history of the universe? Are there methods of investigation that will provide empirical evidence for this type of question? The hierarchical theory of time has answered these questions to the positive; it would be more consistent to reflect this in its epistemology.

(4) An adequate analysis of the noo-temporal world has to go beyond the mere quantitative expansion of the bio-temporal *umwelten.* It has to account for the role of normative fictions—*ideas*—which are not taken from experience, but used to organize, measure and differentiate it. *Reality* is one such necessary fiction. It is a key concept that helps us out of Uexküll's cave. To put it in philosophical terms: An adequate account of human knowledge cannot be based on the empiristic model of deduction from observation alone. It needs to reflect the crucial role of *counterfactional* norms. It is the creativity of those norms that breaks the barriers of received experience and wisdom and propels us into the constitutive insecurities of human existence. Without this structural friction, without counterfactional ideas, there could be no truth, no morals, not even beauty.

Fraser's descriptions of the noo- and sociotemporal world, the world of human culture, contain ample illustrations of these philosophical observations. My argument is that these insights should be more strongly reflected in the methodology of the theory of time. Doing so will bring it closer to its own aim of a rational exploration of time and human existence.

References

Aristotle 1955. *Aristotle XII. Parts of Animals. Movement of Animals. Progression of Animals.* Trans. A. L. Peck and E. L. Forster. Cambridge: Harvard University Press.

Cassirer, Ernst. 1990. *Substanzbegriff und Funktionsbegriff.* Darmstadt: Wissenschaftliche Buchgesellschaft. Translated as *Substance and Function.* Chicago: Open Court, 1923.

———. 1923. *Philosophie der symbolischen Formen. Erster Teil: Die Sprache.* Berlin: Cassirer. Translated as *The Philosophy of Symbolic Forms. Volume One: Language.* New Haven: Yale University Press, 1955.
———. 1956. *Determinism and Indeterminism in Modern Physics.* New Haven: Yale University Press.
Cohen, Hermann. 1907. *Ethik des reinen Willens.* Berlin: Cassirer.
———. 1914. *Logik der reinen Erkenntnis.* Berlin: Cassirer.
Descartes, René: *Meditationes de Prima Philosophia.* Quoted after: Id.: *Oeuvres.* Ed. Charles Adam and Paul Tannery. Vol. 7: 1–90. Paris: Cerf, 1897–1913.
Feyerabend, Paul. 1975. *Against Method.* London: Verso.
Fraser, J. T. 1990. *Of Time, Passion, and Knowledge.* Princeton, NJ: Princeton UP.
———. 1999. *Time, Conflict, and Human Values.* Urbana and Chicago: University of Illinois Press.
———. 2007. *Time and Time Again.* Leiden et al.: Brill.
Gross, Charles G. 1995. Aristotle on the Brain. *The Neuroscientist* 1.4: 245–250.
Hegel, Georg Wilhelm Friedrich. 1986. *Phänomenologie des Geistes.* Frankfurt a. M.: Suhrkamp [Werke; 3].
Heinsohn, Gunnar and Otto Steiger. 2002. *Eigentum, Zins und Geld.* Marburg: Metropolis.
Hönigswald, Richard. 1997. *Grundfragen der Erkenntnistheorie.* Hamburg: Meiner.
Kant, Immanuel. 1787. *Critik der reinen Vernunft.* 2nd ed. (B). Riga: Hartknoch. Quoted after: Kant, Immanuel. 1998. *Kritik der reinen Vernunft. Nach der 1. und 2. Originalausgabe herausgegeben von Jens Timmermann.* Hamburg: Meiner.
Marx, Wolfgang. 1981. *Ästhetische Ideen. Untersuchungen über die Grundlagen einer Theorie der Kunst.* Bonn: Bouvier.
Plato: *Sophistes.* Quoted after: Platon. 1965. *Spätdialoge. Eingeleitet von Olof Gigon. Übertragen von Rudolf Rufener.* Zürich: Artemis.
Uexküll, Jakob von. 1957. A stroll through the worlds of animals and men. In *Instinctive Behavior. The Development of a Modern Concept.* Claire H. Schiller, ed. and trans. 5–80. New York: International Universities Press.

INDEX

Jo Alyson Parker, Paul A. Harris and Christian Steineck (Eds), Time: Limits and Constraints, pp. 367–377

www.ingramcontent.com/pod-product-compliance
Lightning Source LLC
La Vergne TN
LVHW010631300926
849162LV00001B/120
* 9 7 8 9 0 0 4 1 8 5 7 5 3 *